Preface

The subject of the present book is the mechanics of Lorentz transformations which we also label concisely as Lorentz mechanics. This branch of physics is commonly investigated under the title of special relativity or special relativistic mechanics. Our motive for using the title "The Mechanics of Lorentz Transformations" (instead of "Special Relativity" or "Special Relativistic Mechanics" or something like these) and approaching the subject as such is our belief that special relativity is essentially a particular epistemological and philosophical interpretation of the formalism of this branch of physics, and hence to be more objective and less exposed to possible errors and pitfalls of interpretations and non-scientific contents, the subject should be approached and structured according to its formalism and not according to these interpretations and their philosophical and epistemological framework. Various interpretations can then follow but with avoidance of mixing these interpretations with the formalism which is the only experimentally verified part. In this approach, we follow the style of quantum mechanics where the formalism and interpretations are kept apart. So, in this book we outline our vision for the way that should be followed in formulating, structuring and presenting the mechanics of Lorentz transformations in education, academia and research to avoid hijacking the whole subject by any particular philosophical and epistemological framework or being imprisoned by attachments where these frameworks and attachments may not be scientifically verified or verifiable through experiment and observation.

Accordingly, instead of starting from the two postulates of special relativity[1] (i.e. the relativity principle and the constancy of the observed speed of light) and deriving the Lorentz transformations from these postulates, Lorentz mechanics starts from the Lorentz spacetime coordinate transformations as the main postulates of this mechanics where the direct and indirect experimental and observational evidence is the only foundation for these transformations and their logical, mathematical and physical consequences. Our belief is that the postulates of special relativity are essentially philosophical and epistemological rather than scientific statements or at least they have unverified or unverifiable contents and implications. The advantage of postulating the Lorentz transformations to replace the special relativity postulates is that the formalism will then be neutral to any particular philosophical and epistemological framework and hence all the conclusions and implications will be completely based on the formalism and its scientific evidence rather than on philosophical and epistemological frameworks with non-scientific attachments. As a consequence, any experimental evidence according to our approach will establish the bare formalism while it remains neutral towards any particular interpretation. Moreover, more freedom will be gained from this non-association with a particular philosophical and epistemological framework. On the other hand, basing Lorentz mechanics on the special relativity theory and its postulates will inevitably lead the subject towards conclusions

---

[1] The special theory of relativity, which is commonly known as the special relativity theory and represents in our view an epistemological and philosophical interpretation to the mechanics of Lorentz transformations and its formalism, was originated by Poincare and elaborated by Einstein.

and implications that are essentially based on the philosophical and epistemological contents of the framework where the evidence in support of the formalism will be wrongly interpreted as support to the philosophical and epistemological foundations of the theory. Moreover, the whole subject will be imprisoned by the framework of the theory and its attachments leading to the closure of many potential routes of investigation and directing the scientific research in potentially wrong directions.

In this context, we assume that the mechanics of Lorentz transformations is well established and supported by experimental and observational evidence as claimed in the literature of Lorentz mechanics and hence we rule out fundamental errors that may put question marks on certain issues and conclusions. We also rule out damaging prejudice that affects the credibility and authenticity of certain issues and claims that are reported in the literature of this subject with common acceptance. However, we may criticize or assess certain issues or evidence since many of the alleged verifications of the mechanics of Lorentz transformations are controversial in their outcome and explanation. In fact, scientific evidence in general, not only in this part of physics, is not immune to errors, prejudice and wrong interpretation.

Although we adopt a rather different approach to the common approach of presenting and formulating the whole physics of Lorentz transformations in terms of special relativity and its philosophical and epistemological foundations, we still discuss comprehensively special relativity as it represents the dominant view in the literature of Lorentz mechanics. In fact, we dedicate a whole chapter (see § 9), as well as significant parts of other chapters, to the discussion and investigation of special relativity theory and related issues and hence this theory will be present in most parts of the book at least in the background. Moreover, some of the formulations and exercises are based on special relativistic viewpoints to be fair and balanced. We are motivated in this regard by our appreciation to the fact that special relativity is not only considered by the overwhelming majority of the scientific community as the only viable interpretation of Lorentz mechanics but it is even considered to be Lorentz mechanics itself with the denial of the existence of Lorentz mechanics as such. On the other side, we will not hesitate to present views, interpretations and even challenges to special relativity that are not frequently heard in public about this subject which, unlike most other branches of physics, is emotionally charged since it is associated, thanks to education and indoctrination, with Einstein and hence in the eye of many people any opponent view or criticism, or even question mark, is a direct attack against Einstein and his legacy.

The book originates from a collection of personal notes and tutorials about topics and applications related to modern physics and tensor calculus. The book includes extensive sets of exercises which are distributed throughout the book in the end of sections and subsections. The detailed solutions of all these exercises are provided in a separate book (titled: Solutions of Exercises of The Mechanics of Lorentz Transformations). The book is also furnished with many examples and solved problems inside the main text. It also contains a number of good quality graphic illustrations. A rather thorough index that contains all the essential terms of this branch of physics is also added to the book to enable keyword search by the readers and to provide a useful list for the main technical terms of

this subject. Cross referencing is used extensively in the book to connect related parts and help the readers to discover the relationships between various concepts, formulations and sections. These cross references are hyperlinked in the digital versions of the book. We believe that the book can be used as a guiding text or as a reference for a first course on the mechanics of Lorentz transformations or as part of a course on modern physics or tensor calculus or even the special theory of relativity. Regarding the credit for the preparation of the book, this book like its predecessors is totally made by the author including all the graphic illustrations, indexing, typesetting, book cover, and overall design.
Taha Sochi
London, May 2018

Contents

Preface   1

Table of Contents   4

Nomenclature   9

1 Preliminaries   **12**
   1.1 General Background 12
   1.2 Historical Issues and Credits 14
   1.3 General Terminology 16
   1.4 Mathematical Preliminaries 20
   1.5 General Conventions, Notations and Remarks 26
   1.6 Physical Reality and Truth 35
   1.7 Intrinsic and Extrinsic Properties 40
   1.8 Invariance of Physical Laws 41
   1.9 Galilean Transformations 43
     1.9.1 Space and Time Transformations 44
     1.9.2 Velocity Transformations 46
     1.9.3 Velocity Composition 48
     1.9.4 Acceleration and Force Transformations 49
   1.10 Newton's Laws of Motion 51
   1.11 Thought Experiments 54
   1.12 Requirements for Scientific Theories and Facts 55
   1.13 Speed 56
     1.13.1 Speed of Projectile 56
     1.13.2 Speed of Wave 57
   1.14 Speed of Light 59

2 Emergence of Lorentz Mechanics   **62**
   2.1 Classical View of the World 62
   2.2 Galilean Relativity 65
   2.3 Maxwell's Equations and Speed of Light 69
   2.4 Maxwell's Equations and Galilean Transformations 71
   2.5 Light as Wave Phenomenon and Luminiferous Ether 72
   2.6 Michelson-Morley Experiment 73
   2.7 FitzGerald-Lorentz Proposal and Emergence of Lorentz Transformations . 76
   2.8 Poincare Suggestion and Subsequent Developments 78

3 Introduction to Lorentz Mechanics — 80
3.1 Lorentz Mechanics versus Other Mechanics — 80
3.2 Restrictions and Conditions on Lorentz Mechanics — 82
3.3 Space Coordination — 84
3.4 Time Measurement and Synchronization — 85
3.5 Calibration of Space and Time Measurement — 89
3.6 Reference Frame — 89
3.6.1 Construction of Reference Frame — 89
3.6.2 Inertial and non-Inertial Frames — 90
3.6.3 Difference between Inertial and non-Inertial Frames — 93
3.7 Causal Relations — 94
3.8 Speed of Light — 95
3.8.1 Speed of Light as Restricted and Ultimate Speed — 96
3.8.2 Measuring the Speed of Light — 99
3.9 Spacetime in Lorentz Mechanics — 100
3.9.1 Spacetime Diagram and World Line — 102
3.9.2 Light Cone — 107
3.9.3 Spacetime Interval — 112
3.9.4 Invariance of Spacetime Interval — 114

4 Formalism of Lorentz Mechanics — 117
4.1 Physical Quantities — 117
4.1.1 Length — 117
4.1.2 Time Interval — 119
4.1.3 Mass — 123
4.1.4 Velocity — 124
4.1.5 Acceleration — 125
4.1.6 Momentum — 126
4.1.7 Force — 127
4.1.8 Energy — 128
4.1.9 Work — 133
4.2 Physical Transformations — 133
4.2.1 Lorentz Spacetime Coordinate Transformations — 135
4.2.2 Velocity Transformations — 138
4.2.3 Velocity Composition — 144
4.2.4 Acceleration Transformations — 152
4.2.5 Length Transformation — 153
4.2.6 Time Interval Transformation — 153
4.2.7 Mass Transformation — 154
4.2.8 Frequency Transformation and Doppler Shift — 154
4.2.9 Charge Density and Current Density Transformations — 159
4.3 Physical Relations — 160
4.3.1 Newton's Second Law — 160

		4.3.2	Mass-Energy Relation	161

        4.3.2   Mass-Energy Relation 161
        4.3.3   Momentum-Energy Relation 163
        4.3.4   Work-Energy Relation 165
  4.4 Conservation Laws 166
  4.5 Restoring Classical Formulation at Low Speed 170
  4.6 Restrictions at High Speed 172

5 Derivation of Formalism   174
  5.1 Physical Transformations 176
        5.1.1   Lorentz Spacetime Coordinate Transformations 176
        5.1.2   Velocity Transformations 178
        5.1.3   Acceleration Transformations 180
        5.1.4   Length Transformation 183
        5.1.5   Time Interval Transformation 184
        5.1.6   Mass Transformation 186
        5.1.7   Frequency Transformation and Doppler Shift 187
        5.1.8   Charge Density and Current Density Transformations 188
  5.2 Physical Quantities 189
        5.2.1   Momentum 189
        5.2.2   Force 193
        5.2.3   Energy 194
  5.3 Physical Relations 196
        5.3.1   Newton's Second Law 196
        5.3.2   Mass-Energy Relation 197
        5.3.3   Momentum-Energy Relation 206
        5.3.4   Work-Energy Relation 208
  5.4 Conservation Laws 209

6 Tensor Formulation of Lorentz Mechanics   216
  6.1 Preliminaries 216
  6.2 Useful Mathematics 219
  6.3 Minkowski Metric Tensor 221
  6.4 Lorentz Transformations in Matrix and Tensor Form 226
  6.5 Vector, Tensor and Matrix Formulation 228
        6.5.1   Spacetime Position and Displacement 4-Vector 228
        6.5.2   Quadratic Form of Spacetime Interval 229
        6.5.3   Velocity 229
        6.5.4   Acceleration 230
        6.5.5   Momentum 230
        6.5.6   Force and Newton's Second Law 231
        6.5.7   Electromagnetism and Maxwell's Equations 233

7 Consequences and Predictions of Lorentz Mechanics — 240
7.1 Merging of Space and Time into Spacetime — 240
7.2 Length Contraction — 240
7.3 Time Dilation — 241
7.4 Relativity of Simultaneity — 242
7.5 Relativity of Co-positionality — 245
7.6 Equivalence of Mass and Energy — 249

8 Evidence for Lorentz Mechanics — 251
8.1 Success of Lorentz Transformations — 253
8.2 Mass-Energy Equivalence — 253
8.3 Prolongation of Lifetime of Elementary Particles — 253
8.4 Atomic Clock Experiment — 256
8.5 Stellar Aberration — 256

9 Special Relativity — 258
9.1 Characteristic Features of Special Relativity — 258
9.2 Postulates of Special Relativity — 259
9.3 Assessing the Postulates of Special Relativity — 261
9.3.1 Relativity Principle — 261
9.3.2 Invariance of Observed Speed of Light — 263
9.3.3 Overall Assessment of Special Relativity Postulates — 268
9.4 Abolishment of Fundamental Concepts — 269
9.5 Controversies within Special Relativity — 271
9.6 Thought Experiments in Special Relativity — 271
9.6.1 Train Thought Experiment — 271
9.7 Light Clock — 277
9.7.1 Assessing Light Clock — 281

10 Challenges, Criticisms and Controversies — 296
10.1 Twin Paradox — 296
10.1.1 Time Dilation Effect is Apparent — 298
10.1.2 Traveling Twin is Distinguished by being non-Inertial — 298
10.1.3 Traveling Twin has Two Inertial Frames — 301
10.1.4 Calling for General Relativity — 301
10.2 Barn-Pole Paradox — 302
10.3 Other Paradoxes — 305
10.4 Speeds Exceeding c — 306
10.5 Non-Local Reality of Quantum Mechanics — 306

11 Interpretation of Lorentz Mechanics — 308
11.1 Criteria for Acceptable Interpretation — 309
11.2 Essential Elements of Potentially Acceptable Interpretation — 310

12 Appendices · **313**
   12.1 Maxwell's Equations 313
   12.2 Michelson-Morley Experiment 314
   12.3 Invariance of Laws under Galilean and Lorentz Transformations 319
      12.3.1 Laws of Classical Mechanics 319
      12.3.2 Maxwell's Equations 320
      12.3.3 Electromagnetic Wave Equation 320
   12.4 Derivation of Lorentz Spacetime Coordinate Transformations 323
      12.4.1 Special Relativity Method of Derivation 324
      12.4.2 Our Method of Derivation 327

Epilogue   **331**

References   **332**

Index   **333**

Author Notes   **340**

Nomenclature

In the following list, we define the common symbols, notations and abbreviations that are used in the book as a quick reference for the reader.

∇	nabla differential operator in ordinary 3D space
∇h	gradient of scalar h
$\nabla \cdot \mathbf{A}$	divergence of vector $\mathbf{A}$
$\nabla \times \mathbf{A}$	curl of vector $\mathbf{A}$
∇^2	Laplacian operator in ordinary 3D space
$\Box$	nabla operator in Minkowski 4D spacetime
$\Box^2$	d'Alembertian operator in Minkowski 4D spacetime
$\partial_\mu, \partial^\mu$	partial derivative with respect to the μth coordinate
$\delta/\delta\tau$	absolute or intrinsic derivative operator with respect to τ
$\sim$	comparable in size
$'$ (prime)	mark of reference frame in motion relative to a given reference frame
0 (subscript)	proper quantity, e.g. proper length L_0
a	magnitude of acceleration or 1D acceleration
$\mathbf{a}, \mathbf{A}$	acceleration vector in 3D, 4D
$\mathbf{a}, \mathbf{A}$	electromagnetic vector potential in 3D, 4D
a^i, A^μ	acceleration vector or its components in 3D, 4D
a^i, A^μ	electromagnetic vector potential or its components in 3D, 4D
a_x, a_y, a_z	components of 3D acceleration vector in x, y, z directions
a_x, a_y, a_z	components of 3D electromagnetic vector potential in x, y, z directions
B	magnitude of magnetic field
$\mathbf{B}$	magnetic field vector
B_x, B_y, B_z	components of magnetic field vector in x, y, z directions
c	characteristic speed of light in vacuum
$\text{diag}[\cdots]$	diagonal matrix with embraced diagonal elements
$ds, d\sigma$	infinitesimal line element in 3D space, 4D spacetime
E	energy
$\mathbf{E}$	electric field vector
E_0, E_k, E_t	rest, kinetic, total energy
E_x, E_y, E_z	components of electric field vector in x, y, z directions
Eq./Eqs.	Equation/Equations
f	magnitude of force or 1D force
$\mathbf{f}, \mathbf{F}$	force vector in 3D, 4D
f^i, F^μ	3D, 4D force vector or its components
f_x, f_y, f_z	components of 3D force vector in x, y, z directions
g_{ij}, g^{ij}	covariant, contravariant metric tensor of 3D space or its components
$g_{\mu\nu}, g^{\mu\nu}$	covariant, contravariant metric tensor of 4D spacetime or its components
i	imaginary unit

iff	if and only if
j, **J**	electric current density vector in 3D, 4D
j_x, j_y, j_z	components of 3D electric current density vector in x, y, z directions
J^μ	electric current density vector or its components in 4D
k	dragging coefficient
k_w	dragging coefficient of water
L, L_0	length, proper length
L, $[L^\mu_\nu]$	Lorentz matrix
$\mathbf{L}^{-1}$	inverse of Lorentz matrix
L^μ_ν	Lorentz tensor or its components
Ly	light year (distance)
m, m_0	mass, rest mass
m_e, m_n, m_p	mass of electron, neutron, proton
M^μ_ν	inverse of Lorentz tensor or its components
n	refractive index
nD	n-dimensional
O	observer or frame of reference
p	magnitude of momentum or 1D momentum
p, **P**	3D, 4D momentum vector
p^i, P^μ	3D, 4D momentum vector or its components
p_x, p_y, p_z	components of 3D momentum vector in x, y, z directions
Q	electric charge
r	radius
r	position vector in 3D space, i.e (x, y, z) or (x^1, x^2, x^3)
r, θ, ϕ	spherical coordinates of 3D space
s	3D space interval (or length of arc)
S, $[S^{\mu\nu}]$	electromagnetic field strength matrix
$S^{\mu\nu}$	electromagnetic field strength tensor or its components
t, t_0	time, proper time
T, $[T^{\mu\nu}]$	electromagnetic dual field strength matrix
$T^{\mu\nu}$	electromagnetic dual field strength tensor or its components
u	speed or 1D velocity (usually belongs to observed object)
u, **U**	3D, 4D velocity vector
u^i, U^μ	3D, 4D velocity vector or its components
u_x, u_y, u_z	components of 3D velocity vector in x, y, z directions
v	speed or 1D velocity (usually belongs to inertial frame)
v	3D velocity vector
v_x, v_y, v_z	components of 3D velocity vector in x, y, z directions
V	event
V, V_0	volume, proper volume
x	position vector in 4D spacetime, i.e. (x^0, x^1, x^2, x^3) or (x^1, x^2, x^3, x^4)
x^0, x^1, x^2, x^3	spacetime coordinates (x^0 temporal)
x^1, x^2, x^3, x^4	spacetime coordinates (x^4 temporal)

x^i, x^μ	coordinates of 3D space, 4D spacetime
x, y, z	spatial coordinates (normally rectangular Cartesian)
β	speed ratio
γ	Lorentz factor
$\Gamma^\mu_{\nu\omega}$	Christoffel symbol of 2nd kind for 4D spacetime
δ^μ_ν	Kronecker delta tensor in 4D
Δ	finite change
ε_0	permittivity of free space
λ, λ_0	wavelength, proper wavelength
μ_0	permeability of free space
ν, ν_0	frequency, proper frequency
ρ, ρ_0	charge density, proper charge density
ρ, ϕ, z	cylindrical coordinates of 3D space
σ	spacetime interval in 4D
τ	proper time parameter
ϕ	electromagnetic scalar potential
ω	angular speed

Chapter 1
Preliminaries

In this chapter, we discuss a number of background issues about Lorentz mechanics in its historical and scientific context, general concepts and conventions that will be used in the upcoming parts of the book, and some other related issues such as preliminary mathematical requirements.

1.1 General Background

Mechanics is the science of describing and predicting the physical motion of objects through space and time. It has a kinematical aspect where it specializes in describing the motion regardless of its causes and a dynamical aspect where it investigates the causes of motion as well. There are several types and branches of mechanics that make parts of the science of physics, as well as a number of theories that may still require further evidence to be established as facts. For instance, we have classical (or Newtonian) mechanics, quantum mechanics, Lorentz mechanics, and general relativistic mechanics. These different types and branches were developed in different stages of the history of science to address issues and solve problems that could not be addressed and solved by the existing knowledge at that time.

As we will see later in more details, each one of these types and branches is distinguished by its domain of applicability and limits of its use. For example, classical mechanics applies to large objects (i.e. objects comparable in size to human perception) moving at normal speeds which human is familiar with in his everyday life, while quantum mechanics is the mechanics for atomic and subatomic world where the size of objects is beyond the perception of human and hence it represents observations that have not been experienced during his evolution. Similarly, the primary domain of Lorentz mechanics is objects moving at very high speeds (i.e. speeds comparable to the speed of light) as observed from inertial frames (see § 3.6.2). On the other hand, the domain of general relativistic mechanics is gravity and its influence on the physical objects such as matter and light where the unique perceptible effects of this mechanics are supposed to be observable in strong gravitational fields.

As stated in the Preface, the present book is about the mechanics of Lorenz transformations, which for convenience we frequently call "Lorentz mechanics", where we use this term to refer to the branch of mechanics that is based on the Lorentz spacetime coordinate transformations. This mechanics in its general features is the same as the mechanics of special relativity. However, we depart from the common (and almost universal) terminology because of our belief that special relativity is just one possible philosophical and epistemological interpretation of the mechanics of Lorenz transformations and hence the science of this mechanics should be restricted to the formalism of Lorentz transformations

1.1 General Background

and their logical, mathematical and physical consequences without mixing this with any philosophical or epistemological interpretation which may contain false premises or logical errors and inconsistencies.

We also consider in our choice of name another issue that is the credit for inventing and developing this branch of modern physics where there is virtually a general consensus these days that Einstein is the sole contributor (or at least the main contributor) to the creation and development of this part of science. There is also an association of this part of science with the theory of special relativity which is supposed to be created by Einstein, while our view, which is based on reliable historical evidence, is that this part of modern physics was developed by a generation of scientists in the last decades of the 19th century and the early decades of the 20th century with Einstein being just one among many contributors. In fact, Einstein is neither the first nor the last of these contributors and he is not even the main contributor. As we will see, Einstein was preceded and succeeded by a number of original contributors whose role is at least as important individually as his own role. So, labeling this branch of physics as "Lorentz mechanics" instead of "special relativity" serves another purpose that is attributing the credit in a fair way and acknowledging the work of other contributors. This cannot be achieved if we follow the general practice of labeling this mechanics as "special relativity" because of the strong association of this theory with Einstein although we do not believe that Einstein should get the main credit even for originating and developing special relativity which fundamentally is the theory of Poincare.

Accordingly, our method in this book is to adopt a more inclusive approach by presenting opposite views and opinions. We also present the formalism of Lorentz mechanics in a rather different way where we start from the Lorentz spacetime coordinate transformations as postulates instead of starting from the special relativity postulates. Hence, all the main results of the formalism of Lorentz mechanics will be derived from these coordinate transformations with no reference to the special relativity postulates or methods like light clock and thought experiments which are common tools in special relativity. The bare formalism will then be followed by the other aspects of Lorentz mechanics including the commonly proposed epistemological and philosophical interpretations, where in this context the theory of special relativity, which largely represents the interpretation of Poincare to Lorentz mechanics, will be given particular attention due to its dominance in the modern literature of Lorentz mechanics.

We would like to refer to the fact that these interpretations, which are structured within particular philosophical and epistemological frameworks, may contain logical inconsistencies or incorrect philosophical and scientific conclusions, and hence these interpretations may lead to incorrect scientific conclusions and results. However, the validity of the Lorentz mechanics will not be affected by these interpretations according to our approach since the formalism is not based on any particular interpretation. In this regard, we follow a similar approach to the approach that is used in quantum mechanics where the formalism is presented first while the presentation of any philosophical and epistemological interpretations is postponed with obvious understanding that the experimental evidence and validation belong to the formalism and not to the interpretations. Accordingly, the fallacy or collapse of any particular interpretation will not affect the validity of the formalism.

Exercises
1. Define the science of mechanics in a few words.
2. Make a brief comparison between the mechanics of Lorentz transformations and the following branches and theories of mechanics: Newtonian mechanics, quantum mechanics and general relativistic mechanics.
3. Why we think it is better to use expressions like "mechanics of Lorentz transformations" or "Lorentz mechanics" instead of the common expressions like "special relativity" or "special relativistic mechanics" to refer to the part of mechanics that is based on the Lorentz transformations?
4. Compare between the approach followed in the construction and presentation of quantum mechanics in the common textbooks of physics and the approach followed in the construction and presentation of Lorentz mechanics in these texts to see if the two subjects are treated equally.

1.2 Historical Issues and Credits

Based on the available historical records, the mechanics of Lorentz transformations was developed over three or four decades in the late 19th century and the early 20th century by a number of scientists such as FitzGerald, Lorentz, Larmor, Voigt, Poincare, Einstein, Planck, Laue, Minkowski, and Sommerfeld. So, despite the common belief that Lorentz mechanics is the brainchild of Einstein alone or at least he is the main contributor, there are many scientists, before and after Einstein, who contributed to the development of this branch of physics. In fact, Einstein is not only just one among many contributors but he is not even the main contributor. This bias in favor of Einstein started after the alleged success of the 1919 solar eclipse expedition in proving the prediction of the general theory of relativity about the bending of light by the gravity of the Sun. Reliable historical records show that prior to this questionable endorsement of general relativity by Eddington and his team, the theory of Lorentz mechanics was generally attributed to Lorentz and Poincare and not to Einstein. This may be exemplified in labeling this theory by Klein in 1910 as the invariant theory of the Lorentz group which clearly associates this part of mechanics with the work of Lorentz and his transformations.

In the following paragraph we outline some key stages in the development of Lorentz mechanics in its early days for the sake of completeness and to support the aforementioned claim about the credit of creating and developing this branch of physics. However, these should be seen as only gross examples and hence they are not comprehensive or meticulous in any way. There are many other original contributions by other scholars which are not presented here; moreover the given contributions are just outlines and hence they lack many important details. The interested reader should refer to specialized books and articles on the history of science and Lorentz mechanics in particular.

FitzGerald and Lorentz proposed the effect of length contraction as a fix for the null

1.2 Historical Issues and Credits

result of the Michelson-Morley experiment.[2] This proposal resulted into the subsequent development by Lorentz of what will be known later as Lorentz transformations of spacetime coordinates. Larmor and Voigt were also active participants in the development of these transformations as well as some other elements of Lorentz mechanics in its early days. Poincare suggested the essence of what will be called subsequently the special theory of relativity which is a philosophical and epistemological interpretation of these transformations and their physical and logical consequences. Einstein adopted Poincare interpretation and hence he derived the Lorentz transformations from the two postulates of special relativity, which are the principle of relativity and the constancy of the observed speed of light (refer to § 9.2 and § 12.4.1), with some elaboration on the consequences of Lorentz transformations and their interpretation. Planck proposed the definition of force and the Lagrangian formulation according to the principles of Lorentz mechanics. Minkowski proposed the mathematical formulation of the 4D spacetime conceptualization of Lorentz mechanics which is commonly used in modern times.[3] This 4D approach has eventually propagated into the tensor formulation of Lorentz mechanics (see § 6) and several other disciplines. Sommerfeld added some contributions and elaborations to the Minkowski 4D spacetime approach. Laue proposed the mechanical interpretation of the Fresnel drag as a consequence of the principles of Lorentz mechanics. There are many other contributions to the main body of Lorenz mechanics as well as to other issues that are related to this mechanics, like the equivalence between mass and energy and the adaptation of quantum mechanics to the framework of Lorentz mechanics, where several scholars like Pauli, Lenard and Dirac contributed to the development of original ideas and techniques in this field. In fact, the development of Lorentz mechanics and the hot discussion about it and its significance and interpretation, as well as many controversies that surround it, are still going on to these days. This highlights the importance of this branch of physics and its exclusive status among all scientific disciplines.

Solved Problems

1. What we mean by saying: the mechanics of Lorentz transformations was developed over three or four decades in the late 19th century and the early 20th century?
   Answer: We mean by this the development of the main body of Lorentz mechanics; otherwise many issues that contributed to the emergence and rise of Lorentz mechanics (such as the relativity principle) go back to the early days of renaissance and some of these issues go back even further where their deep roots can be detected in the ancient

---

[2] It may be more appropriate to start from Maxwell and his equations which triggered the subsequent development of Lorentz mechanics. Moreover, these equations may be regarded as the first Lorentzian formulation of a physical theory due to their compatibility with the Lorentz transformations. However, starting from FitzGerald and Lorentz seems more natural and specific to the development of Lorentz mechanics due to the direct nature of their contributions. We should also give credit to the experimental work of Michelson and Morley and their alike for initiating the debate about fundamental issues in physics that led to the emergence and rise of Lorentz mechanics.

[3] In fact, this merge of space and time into spacetime seems to be proposed first by Poincare (and may be even before Poincare) although Minkowski put this in a more elegant mathematical form and in a more explicit fashion.

1.3 General Terminology 16

science and philosophy of the middle ages and old civilizations. Also, the development of Lorentz mechanics continued beyond the aforementioned few decades through elaboration, extension, application and incorporation in other branches of science. In fact, the development of Lorentz mechanics in its wider sense is still going on even in these days.

Exercises
1. Name a number of physicists who have contributed to the development of Lorentz mechanics in its early days.
2. When the credit for the creation and development of Lorentz mechanics started to shift from Lorentz, Poincare and other major contributors to Einstein?
3. Why we attribute the main credit for the special relativity interpretation to Poincare?

1.3 General Terminology

In this section, we present brief definitions of a number of terms and concepts that will be used frequently in the forthcoming parts of the book. These terms are also in common use in the literature of this subject. We note that these definitions are elementary and some of which may be revised or elaborated further later in the book.

Massive or **material** object is a physical object that possesses finite mass (i.e. $m > 0$) such as a particle of sand, while **massless** object is a physical object with no mass (i.e. $m = 0$) such as a photon. A **uniform motion** is a motion with constant velocity, i.e. constant speed and constant direction. A **free particle** is a massive object that is not under the influence of a net external force. A free particle can be stationary or moving uniformly in space with constant velocity. An **observer** is someone who observes physical events taking place in the physical world and can take measurements of physical quantities such as time, space coordinates, mass and force. An observer needs a frame of reference (see next) to take quantitative measurements of the observed physical quantities that are related to space and time such as position and velocity.

A **reference frame** or a **frame of reference** is a combination of a spatial coordinate system to identify points in space and a temporal mechanism to identify points in time. Since a reference frame is inevitably associated with an observer who is using the frame to observe and record events taking place in space and time, the terms "reference frame" and "observer" are often used interchangeably.[4] A **rest frame** (or **proper frame**) of an object is a frame of reference in which the object is at rest and not in motion relative to this frame. There is an infinite number of rest frames for any particular object or observer where these frames are obtained from each other by static transformations like translation of the origin of coordinates, rotation and reflection of the spatial coordinate axes (and possibly other types of transformation like scaling and shearing) as well as translation of the origin of time and scaling of the time unit. However, all these frames can be reduced to a single frame by these static transformations. A **proper** quantity (e.g. proper length

---
[4] In fact, observer in this context means a global observer who is present everywhere in his frame all the time and hence he is not localized into a particular position in the frame or a particular interval of time.

1.3 General Terminology

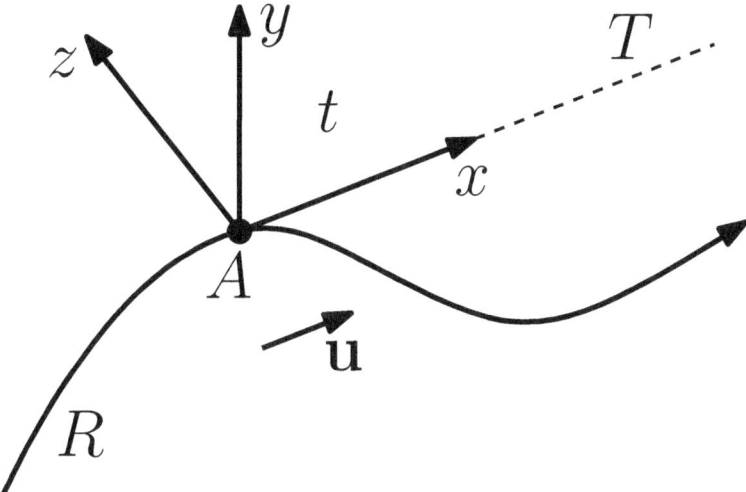

Figure 1: Instantaneous rest frame $xyzt$ of an object A in its trajectory R at time t_g where this frame is moving with constant velocity **u**, which is the instantaneous velocity of A, along the tangent T of the trajectory of A at t_g.

and proper time) of an object is the quantity as measured in the rest frame of the object, while an **improper** quantity is the quantity as measured by an observer who is in a state of relative motion with respect to the rest frame of the object. Accordingly, a proper quantity may be described as rest quantity since it is measured in a state of rest and hence we have, for example, rest mass and rest energy which are proper mass and proper energy. **Inertial frame** is a frame in which Newton's laws of motion are valid,[5] while **non-inertial** or **accelerating** frame is a frame in which Newton's laws of motion are not valid. Inertial frames may also be called Galilean reference frames. **Instantaneous rest frame** of a physical object A at a given instant of time t_g is an inertial frame whose origin is at the position of A and it is moving uniformly with the same velocity of A at t_g, i.e. it is moving with a constant speed along the tangent to the trajectory of A at t_g (see Figure 1). This definition should be extended to include instantaneous rest frame of non-inertial frames.

Spacetime is a combination of space and time in which objects do exist and physical events take place. In Lorentz mechanics, spacetime is not seen as space plus time but as a single manifold in which temporal and spatial dimensions are treated equally as dimensions of a 4D manifold. They are also intertwined in a way that the spatial and temporal coordinates are mixed in their transformations between inertial frames such that a space or time coordinate in one frame is expressed in terms of both space and

---

[5] In fact, this is a sufficient and necessary condition for the frame to be inertial. Also, this is equivalent (at least empirically) to holding Newton's first law of motion and hence inertial frames are commonly defined as the frames in which the law of inertia holds true. We also note that since all the laws of motion in classical mechanics are derived directly or indirectly from Newton's three laws then this means that inertial frames are also characterized by the fact that all the laws of classical mechanics hold true in these frames.

1.3 General Terminology

time coordinates of the other frame (see for example Eqs. 90 and 93). For dimensional consistency,[6] the temporal coordinate of spacetime is usually taken as ct (which involves the characteristic speed of light c as well as t which symbolizes time) rather than t so that all the coordinates of the spacetime manifold have the same physical dimension of length. An **event** is a physical occurrence that takes place in spacetime. An event is usually represented by a single point in the 4D spacetime manifold to identify its temporal and spatial coordinates, and hence an event is usually identified by four real numbers which are one temporal coordinate and three spatial coordinates. Accordingly, the presence of a physical object in spacetime (which can be seen as a continuous series of events representing its existence in spacetime) is depicted by a continuous curve in spacetime.

Minkowski space, which is also known as Minkowski spacetime, is the 4D manifold of spacetime that consists of one temporal dimension and three spatial dimensions. Accordingly, a 4D manifold whose points represent events is also called an **event space**, as well as **Minkowski space**. As indicated earlier, the points of an event space are usually identified by 4 coordinates x^μ ($\mu = 0, 1, 2, 3$) where $x^0 \equiv ct$ while $(x^1, x^2, x^3) \equiv (x, y, z)$ with c being the characteristic speed of light in free space, t is time and x, y, z are the three spatial coordinates of a rectangular Cartesian system (also see § 1.5 about other identifications). **World line** is a continuous curve in spacetime and hence it represents the trajectory of a physical object or a continuous series of correlated events in spacetime. Accordingly, the points of world line represent distinct events in spacetime. World line, for example, can represent the path in spacetime of an object which is at rest or moving uniformly or accelerating under the influence of a net external force, e.g. the trajectory in spacetime of a stone thrown by a warrior from its point of release to its point of impact. Similarly, world line can represent the trace in spacetime of a sequence of correlated events such as the interaction between a number of objects through gravitational or magnetic forces.

Two events, $V_1(\mathbf{x}_1)$ and $V_2(\mathbf{x}_2)$, in an event space are described as:
1. **Simultaneous** when they occur at the same instant of time and hence they have identical temporal coordinate, i.e. $x_1^0 = x_2^0$.
2. **Co-positional** when they occur at the same location in space and hence they have identical spatial coordinates, i.e. $x_1^i = x_2^i$ ($i = 1, 2, 3$).
3. **Identical** when they occur at the same time and location and hence they have identical temporal and spatial coordinates, i.e. $x_1^\mu = x_2^\mu$ ($\mu = 0, 1, 2, 3$).
4. **Anti-identical**[7] when they occur at different times and different locations, i.e. $x_1^0 \ne x_2^0$ and at least one of the following conditions is true: $x_1^1 \ne x_2^1$ or $x_1^2 \ne x_2^2$ or $x_1^3 \ne x_2^3$.

We generally consider these four categories as mutually exclusive and hence identical events cannot be described as simultaneous although they satisfy the condition $x_1^0 = x_2^0$. However, there are some exceptions to this rule in the book where the meaning is obvious (i.e. "simultaneous" and "co-positional" are used in their generic sense to indicate same time and same place regardless of being so exclusively).

The **space interval** Δs between two events, $V_1(\mathbf{r}_1)$ and $V_2(\mathbf{r}_2)$, in an event space is a

---
[6] There may also be other reasons for this like elegance in form and symmetry.
[7] We carved this jargon (which we did not find in the literature) to represent this concept.

1.3 General Terminology

real number given by:

$$\Delta s = \sqrt{(\Delta x^1)^2 + (\Delta x^2)^2 + (\Delta x^3)^2} = \sqrt{\Delta x^i \Delta x^i} \tag{1}$$

where $\Delta x^i = x_2^i - x_1^i$ and $i = 1, 2, 3$ with the summation convention being used in the last equality (with employed Cartesian system). Similarly, the **spacetime interval** $\Delta \sigma$ between two events, $V_1(\mathbf{x}_1)$ and $V_2(\mathbf{x}_2)$ is defined as:

$$\Delta \sigma = \sqrt{(\Delta x^0)^2 - (\Delta x^1)^2 - (\Delta x^2)^2 - (\Delta x^3)^2} \tag{2}$$

where $\Delta x^\mu = x_2^\mu - x_1^\mu$ and $\mu = 0, 1, 2, 3$. We note that the spacetime interval as defined above is not necessarily real. Hence, it may be defined mathematically as above but with the modulus sign to make it necessarily real. Spacetime interval may also be defined as:

$$\Delta \sigma = \sqrt{(\mathrm{i}\Delta x^0)^2 + (\Delta x^1)^2 + (\Delta x^2)^2 + (\Delta x^3)^2} \tag{3}$$

where i is the imaginary unit. This is the same as Eq. 2 but with the reversal of the sign of all terms inside the square root. Now, if we define the temporal coordinate to be $x^0 = \mathrm{i}ct$ and we employ the summation convention, then we can write the last form of spacetime interval compactly as:[8]

$$\Delta \sigma = \sqrt{\Delta x^\mu \Delta x^\mu} \tag{4}$$

where $\Delta x^\mu = x_2^\mu - x_1^\mu$ and $\mu = 0, 1, 2, 3$ with summation over μ. A quantity that is intimately related to spacetime interval is the **proper time parameter** τ which will be investigated in detail later in the book (see § 6.1). As we will see, the proper time parameter τ, like the spacetime interval, is invariant under the Lorentz transformations. We note that the spacetime interval may be defined by one of the above expressions but in an infinitesimal form, i.e. $d\sigma$.[9] These issues will be investigated in more details later (see for example § 3.9.3).

A **relation** or a law is described as **invariant** if a group of observers, characterized by a particular attribute and between whom certain transformations apply, observe the same form of the relation. A physical **quantity** is described as **invariant** if the aforementioned group of observers obtain the same value when they measure this quantity. So, we have form-invariance that applies to the laws of physics and value-invariance that applies to physical quantities. For example, in classical mechanics Newton's laws are form-invariant under the Galilean transformations and mass is value-invariant under these transformations. As we will see (refer to § 1.8), the invariance of physical laws is a fundamental principle of physics and science in general.[10] We finally remark that we frequently use the term "Lorentz invariant"[11] as an attribute to certain physical laws or quantities (e.g. the

---

[8] More rigorous definitions and restrictions related to spacetime interval and the rules of indices will be given later in the book.

[9] This also applies to the proper time parameter τ and space interval s where they are defined in an infinitesimal form as $d\tau$ and ds respectively.

[10] In fact, it belongs more appropriately to the philosophy of science and its epistemological foundations and principles.

[11] This is commonly known in the literature as "Lorentz covariant", but we deliberately avoid this term to prevent any confusion with "covariant" as a tensor attribute, i.e. opposite to "contravariant".

electromagnetic wave equation and the spacetime interval are Lorentz invariant) to mean that the concerned law or quantity keeps its form or value under the Lorentz spacetime coordinate transformations. This terminology may also be used with the Galilean transformations and hence we may describe Newton's laws or mass, for instance, as Galilean invariant.

Exercises

1. Briefly define the following terms: massive object, massless object, uniform motion, inertial frame, rest frame, proper quantity, Minkowski space, world line, event, simultaneous events, co-positional events, space interval, and spacetime interval.
2. What is the difference between coordinate system and frame of reference?
3. What is the difference between "space and time" and "spacetime"?
4. List a number of transformations between different frames of reference which are at rest relative to each other.
5. List a number of differences between different reference frames which are in a state of motion relative to each other.
6. Which of the differences that we considered in the previous two exercises do not affect the status of inertiality (i.e. the state of being inertial or non-inertial) between two reference frames?

1.4 Mathematical Preliminaries

In the literature of Lorentz mechanics, the symbols β and γ are among the most used symbols where they are defined as follows:[12]

$$\beta = \frac{v}{c} \qquad \gamma = \frac{1}{\sqrt{1-\beta^2}} \qquad (5)$$

where β is the speed ratio, v is the speed of an object or a frame as measured from an inertial frame, c is the characteristic speed of light in vacuum, and γ is the Lorentz factor. The relation between γ and β is depicted in Figure 2. As we see, $\gamma = 1$ at $\beta = 0$, i.e. when $v = 0$. Moreover, γ stays very close to unity at comparatively low speeds. i.e. speeds much lower than c (or $v \ll c$). However, γ shoots up sharply as β approaches unity from below where the line $\beta = 1$ is a vertical asymptote to the γ curve when β becomes close to unity, i.e. $\beta \simeq 1$. As seen, both these factors are dimensionless and hence they are numbers without units.

In this context, we find it useful to list a number of identities (including one approximation) related to β and γ. These identities are either used in the forthcoming derivations of Lorentz mechanics formulae or used in the solutions of solved problems and exercises. These identities are:

$$\gamma^2 - 1 = \gamma^2\beta^2 = \gamma^2\frac{v^2}{c^2} \qquad (6)$$

---

[12] We note that v in the definition of β stands for speed in general and hence β may also be given as $\beta = u/c$.

1.4 Mathematical Preliminaries

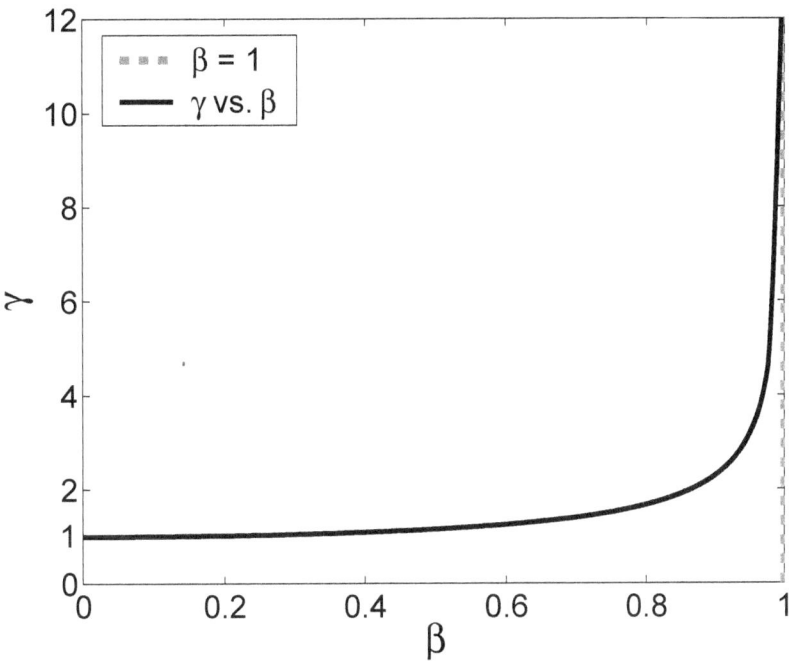

Figure 2: The curve representing Lorentz factor γ as a function of speed ratio β (solid line) and the vertical line $\beta = 1$ (dashed line) which is an asymptote to the γ curve as β approaches unity (i.e. $v \to c$) from below.

$$\gamma^2 - \gamma^2 \beta^2 = 1 \tag{7}$$

$$\frac{d\gamma}{dt} = \frac{\beta \frac{d\beta}{dt}}{(1-\beta^2)^{3/2}} = \gamma^3 \beta \frac{d\beta}{dt} \tag{8}$$

$$d\left[\frac{1}{(1-\beta^2)^{1/2}}\right] = \frac{\beta \, d\beta}{(1-\beta^2)^{3/2}} \tag{9}$$

$$1 - \frac{1}{\gamma^2} = \beta^2 \tag{10}$$

$$\gamma - \frac{1}{\gamma} = \gamma \beta^2 = \gamma \left(\frac{v}{c}\right)^2 \tag{11}$$

$$v \frac{dv}{dt} = c^2 \beta \frac{d\beta}{dt} \tag{12}$$

$$\frac{1}{\sqrt{1-\beta^2}} = 1 + \frac{1}{2}\beta^2 + \frac{3}{8}\beta^4 + \cdots \tag{13}$$

$$\gamma^2 - \gamma \simeq \frac{\beta^2}{2} \quad (\beta \ll 1) \tag{14}$$

$$\gamma \beta = \sqrt{\gamma^2 - 1} \tag{15}$$

$$\int v \, d(\gamma v) = c^2 \gamma \tag{16}$$

1.4 Mathematical Preliminaries

We note that some of these formulae are just variants of other formulae but they are included as separate formulae for the convenience of the readers. Anyway, their presentation and derivation may be useful as a mathematical exercise to improve the skills in recognizing and managing the mathematical expressions that involve β and γ which are present everywhere in the mathematics of Lorentz mechanics. In the following we derive these formulae. However, before that we should remark that some of these derivations are based on treating c as constant which may be seen by some as equivalent to the second postulate of special relativity. However, this is not the case because c is the *characteristic* speed of light which is a constant regardless of special relativity and its postulates. In fact, the second postulate of special relativity is about the constancy of the *observed* speed of light. The reader is referred to § 1.13, § 1.14 and § 9.2 for more details about this issue.

- **Derivation of Eq. 6**:

$$\begin{align}
\gamma^2 - 1 &= \frac{1}{1-\beta^2} - 1 \tag{17}\\
&= \frac{1}{1-\beta^2} - \frac{1-\beta^2}{1-\beta^2}\\
&= \frac{\beta^2}{1-\beta^2}\\
&= \gamma^2 \beta^2\\
&= \gamma^2 \frac{v^2}{c^2}
\end{align}$$

- **Derivation of Eq. 7**: this is another form of Eq. 6. However, as an exercise we derive it again using a slightly different method:

$$\begin{align}
\gamma^2 - \gamma^2\beta^2 &= \frac{1}{1-\beta^2} - \frac{\beta^2}{1-\beta^2} \tag{18}\\
&= \frac{1-\beta^2}{1-\beta^2}\\
&= 1
\end{align}$$

- **Derivation of Eq. 8**:

$$\begin{align}
\frac{d\gamma}{dt} &= \frac{d}{dt}\left(\frac{1}{\sqrt{1-\beta^2}}\right) \tag{19}\\
&= \frac{d}{dt}\left(1-\beta^2\right)^{-1/2}\\
&= -\frac{1}{2}\left(1-\beta^2\right)^{-3/2}\left(-2\beta\frac{d\beta}{dt}\right)\\
&= \left(1-\beta^2\right)^{-3/2}\beta\frac{d\beta}{dt}\\
&= \frac{1}{\left(1-\beta^2\right)^{3/2}}\beta\frac{d\beta}{dt}
\end{align}$$

1.4 Mathematical Preliminaries

$$
\begin{aligned}
&= \left(\frac{1}{\sqrt{1-\beta^2}}\right)^3 \beta \frac{d\beta}{dt} \\
&= \gamma^3 \beta \frac{d\beta}{dt}
\end{aligned}
$$

- **Derivation of Eq. 9**: this formula is the differential form of the previous formula (i.e. Eq. 8) which is in a derivative form and hence the derivation of this formula is similar to the derivation of Eq. 8.
- **Derivation of Eq. 10**:

$$ 1 - \frac{1}{\gamma^2} = 1 - \left(1 - \beta^2\right) = \beta^2 \tag{20} $$

- **Derivation of Eq. 11**: this formula can be obtained from Eq. 10 by multiplying both sides with γ.
- **Derivation of Eq. 12**:

$$
\begin{aligned}
v\frac{dv}{dt} &= c\frac{v}{c}\frac{dv}{dt} \\
&= c\beta\frac{dv}{dt} \\
&= c^2\beta\frac{d}{dt}\left(\frac{v}{c}\right) \\
&= c^2\beta\frac{d\beta}{dt}
\end{aligned} \tag{21}
$$

- **Derivation of Eq. 13**: this is a standard binomial expansion or a Taylor (or Maclaurin) series expansion and hence it can be found in any standard mathematical textbook on power series. It can also be derived easily using the standard methods of binomial expansion or power series.
- **Derivation of Eq. 14**: the Taylor series expansions of γ^2 and γ are:

$$ \gamma^2 = \frac{1}{1-\beta^2} = 1 + \beta^2 + \beta^4 + \cdots \tag{22} $$

$$ \gamma = \frac{1}{\sqrt{1-\beta^2}} = 1 + \frac{1}{2}\beta^2 + \frac{3}{8}\beta^4 + \cdots \tag{23} $$

where $|\beta| < 1$. Hence, up to the quadratic term, which is a good approximation when $\beta \ll 1$, we have:

$$ \gamma^2 - \gamma = \left(1 + \beta^2 + \beta^4 + \cdots\right) - \left(1 + \frac{1}{2}\beta^2 + \frac{3}{8}\beta^4 + \cdots\right) \simeq \frac{\beta^2}{2} \tag{24} $$

- **Derivation of Eq. 15**:

$$ \gamma\beta = \sqrt{\gamma^2\beta^2} \tag{25} $$

1.4 Mathematical Preliminaries

$$= \sqrt{\frac{\beta^2}{1-\beta^2}}$$

$$= \sqrt{\frac{1-1+\beta^2}{1-\beta^2}}$$

$$= \sqrt{\frac{1}{1-\beta^2} - \frac{1-\beta^2}{1-\beta^2}}$$

$$= \sqrt{\gamma^2 - 1}$$

- **Derivation of Eq. 16**: using the method of integration by parts, we have:

$$\int v\, d(\gamma v) = \int v\, d\left(\frac{v}{\sqrt{1-(v/c)^2}}\right)$$

$$= \frac{v^2}{\sqrt{1-(v/c)^2}} - \int \frac{v\, dv}{\sqrt{1-(v/c)^2}}$$

$$= \frac{v^2}{\sqrt{1-(v/c)^2}} - \int \left(-\frac{c^2}{2}\right) \frac{-2\frac{(v/c)}{c} dv}{\sqrt{1-(v/c)^2}}$$

$$= \frac{v^2}{\sqrt{1-(v/c)^2}} - \int \left(-\frac{c^2}{2}\right) \frac{dw}{\sqrt{w}}$$

$$= \frac{v^2}{\sqrt{1-(v/c)^2}} - \int \left(-\frac{c^2}{2}\right) w^{-1/2} dw$$

$$= \frac{v^2}{\sqrt{1-(v/c)^2}} - \left(-\frac{c^2}{2}\right) 2w^{1/2}$$

$$= \frac{v^2}{\sqrt{1-(v/c)^2}} + c^2 \sqrt{1-(v/c)^2}$$

$$= \frac{v^2}{\sqrt{1-(v/c)^2}} + c^2 \frac{[1-(v/c)^2]}{\sqrt{1-(v/c)^2}}$$

$$= \frac{v^2}{\sqrt{1-(v/c)^2}} + \frac{c^2 - v^2}{\sqrt{1-(v/c)^2}}$$

$$= \frac{c^2}{\sqrt{1-(v/c)^2}}$$

$$= c^2 \gamma$$

1.4 Mathematical Preliminaries

where an arbitrary constant of integration may also be added.

Solved Problems

1. Discuss the cases in which classical mechanics can be regarded as a good approximation to Lorentz mechanics and the cases in which it cannot, and correlate this to the numerical value of the Lorentz factor γ and the speed ratio β.
   Answer: There is no such sharp line separating the two regions because what is good approximation and what is bad approximation depends on the objectives and the required level of accuracy that should be achieved. However, there are two extreme cases where the judgment about the good and bad is obvious. The first case is when β is very close to zero which is equivalent to having γ very close to 1, and hence for all legitimate practical purposes γ can be seen as equal to 1, where classical mechanics can replace Lorentz mechanics as a good approximation. The second of these cases is when β is very close to 1 from below where classical mechanics cannot replace Lorentz mechanics as a valid approximation. Apart from these two extreme cases, the situation should be assessed within its proper context and considerations. Some rules of thumb about this issue may be found in the literature of Lorentz mechanics, e.g. restricting the region of classical mechanics as a legitimate approximation to be below $\beta = 0.1$. However, these alleged rules may apply under specific circumstances or in certain contexts but they cannot be accepted unconditionally, as explained above.

2. Find the first five terms of the power series of γ.
   Answer: These can be obtained from the method of power series expansion or from standard mathematical textbooks:
   $$\frac{1}{\sqrt{1-\beta^2}} = 1 + \frac{1}{2}\beta^2 + \frac{3}{8}\beta^4 + \frac{5}{16}\beta^6 + \frac{35}{128}\beta^8 + \cdots$$

Exercises

1. Define β and γ as used in the literature of Lorentz mechanics.
2. Show that:
$$\gamma^2 = \gamma^2\beta^2 + 1$$
3. Show that:
$$\gamma + 1 = \frac{\gamma^2\beta^2}{\gamma - 1} \qquad (\gamma \neq 1)$$
4. Find the relative error in using the approximation:
$$\gamma \simeq 1 + \frac{1}{2}\beta^2$$
   when $v = 0.1c$ and hence assess the reliability of this approximation.
5. Repeat the previous question with the approximation:
$$\gamma^2 - \gamma \simeq \frac{\beta^2}{2}$$

6. Plot the Lorentz γ factor as a function of the speed ratio β and discuss the distinct features of this plot and the physical significance of these features on the relation between classical and Lorentz mechanics and on the issue of speed restrictions in Lorentz mechanics.

1.5 General Conventions, Notations and Remarks

In the following points we provide some notes about the conventions and notations that are used in this book; many of which are also used in other books of Lorentz mechanics. We also include some general remarks that are needed for future investigation.

• Following the convention of several authors, the Latin indices range over $1, 2, 3$ while the Greek indices range over $0, 1, 2, 3$ or over $1, 2, 3, 4$.[13] The Latin indices usually represent the three spatial coordinates while the Greek indices represent the four spacetime coordinates (one temporal and three spatial). For example, (x^0, x^1, x^2, x^3) is used to represent the spacetime coordinates of an event where the first coordinate (i.e. x^0) represents the temporal coordinate ct while the last three represent the 3D spatial coordinates. Alternatively, (x^1, x^2, x^3, x^4) is used to represent the spacetime coordinates where the last coordinate (i.e. x^4) represents the temporal coordinate ct while the first three represent the 3D spatial coordinates. Both these notations are common in the literature of Lorentz mechanics. The latter convection may be seen as more appropriate for tensor formulation where tensor indices usually start from 1 rather than 0 while the first may be more common in the physically oriented textbooks of Lorentz mechanics. In this book, we generally use the first notation even in tensor formulation (see § 6).

• The term "coordinates" in this book may be used to represent the spatial variables exclusively or with the inclusion of the temporal variable; the meaning should be obvious from the context. Also, we use "temporal coordinate" commonly to refer to ct although it may be used occasionally to refer to t.

• We generally employ the summation convention in the tensor formulations (mainly in § 6) and hence a twice-repeated index like i in $dx_i dx^i$ means summing over the range of i. Hence, if the range of i is $1, 2, 3$ and the range of μ is $1, 2, 3, 4$ then we have:

$$dx_i dx^i \equiv \sum_{i=1}^{3} dx_i dx^i = dx_1 dx^1 + dx_2 dx^2 + dx_3 dx^3 \qquad (26)$$

$$dx_\mu dx^\mu \equiv \sum_{\mu=1}^{4} dx_\mu dx^\mu = dx_1 dx^1 + dx_2 dx^2 + dx_3 dx^3 + dx_4 dx^4 \qquad (27)$$

We note that the summation convention follows the above convention about the range of Latin and Greek indices as seen in the last two examples.

• We note that "light" in the literature of Lorentz mechanics (including this book) includes all types of electromagnetic radiation and is not restricted to the visible part of the electromagnetic spectrum because what is important in Lorentz mechanics is the speed

---
[13] Some authors reverse the convention about the use of Latin and Greek indices.

1.5 General Conventions, Notations and Remarks

of propagation in free space which is the same for all types of electromagnetic radiation regardless of their frequency and wavelength and hence they are all treated equally in this regard although they have different speeds in transparent material media such as air, water and glass.

• We generally use "classical mechanics" to refer to the science of mechanics as represented by the Newtonian formulation prior to the emergence of Lorentz mechanics.[14] However, Lorentz mechanics may also be considered in some physics texts as part of classical mechanics in contrast to quantum mechanics and other related branches and theories of modern physics.

• As we will see, Lorentz mechanics is generally concerned with inertial frames of reference which are in a state of relative motion with respect to each other. To simplify the presentation and to ease the derivation and formulation of Lorentz mechanics, a standard setting between two inertial frames which are in a state of relative uniform motion is commonly used in the literature of Lorentz mechanics. Following this general practice, most of the forthcoming formulations of Lorentz mechanics in this book are based on using this standard setting. It is important to note that the use of this setting does not affect the generality[15] of the formulation or the validity of the derived results in more complex frame settings although some modifications in the employed notation and mathematical techniques (such as using vector and tensor symbolism and methods) may be required. We should also remark that an important beneficial aspect for the use of standard setting is that many vector quantities and formulae will be expressed more simply in a scalar-like form.

According to the standard setting (refer to Figure 3), two inertial frames, O and O', that employ in their spatial identification rectangular Cartesian coordinate systems and they are in a relative uniform translational motion with respect to each other, are in a state of standard setting (or standard configuration) if they satisfy the following conditions:

1. Their coordinate systems (unprimed and primed) are positioned and oriented in such a way that they have a common x-x' axis[16] and parallel y-y' and z-z' axes throughout their relative motion where all the corresponding coordinate axes have the same sense of orientation, i.e. the x-x' axes point in the same direction and similarly for the y-y' and z-z' axes.

2. The relative translational motion between the two frames is presumed to take place only along the common x-x' axis with a constant velocity v (i.e. v can be positive or negative) while they remain in a state of relative rest in the y-y' and z-z' directions.

3. The origins of the two coordinate systems are supposed to coincide at the start of time in both frames (i.e. $x = 0 = x'$ at $t = 0 = t'$)[17] and the y-y' and z-z' axes become

---

[14] In fact, we should exclude the Newtonian gravity theory since in this book we have no interest in the physics of gravity (refer to the solved problems).

[15] The generality may also be guaranteed by the tensor formulation where the validity of the formulation is not affected by any linear transformation between the standard setting and any other setting.

[16] "Common" here means overlapping since the x and x' axes slide against each other during the perpetual relative motion and hence they are not identical in this sense.

[17] In the equation $x = 0 = x'$ and its alike, x represents the x coordinate (as seen in O frame) of the

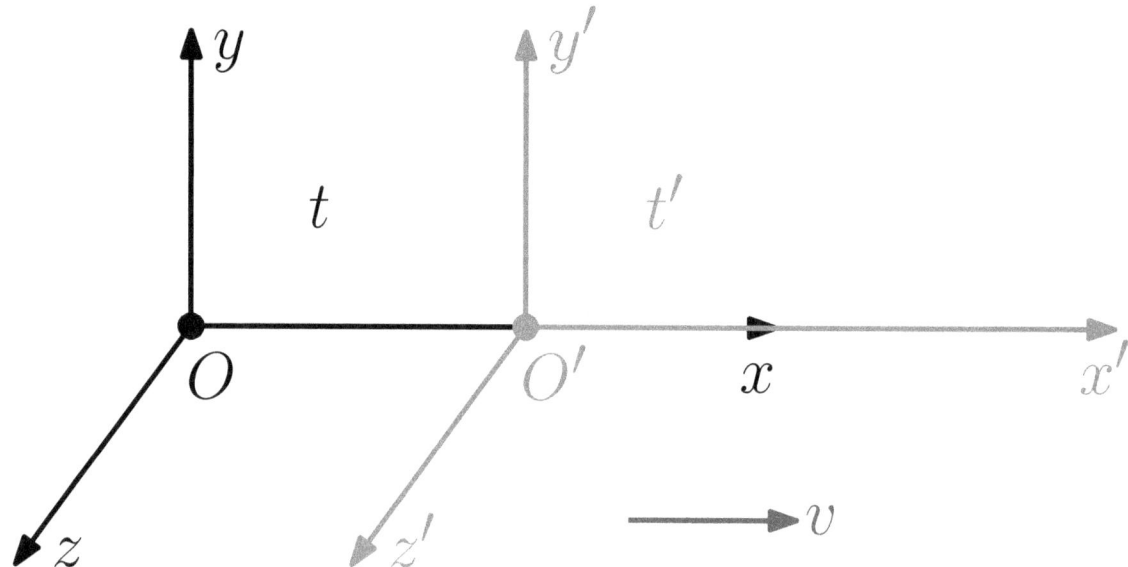

Figure 3: Two inertial frames, O and O', in a state of standard setting where the primed frame is moving to the right relative to the unprimed frame with a constant velocity v along the common x-x' axis.

    identical at this moment. In brief, at time 0 the two frames become identical as they have identical coordinate systems and identical times.
4. The two frames use the same scale for time and the same scale for coordinate axes. Accordingly, if these frames are at rest with respect to each other they share the same unit of time and length and hence they have identical measurements of time intervals and distances.
5. It should be obvious that since the frames are assumed to be inertial, then there is no relative rotational motion of any kind (uniform or non-uniform) between the frames because any rotational motion means that at least one of the frames is not inertial (refer to § 3.6.2).[18]

Based on the above description, the motion of two frames which are in a state of standard setting is essentially one-dimensional since it takes place only along the x-x' dimension, even though it is occurring in a 3D space. Because of this, the y-y' and z-z' coordinates of any event will be the same in both frames at all times and hence they can be ignored or manipulated easily.

    As indicated above, the use of this standard setting will result in major simplifications

---

    origin of O' while x' represents the x' coordinate (as seen in O' frame) of the origin of O. In fact, only one of these conditions is needed. We should similarly have $y = 0 = y'$ and $z = 0 = z'$ but this applies at all times and not only at $t = 0 = t'$.

[18] In fact, we should say "there is no rotational motion" whether relative or not where this in our view is an implicit reference to the absolute frame and hence it is a confession of the existence of such a frame by any one who accepts this statement. In other words, even if there is no relative rotational motion between the two frames, the frames could still be non-inertial when they have a rotational motion relative to the absolute frame.

1.5 General Conventions, Notations and Remarks

to the formulation of Lorentz mechanics where many vector quantities can be treated like scalars since they are essentially one-dimensional. Nevertheless, this setting will not affect the generality of any formulation since the motion is actually occurring in a 3D space and the choice of the coordinate systems is rather arbitrary so the coordinate system of any frame can be translated, rotated, reflected and scaled (if some or all of these transformations are needed at all) to be in the configuration of this standard setting. Similarly, the temporal coordinate can also be translated and scaled to satisfy the conditions of standard setting. As a consequence, in most occasions we will not need to boldface the symbols or use similar sophisticated notational techniques to refer to vector quantities as the formulation will be based only on one component which corresponds to the x dimension.

• The state of standard setting, as described in the last point, between two frames of reference may be extended to include more than two frames where the added frames in their relation to each one of the other frames are subject to the same conditions that the two original frames are assumed to be subject to, i.e. uniform translational motion restricted to the x dimension with the origin of time being unified when all the coordinate systems coincide (that is $x = x' = x'' = \cdots = 0$ at $t = t' = t'' = \cdots = 0$) etc. However, in this book we do not need such an extension to more than two frames.

• The subject of Lorentz mechanics is the physical laws as observed from inertial frames.[19] Hence, it may be said frequently in this book and other books that Lorentz mechanics is about inertial frames and inertial observers. However, we would like to clarify an important issue that is: being about inertial frames means that Lorenz mechanics is about the physical laws as observed from inertial frames but it does not mean that the observed objects and phenomena in these laws should be "inertial". So, for example we can observe accelerating objects from an inertial frame and we are still within the applicability domain of Lorentz mechanics.

• Based on the previous points, the standard setting between two inertial frames is usually extended to a "standard setting" between an inertial frame and the observed phenomena. Accordingly, when we (as inertial observers in the domain of Lorentz mechanics) observe a physical phenomenon we commonly set our inertial frame in a state of "standard setting" with the observed phenomenon by choosing (or positioning and orienting) our frame in such a way that the principal vector quantities in the observed phenomenon become one-dimensional and hence they can be represented by a single component. For example, we may observe from our inertial frame a massive object that is under the influence of an external constant force and hence it is accelerating. So, to simplify the situation we orient our inertial frame in such a way that the force and acceleration will be in the x direction only and hence the main vector quantities (i.e. force and acceleration) will be along the x axis of our frame. Therefore, they can be fully represented by a single vector component and hence they are treated like scalar quantities. Again, this choice of "standard setting" has the benefits and advantages of ordinary standard setting between inertial frames by reducing vector quantities to scalar-like quantities and hence treating

---

[19] In fact, the essence of Lorentz mechanics is the transformation of space and time between different inertial frames and hence it is a theory about space and time and how they enter in the fabric of the physical laws according to the view of inertial observers.

1.5 General Conventions, Notations and Remarks

them like scalars. Moreover, this "standard setting" usually does not affect the generality of the physical arguments or the derived mathematical formulae.[20]

- A point related to the standard setting is that: for brevity we sometimes speak about x coordinate or x direction or x dimension and we mean the common x-x' coordinate or direction or dimension, i.e. x is not used as opposite to x'. This also applies to y and z coordinates or directions or dimensions where they are used generically not as opposite to their corresponding primed symbols.

- As indicated earlier, we frequently use the label "Lorentz mechanics" to mean the mechanics that is based on the Lorentz transformations (or what we call "the mechanics of Lorentz transformations" which is the title of this book). This branch of physics is almost universally known as special relativity theory. This change of label is inline with our approach of structuring and presenting Lorentz mechanics in its pure formalism independent of any particular philosophical or epistemological interpretation or unverified or unverifiable assumptions or postulates. Accordingly, we commonly use "Lorentz mechanics" to replace the commonly used "special relativity" and "relativistic" attributes. For example, we talk about spacetime in Lorentz mechanics instead of spacetime in special relativity and talk about Lorentz mechanics energy or Lorentz mechanics momentum instead of relativistic energy or relativistic momentum. Similarly, we may use "Lorentzian" description as a substitute for "relativistic" for the same purpose. However, we still use the "relativistic" label in some places where the situation belongs to the special relativity theory.

- We usually use zero as a subscript to label proper quantities, e.g. L is improper length or length in general and L_0 is proper length specifically. However, a zero subscript may also be used occasionally for other purposes, e.g. as part of the commonly used symbols that represent the permittivity and permeability of free space, i.e. ε_0 and μ_0. The use of zero as a subscript to label proper quantities should also be distinguished from its use as a superscript index (and even subscript index sometimes) to mark the temporal coordinate and variables of spacetime. In all cases, the meaning should be obvious from the context if not explicit explanation is given.

- For convenience and to familiarize the reader with different conventions and notations, all of which are in common use in the literature of Lorentz mechanics, we use different equivalent notations. The following are some examples:

1. We use both x, y, z and x^1, x^2, x^2 to represent the spatial coordinates which are normally orthonormal Cartesian.
2. We use both t and ct to represent the temporal variable (or temporal coordinate) although the first is usually used in the context of classical mechanics where space and time are separate entities and they are not merged in a single spacetime manifold while the second is usually used in Lorentz mechanics particularly in its tensor formulation.
3. We use both x^0, x^1, x^2, x^3 and x^1, x^2, x^3, x^4 to represent the spacetime coordinates where

---

[20] We note that the physical essence of Lorentz mechanics (which is the objective of this book) can be properly and entirely presented by the use of this simplified scalar-like approach. The use of more sophisticated vector notation and techniques at this introductory level is no more than mathematical elaboration that complicates things and may obscure and damage the main objectives.

1.5 General Conventions, Notations and Remarks 31

the temporal variable is represented by x^0 in the first notation and by x^4 in the second notation. These two notations will be distinguished by the use of x^0 or x^4 for the temporal coordinate since the other coordinates (i.e. x^1, x^2, x^3) are spatial in both notations. We also note that both x^0 and x^4 in these notations stand for ct.

We note that using different notations for the same concepts and formalism is not only for the purpose of convenience or to familiarize the reader with different notations which are in common use in the literature, but it may also be for certain advantages in each notation depending on the context where some notations may be more advantageous to use than others in a particular case. In some cases, this multiple use of notation can be necessary because certain mathematical techniques require certain notation and symbolism. This is particularly true in tensor formulation where some types of notation (e.g. x^1, x^2, x^3 instead of x, y, z) are needed for the application of tensor techniques.

- "Space" may be used to label the ordinary "spatial" space, which is usually a 3D Euclidean space. It may also be used to mean "manifold" in the context of talking about spacetime manifold which includes both temporal and spatial dimensions. The meaning should be obvious from the context in the absence of explicit clarification.

- As indicated earlier, "coordinate system" is used to mean an abstract device to locate points in space, while "frame" is used to mean an abstract device to locate points in spacetime. This is based on the definition of reference frame as a coordinate system with a time measuring mechanism. However, in the literature of Lorentz mechanics these terms may be used interchangeably by some authors.

- Since in classical mechanics space and time are regarded as two separate independent entities while in Lorentz mechanics space and time are supposed to be entangled in a "spacetime" manifold, we use expressions like "space coordinates and time" in the context of classical mechanics while we use expressions like "spacetime coordinates" in the context of Lorentz mechanics.

- The characteristic speed of light in vacuum is based on the definition of meter and hence it is given exactly as $c = 299792458$ m/s.[21] However, in the numerical calculations in the solution of solved problems and exercises we generally use an approximate value of $c \simeq 3 \times 10^8$ m/s.

- We may describe the constant c as restricted and ultimate speed (according to the special relativistic view) to mean that nothing other than light, in its extended sense that includes all types of electromagnetic radiation, can reach this speed and hence c is restricted to light, moreover nothing, including light, can exceed this speed and hence c is the ultimate value for any physical speed.[22]

- In the literature of Lorentz mechanics c may stand for the above constant and may stand for the observed speed of light in vacuum. Although these are generally considered to be the same, due to the general acceptance of the universality of the speed of light according to special relativity, they are not the same in principle (or conceptually) since

---

[21] In fact, this is a definition of the meter more than a measurement of the characteristic speed of light.
[22] "Restricted" may also mean that the speed of light must be c and hence it cannot be less than or greater than c. So, "restricted and ultimate speed" could mean that the speed of all massive objects must be less than c while the speed of light (and all massless objects) must be c.

1.5 General Conventions, Notations and Remarks

the observed speed of light can be frame dependent if we do not adopt the postulates of special relativity. To avoid confusion, with keeping the freedom of considering all the possibilities about the speed of light without committing ourselves to any particular interpretation, we label c as the *characteristic* speed of light (refer to § 1.13 and § 1.14) and hence any variation in the observed speed of light due to potential frame dependency (assuming such a dependency is physically viable) will not be absorbed within c but will be represented by another symbol like v. Hence, if the *observed* speed of light is really frame dependent, then this speed will be expressed as $c \pm v$ where v is the frame-dependent part of the observed speed while c is the *characteristic* speed which is equal to the above given constant.

• "Frame" and "observer" may be used interchangeably in this book. In fact, this is based on what we indicated earlier that is the observer in this context means a global observer who is present everywhere in his frame all the time and hence he is not localized into a particular position in the frame or a particular period of time, and therefore he is identical to the frame in this sense. However, we should note that we may also use "observer" to refer to a spatially- and temporally-localized observer in a given frame, as discussed in the context of the difference between the simultaneity of occurrence and the simultaneity of observation (see § 7.4 and § 9.6.1). Also, when we use an expression like "observer in a frame" or "observer of a frame" it should mean that the frame is the rest frame of that observer.

• We may use the adjective "experimental" (e.g. experimental data or evidence) to mean something obtained by direct observation to Nature whether the observed physical occurrence takes place with or without human intervention and arrangement, and hence it generally includes observational as well as experimental phenomena. Accordingly, astronomical observations are experimental data or evidence in this sense although they are not obtained from a physical occurrence that is set and planned by the observer. Similarly, "experiment" may be used to include both experiment and observation. We may also use "observational" or "observation" to refer to both these types of observation.

• The dimensional physical quantities, such as mass and velocity, are generally assumed to be in standard SI units (Système Internationale d'Unités) unless it is stated otherwise. Hence, in the solved problems and exercises, as well as in their solutions, these quantities are commonly given with no units for the purpose of simplicity.

• Two reference frames in their relation to each other can be in a state of relative rest, or in a state of relative uniform motion, or in a state of relative acceleration. Since the first two states (i.e. rest and uniform motion) are mechanically equivalent[23] (i.e. the same laws of mechanics apply to two frames which are at rest or in uniform motion relative to each other), we may combine these two states as "uniform motion" to be concise where the state of rest is considered as a special case of uniform motion corresponding to vanishing

---

[23] To be more precise, we should say "inertially equivalent" (i.e. having the same inertiality status) because the equality between mechanical equivalence and inertial equivalence in the general case should require extra effort. However, since our main focus in Lorentz mechanics is inertial frames and their distinction from non-inertial frames, then the above should be sufficient (and even advantageous since it is easier to explain and present).

1.5 General Conventions, Notations and Remarks 33

velocity. Although it is also possible to consider the state of rest as a special case of acceleration corresponding to vanishing acceleration, there is no advantage in doing so since the states of rest and acceleration (i.e. non-vanishing acceleration) do not share the same mechanical status by having the same physical laws. Moreover, vanishing acceleration does not necessarily imply being at rest.

• We generally use "special relativists" to mean the followers of special relativity and the believers in the correctness of this theory.

• We think the following simple rule is very useful and hence it should be memorized and remembered because it will reduce confusion and speed up understanding, that is: the Lorentz γ factor is always greater than or equal to 1 (i.e. $\gamma \geq 1$). Hence, multiplication by γ magnifies the quantity while dividing by γ shrinks the quantity. So, when we see $L = \gamma L'$ we should immediately realize that $L \geq L'$ and when we see $\Delta t = \Delta t'/\gamma$ we should immediately realize that $\Delta t \leq \Delta t'$.

• There is always an implicit assumption that a physical quantity of an object, like length, is the same as seen in any rest frame. So, when we have two inertial observers in relative motion and we have a stick which is at rest in the frame of one of these observers, then although the two observers disagree on the length of the stick while they are moving (since its length will be measured in its rest frame longer than its length as measured in the moving frame according to length contraction effect), the two observers should agree on its length when they gather, i.e. when they stop and hence they are not in a state of relative motion any more. In fact, this assumption is based on the principles of physical reality and truth, which are investigated in § 1.6, as well as on using identical measuring equipment and physical units in both frames.

• As indicated earlier, because of the employment of the previously described standard setting, in most formulations of Lorentz mechanics we use non-bold symbols to represent vector quantities like v for velocity and p for momentum. The justification of this simpler notation, which is also used by many authors, is that the motion is actually one-dimensional and hence the magnitude and sign are sufficient to provide the full information. However, for simplicity these same symbols may also be used to represent the magnitude of these 1D vectors and hence they are positive scalars. For example, v can be used as a vector (representing magnitude and direction which is indicated by its sign) and hence it is labeled as velocity, and can be used as a scalar (representing magnitude only which is positive) and hence it is labeled as speed. Similarly, p can represent momentum or magnitude of momentum. The meaning should be obvious from the context if it is not stated explicitly.

• As stated earlier, we generally assume that the mechanics of Lorentz transformations is well established and supported by experimental and observational evidence as claimed in the literature of Lorentz mechanics, and hence in principle we rule out fundamental errors and prejudice which may put question marks on certain issues and conclusions. This does not mean that we are in a position to endorse these claims or we necessarily believe in the claimed evidence but it means that the reported contents in this book are based on the presumption of the validity of these claims in general. In this regard, we should declare our reservation about a number of claimed evidence and experimental support to Lorentz

1.5 General Conventions, Notations and Remarks

mechanics. This is because of the strong tendency in the scientific community to support special relativity and reject any claimed violations, and hence there is always a considerable possibility of error in the experimental procedure or analysis and interpretation, as well as intentional or unintentional bias.

- In many places in this book, we use the term "absolute frame" to mean the rest frame of absolute space or ether whichever is more appropriate in the particular context. If we admit the generally-accepted understanding that ether in classical physics is at rest in the frame of absolute space, then the two are equivalent because they both share the same frame which is the absolute frame. It is obvious that the concept of absolute frame is incompatible with special relativity because such a frame does not exist in this theory and hence any assumption of an absolute frame means the rejection of special relativity or at least some of its postulates and foundations. We should also note that the concept of absolute frame implies an absolute time which is the time of the frame of the absolute space.

- The statements and details in this book that are directly related to the technical aspects of Lorentz mechanics are of two main types: those related to the formalism and those related to the interpretation. We note that in our presentation of the formalism (i.e. first type), we are not embracing any particular interpretation although the language or the tone may suggest a particular interpretation. In brief, to ease the presentation and to make the content more comprehensible to the reader, we will use any convenient language and method of presentation without adopting any particular interpretation although the used language and method may suggest a particular interpretation.[24] Accordingly, anything related to the interpretation should be obtained only from those parts of the book that are designated and dedicated to the interpretation, e.g. § 9 and § 11, or from the statements that are explicit in this regard.

Solved Problems

1. Which phenomenon of light is caused by the dependency of the speed of light in material media, like glass and water, on frequency?
   Answer: It is dispersion. The decomposition of white light into its colored components by light prism is based on this phenomenon.

2. Elaborate more on the meaning of classical mechanics in this book as contrasted with Lorentz mechanics.
   Answer: We may describe classical mechanics in this context as "Newtonian mechanics" which incorporates Newton's three laws of motion (see § 1.10) and their derived consequences as well as their philosophical and epistemological framework such as basic assumptions about the nature of space and time (see § 2.1) and how they are transformed (see § 1.9).[25] However, we exclude other parts of Newtonian mechanics, such as the law of gravity, which have no counterpart in the concepts and formalism of Lorentz

---

[24] In fact, in some parts a special relativistic tone may be felt due to its common use in the literature and our desire to avoid departure from this common approach (at least in the presentation).

[25] In fact, the rules of transformation of space and time (as represented by the Galilean in classical mechanics and by the Lorentzian in Lorentz mechanics) represent the main distinctive feature when we contrast the two mechanics. Although these transformations are physical in nature, they have deep

mechanics.

Exercises
1. Why "light" in the literature of Lorentz mechanics includes all types of electromagnetic radiation?
2. Discuss the difference between the *characteristic* and *observed* speed of light in free space.
3. Expand the following expressions using the adopted conventions in this book:

$$u_i u^i \qquad cA^\mu b_\mu \qquad \mathbf{r} \qquad \mathbf{x}$$

4. Explain the standard setting between two inertial frames illustrating your descriptive explanation by a simple sketch.
5. What are the main advantages of using the standard setting?
6. Justify the common practice in the literature of Lorentz mechanics of using scalar symbols to represent vector quantities like velocity and momentum.
7. Why symbols like v may be described as speed sometimes and as velocity in others?
8. In a state of standard setting between two inertial frames: (a) How many numbers associate any given event to fully identify its presence in spacetime? (b) How many of these numbers are independent? (c) How are these numbers related? (d) What is the physical quantity that distinguishes these frames?
9. Classify the states of two reference frames, whether inertial or non-inertial, in their relation to each other from the perspective of their relative motion. Which of these states represent the same inertiality status of the two frames, i.e. both frames are necessarily inertial or necessarily non-inertial?
10. Give an example of the tendency in the scientific circles to support certain theories and individuals which may cast a shadow on the validity of the claimed evidence in support of Lorentz mechanics especially in its relativistic version.

1.6 Physical Reality and Truth

This issue is one of the very important foundations of physics and its philosophy and at the heart of any scientific theory. We may start this investigation by asking the following question: is there a single truth representing a single physical reality which the observer will be correct, by having the truth, if he obtained an exact reflection or exact image of this reality in his mind? Or the truth, and perhaps even the physical reality itself if it does exist at all, depends on the observer and hence we could have as many truths, or even physical realities, as the number of observers or at least we could have several different truths all of which are correct reflection of the physical reality or realities?

Obviously this is not a scientific issue or a question that can be inspected, verified or falsified by experimental or observational means. However, this issue, which belongs to the philosophy of science, has particular importance in a number of branches of modern physics notably Lorentz mechanics and quantum mechanics where the physical facts are

---

philosophical and epistemological roots.

1.6 Physical Reality and Truth 36

strongly mixed with philosophical and epistemological issues because these branches of physics state these facts in terms of frames of reference and as seen by observers. This is obvious in Lorentz mechanics where the notions of frame of reference and observer enter in the conceptualization and formulation of the physics and scientific facts and hence they can enter in the definition and determination of spatial coordinates, time, length, and mass among other physical properties and quantities. This is also the case in quantum mechanics where the presence of the observer is supposed to have a direct impact on the outcome of the experiment and hence this presence may determine the physical facts or even the reality.

In fact, the common concepts about physical reality and truth are based on the following fundamental assumptions or principles:

1. The existence of a real world beyond and outside the observer where the reality of this world is independent of the observer. This may be labeled as the principle of existence of physical reality.
2. This reality is unique and hence we have only one physical reality. Accordingly, we cannot have two conflicting realities, e.g. a stick whose real length is 1 m and 0.5 m at the same time. This may be labeled as the principle of uniqueness of physical reality.
3. The truth is unique and hence we have only one truth which represents the honest reflection of the unique physical reality. Accordingly, we cannot have two conflicting truths, e.g. a truth that the real length of a given stick is 1 m and another truth that the real length of the same stick under the same circumstances is 0.5 m. This may be labeled as the principle of uniqueness of truth.

In this regard, we should distinguish between the position of classical physics about the issues of physical reality and truth and the position of modern physics. While classical physics has very clear position about these issues where it embraces a very realistic stand, and hence it bases all its formulations and interpretations on the principles of the existence and uniqueness of physical reality and the uniqueness of truth, the position of some branches of modern physics is not clear about these issues, or at least some of these issues, and hence some branches of modern physics, notably Lorentz mechanics and quantum mechanics, may adopt unconsciously, or even consciously, a non-realistic position where the reality may be seen as non-unique, if it does exist at all, or the truth may depend on the observer and hence we could have multiple truths. Generally, such an attitude or view is not declared explicitly or expressed formally within the formal scientific structure of these branches, but it usually arises as a possibility amid the confusion in the interpretation and the lack of rigor in the definition of fundamental concepts that form part of the formalism. However, although this non-realistic approach is generally contained within the interpretation, the confusion and lack of clarity in the definition of these concepts usually propagate to include even the formalism itself and hence some of these non-realistic interpretations are seen by many as part of the physics itself and its formal and mathematical structure.

Let have some simplifying examples which may not reflect the exact nature of these issues but will certainly help in clarifying the situation as a first step. When we look at a piece of "white" paper directly we see it white but when we look at it through an optical filter that passes only the yellow part of the spectrum we see it yellow. So, does the "color"

1.6 Physical Reality and Truth

property, or an equivalent property that underlies the color property, exist and if so is it unique and hence the paper is in reality either white or yellow. Consequently, are both visions correct and true or one, at least, should be false and misleading and hence we should make corrections to obtain the truth. For example, we may claim that the paper is white in the absence of filter and yellow in its presence. We may also claim that the yellow is not the true color because of the presence of the optical filter and hence when we see the paper yellow we should take account of the absorbed non-yellow parts of the spectrum and hence conclude that the paper is white. We can add other examples of this type such as the apparent fracture of a partly immersed stick in a pool of water. In fact, despite their similar nature the latter example seems less controversial than the former[26] and hence the common belief is that the fracture is just an illusion (instead of being straight outside water and fractured inside water) and therefore the stick is not fractured during its submergence in water.[27]

Now, let have another example of different nature, this time from our subject which is Lorenz mechanics, and analyze it more fundamentally from this perspective. This example is length contraction.[28] When a given stick is measured as 1 m long in its rest frame and as 0.5 m long from a moving frame, what is the real length of this stick? In answering this question we may take one of the following stands (among other possible stands):

1. It could be a matter of definition and convention and hence its real length is 1 m in its rest frame and its real length is 0.5 m in the moving frame with no conflict between these realities because they represent different circumstances due to the difference in observers. However, this is obviously in conflict with our intuition of a unique physical reality which is independent in its existence from the observer. In fact, the conflict with intuition as a reason for rejection is not based on a puritanical philosophical realism but it is based on real pragmatic reasons because the assumption of a unique reality will make the world simpler and hence make our conceptual and practical adaptation to this world much easier. In fact, this is the reason why we should assume that some of our observations are wrong and hence we apply corrections to these observations, i.e. a unique world is more consistent and easier environment to understand and live in than a non-unique world which will be very confusing. Otherwise, why we should assume that the apparently broken immersed stick is not really broken, and hence we apply correction to our vision, if all our observations should be accepted on the background that they represent multiple realities. We obviously discriminate between our observations to have a more consistent vision of the world through this assumption of a unique physical reality. So, the purpose of these corrections is to make our observations more homogeneous and consistent. This equally applies to our scientific observations, as to our daily life observations, because science is no more than an organized collection

---

[26] This may be related to human psychology which can easily propagate to science and philosophy.
[27] In fact, the given explanations to these examples are intentionally very simple and hence more elaborate explanations can be proposed. However, this will not introduce fundamental changes onto the philosophical and epistemological issues that are involved in these examples and are meant to be highlighted by them.
[28] The example and analysis can be repeated on time dilation with minor changes.

1.6 Physical Reality and Truth

of concepts and rules that are based on systematically collected observations and hence science is no more than an improved version of our daily observations and rules. In fact, the fundamental purpose of science is not different from the fundamental purpose of our ideas and practices in daily life; this purpose is to achieve an optimal adaptation to the physical world in the most extensive sense of adaptation.

2. We may improve our stand by saying although the reality is unique the truth is not unique and hence the truth is that the stick is 1 m long in the rest frame and 0.5 m in the moving frame. Regarding the supposedly unique reality which is presumably reflected in our multiple truths, it is irrelevant because all we can reach is our observations and conceptions about this reality. However, this view can be challenged by the same reasons as the previous view because the uniqueness of reality has no value if the truth that represents this reality is not unique as well. This non-uniqueness of truth does not belong to two different observers only but it also belongs to each individual observer because each one of these observers knows that the length in his frame is not the same as the length in the other frame and hence he needs a rational explanation to make sense of this difference and obtain a unique and consistent truth.[29] We also note on these two views (i.e. this and the previous) that they may be ambiguous, like the next view, about the meaning of reality and truth because what it means, in terms of the observed physical consequences and effects, to have multiple realities or truths?

3. We may adopt a more ambiguous view by saying: the reality and truth are unique and hence the stick in reality is 1 m long but the measured length of 0.5 m in the moving frame is apparent. However, this view is rather ambiguous because the meaning of "apparent" in this view is not clear. Does apparent mean that the 0.5 m observation is wrong like the broken stick and hence we should apply a correction to have the correct length and therefore for all realistic physical purposes the stick is actually 1 m long even in the moving frame. As we will see later (refer for example to § 10.1.1 and § 11), this interpretation of apparent is not consistent with other facts about length contraction which are based on treating this effect as a real effect with real physical consequences which are observable in the moving frame. Alternatively, we may interpret apparent as something that is real with real physical consequences but it is not the original reality or truth which is represented by the measurement of the rest frame of 1 m. In fact, we can take this as a convention by associating the reality with the measurement of the rest frame or what we call proper value. But this interpretation of apparent will devoid this answer from any original substance and reduce the issue to be a matter of labeling and verbal distinction and convention because apparent will be just a label for a secondary reality or truth and hence we return to the view of non-uniqueness of reality or truth where this reality or truth is represented, conventionally, by a primary form of 1 m long and a secondary form of 0.5 m long. In brief, although we can accept any label like "apparent" or "secondary truth" we need a clear and useful definition to this label in terms of the observed physical consequences and if these physical consequences, which may be different in the two frames, have real and tangible effects in both frames.

---

[29] In fact, we can even consider a third observer (e.g. the reader) who needs to make sense of this difference.

1.6 Physical Reality and Truth

4. We may also adopt a rather strange view that is the stick is in reality 0.5 m long although its apparent length is 1 m. As we will see later, although this view seems very strange and illogical, it may be the more realistic and consistent view if we adopt a certain framework for our physical theory and certain assumptions about the state of rest and motion of the stick. In brief, let **assume** that we have an absolute frame of reference (say the frame of absolute space or ether) where length contraction will be suffered by any object that is in relative motion with respect to this frame. Also, let **agree**, as a matter of convention, that the reality is the state of the object as observed from this absolute frame, so if a stick, whether at rest or moving, is measured as 1 m long by an observer who is at rest in the absolute frame then its real length is 1 m, while if this same stick, whether at rest or moving, is measured in another occasion as 0.5 m long by an observer who is at rest in the absolute frame then its real length is 0.5 m. Now, if we accept this **assumption** and **agreement** then the explanation can be given as follows. If we assume further that the stick is in a state of relative motion with respect to the absolute frame and it is measured from the absolute frame as 0.5 m and measured in its rest frame as 1 m then its real length should be 0.5 m, according to the convention, and its apparent length should be 1 m because in its rest frame even the measuring rod that is used to measure its length will suffer the same length contraction effect that the measured stick has suffered due to its relative motion with respect to the absolute frame. Although this case (i.e. a stick in relative motion with respect to the absolute frame and an observer at rest in this absolute frame) is not sufficiently general since it is only a special case, it gives a clue about the course of generalization. In summary, when both the stick and the observer are in relative motion with respect to the absolute frame and they are also in relative motion with respect to each other, we calibrate our observations to the observation of a standard observer who is at rest with respect to the absolute frame. Although this interpretation seems more consistent and logical so far, its physical consequences still require evidence. Moreover, this may be seen as another form of multiple reality or multiple truth. However, if we accept the absolute frame as the most fundamental physical reality then the above convention will have deeper roots in our intuition and evolutionary path. After all, all our principles and concepts about the physical reality and truth are no more than products of our intellect, so they are not less original or real than any scientific theory or realistic form of knowledge. More details about this view, which will be proposed as a potential interpretation to Lorentz mechanics, will be given in § 11.

Finally, we should remark that for any scientific theory to be qualified as such, it should comply to some extent with the basic principles of physical reality and truth; otherwise it will be either a piece of nonsense or another type of intellectual activities of mankind such as philosophy or literature or religion. Another remark is that quantum mechanics, or at least some of its interpretations, may not be fully compliant with the principles of physical reality and truth. However, because quantum mechanics is not the subject of the present book, we do not discuss this issue any further. In contrast, Lorentz mechanics, unlike quantum mechanics, is supposed to be based, according to the common understanding and general consensus, on total objectivity of the physical reality, determinism and preciseness as well as other embedded principles of physical reality and truth and hence it is supposed to be fully compliant with the principles of reality and truth. However, we will see later that this may not be the case at least with some interpretations and views where the possibility of having multiple realities and truths seems real.

Solved Problems

1. Can we add more principles of reality and truth?
   Answer: Yes, we can. For example, we can add another principle of reality and truth whose essence is the total independence of the existence of reality from the observer and hence the reality does exist in a completely determined form regardless of the observer and the process of observation. However, this rule is embedded here in one of the other given rules not as an independent rule. This rule may have particular importance in the interpretation of quantum mechanics but not in Lorentz mechanics and hence we do not need to give it more attention or emphasis.
2. Based on the above discussion in the text, make a clear distinction in the meaning of "apparent".
   Answer: We can distinguish two main meanings of "apparent":
   • As opposite to real and hence it should be corrected for like the apparent fracture of the immersed stick.
   • As opposite to absolute and hence the effect is real as seen from the absolute frame like the speed of light which is apparently the same in all inertial frames although this invariance is apparent in this sense due to the spacetime contraction which is caused by the motion relative to the absolute frame. More clarifications about these issues will be given later.

Exercises

1. What are the three principles of reality and truth which are embedded in realistic philosophies and sciences?
2. Briefly discuss realism in the old and modern physics.
3. Discuss briefly the general stand of quantum mechanics and Lorentz mechanics about physical reality and truth.
4. Discuss, giving some examples, the issue of instinctive corrections that we apply unconsciously in our daily life and the relation of this to the issues of reality, truth, adaptation and biological evolution.
5. Briefly examine the claim that modern science is not subject to the same rules as daily life and the necessity of adaptation, and hence modern science may not need to be based on the principles of reality and truth.

1.7 Intrinsic and Extrinsic Properties

An intrinsic property of a physical object is a property that belongs to the object in itself regardless of any observer (or rather particular observer), while an extrinsic property is a property that belongs to the object but with respect to an outside observer and hence it is observer-dependent. An example of intrinsic property is the mass of a material object according to classical mechanics where mass is an inherent property, while an example of an extrinsic property is its kinetic energy which depends on its speed and hence on the observer and his frame of reference. Accordingly, an intrinsic property is unique and hence if the mass of an object (according to classical mechanics) is 1 kg it can only be 1 kg for any observer, while an extrinsic property is not unique since it depends on the observer and hence the kinetic energy of an object can be 0 J for one observer, 10 J for another observer and 100 J for a third observer.

In fact, intrinsic and extrinsic properties can be labeled as absolute and relative properties but we want to avoid any association of these properties with the relativity principle and the relativity theory. We should also note that the issue of being intrinsic or extrinsic has a strong link to the issues of physical reality and truth as discussed in § 1.6. We should remark that any property must be either intrinsic or extrinsic. Hence, a property cannot be intrinsic and extrinsic at the same time because it is either observer-dependent or not. Similarly, a property cannot be neither intrinsic nor extrinsic for the same reason.

Exercises
1. What is the difference between intrinsic and extrinsic properties of a physical object? Give examples of each.
2. Is the acceleration of non-inertial frames an intrinsic or an extrinsic property according to classical mechanics?
3. Are the extrinsic properties examples of multiple reality or multiple truth?

1.8 Invariance of Physical Laws

In gross terms, the principle of invariance of physical laws means that the laws of Nature should be independent of observers, coordinate systems and reference frames. In more technical terms, invariance of a physical law means that the law will take the same form when observed by different observers, or alternatively the law will take the same form under certain coordinate transformations.[30] For example, if we should have a physical law that states "force equals mass times acceleration", it should be so (if it is really a physical law) not only for a particular observer but for different observers where these observers are linked to each other through a certain set of transformations. In this regard we should draw the attention to the following important aspects:
1. The essence of this invariance principle is the invariance of the form of the law and not necessarily the invariance of the physical properties that enter in the definition and formulation of the law and hence two different observers may measure different momenta but they should have the same form of Newton's second law (i.e. "force equals rate of change of momentum") if this should be a law of physics for these observers.[31] Yes,

---

[30] These "coordinate transformations" should be interpreted in a broad sense to include temporal variables.

[31] They should also agree on the fundamental principles of physics like the conservation of energy and momentum as these principles can be put in a given qualitative or quantitative form (e.g. $\Delta E = 0$ and $\Delta \mathbf{p} = \mathbf{0}$) associated with a given set of conditions such as being a closed system or not under the influence of an external force.

1.8 Invariance of Physical Laws

some physical properties, like mass in classical mechanics, may also have this invariance attribute but this generally is independent of the invariance of the law itself (refer to § 1.3).

2. At the root of the above definition is the applicability of certain transformations under which the form of the physical law remains the same when these transformations are applied. For example, in classical mechanics where the Galilean transformations of space coordinates and time apply to inertial observers, when a physical law is determined to have a certain form by an inertial observer O, the Galilean transformations between this observer and another inertial observer O' should produce the same form for the observer O'. Similarly, in Lorentz mechanics a physical law should take the same form in two inertial frames when the Lorentz transformations of spacetime coordinates are employed to transform the law from one frame to the other. The reader is referred to the upcoming parts of the book (e.g. § 1.9.2 and § 12.3) for specific examples.

3. A physical rule may be invariant under certain type of transformations but not under another type of transformations and hence it could be seen as a "conditional law", i.e. it is a law with respect to the first type of transformations but not with respect to the second type. An obvious example is Maxwell's equations of electromagnetism which are invariant under the Lorentz transformations but not under the Galilean transformations.

4. Similarly, some laws of physics may be laws for certain categories of observers and frames of reference but not for other categories. An obvious example is Newton's laws of motion which hold only in inertial frames and hence they are invariant physical laws only in these frames since any transformation (whether Galilean or Lorentzian) from one inertial frame to another inertial frame will not affect their validity as both observers in these inertial frames will see the same form of these laws, e.g. both will verify the law of inertia. Accordingly, Newton's laws are physical laws for inertial observers but they are not physical laws for accelerating observers.

In fact, the principle of invariance of physical laws is a fundamental pillar of science, old and modern. This principle is based on an issue related to the philosophy of science that is if science is to be objective, predictive and of common practical value it should represent observations that do not depend on the individual observers and their frames of reference, and hence all observers (or all observers of a certain category such as inertial) should have a unified view of the world by having a common formal description of the physical phenomena although these views may differ in some details. In brief, if any physical principle or rule is to be qualified as a law, it should be sufficiently general and apply to a large group with some common features and hence it should be invariant for at least a certain type of observers or categories. Physics, and science in general, have no value if everybody has his own version of the laws which cannot be shared with others since these "laws" are not sufficiently general to be laws of common value. In fact, these different versions and individual views cannot even be labeled as laws due to the lack of generality because of their dependence on the individual observers. Although the principle of invariance may sound like a strict rule it is not; in fact there are many details and controversial issues about this principle, its interpretation, restrictions and limitations. Further details and examples about the invariance of physical laws are given in the upcoming parts of the book (see for example § 1.9.2, § 5.4 and § 12.3).

Solved Problems
1. What is the relation between the principle of invariance of physical laws and tensor calculus?
   Answer: Tensors are mathematical objects that are invariant under certain transformations. Hence, tensor calculus is very useful tool in formulating the laws of physics in invariant forms.
2. Contemplate on possible link between the principle of invariance of physical laws and the principles of reality and truth which we discussed earlier in § 1.6.
   Answer: The reader should not fail to notice the strong link between the principle of invariance of physical laws and the principles of reality and truth, which we discussed in § 1.6, where according to the former the observers in different frames should obtain a unified vision of the physical world to be consistent and uniform while according to the latter each individual observer should obtain a unique and self-consistent vision. This means that consistency of vision is not only useful and required by any individual observer for himself but even between different observers to have a common view through which they can interact and communicate. This should serve the adaptation objective since a consistent and common vision for a given group of observers is as important as a self-consistent vision for each individual observer. The principle of invariance of physical laws may also be seen as an example of the uniqueness of reality and truth which are shared by different observers.

Exercises
1. State briefly the principle of invariance of physical laws giving some examples.
2. Analyze the essence and roots of the principle of invariance of physical laws and discuss its significance.
3. State briefly the main assumptions and conditions for the application of the principle of invariance of physical laws.
4. Make a clear distinction between being value-invariant and being form-invariant.
5. Make a clear distinction between being value-invariant and being constant.
6. Assume that a physical quantity is invariant across all inertial frames and it is conserved (i.e. constant) in a particular inertial frame. What should you conclude?

1.9 Galilean Transformations

The Galilean transformations are a set of mathematical relations that transform space coordinates and time of a given inertial frame of reference O_1 to the space coordinates and time of another inertial frame of reference O_2 which is in a state of rest or uniform translational motion relative to O_1. These are the main transformations; other subsidiary or derived transformations, such as velocity and acceleration transformations, which are based on the main transformations of space coordinates and time follow by applying the common rules and definitions of physics, as we will see.

In the following subsections we outline the main Galilean transformations and some of their derived transformations where we assume that these transformations take place between two inertial frames, $O(x, y, z, t)$ and $O'(x', y', z', t')$, with the unprimed and primed variables being representing the space coordinates and time of the unprimed and primed frames, while the two frames are in a state of standard setting as described in § 1.5. Hence, the coordinate axes in the two frames of reference are assumed to be aligned correspondingly (i.e. x is parallel to x', y to y' and z to z') with the x' axis being coincident with the x axis and O' being observed by O to move with a constant velocity v along the orientation of the common x-x' axis.[32] Moreover, the two origins of coordinates in the two frames are assumed to coincide at $t = t' = 0$, i.e. $x(t = 0) = x'(t' = 0) = 0$ and hence the two frames become identical at this instant of time.

As we will see (refer to § 12.3), the Galilean transformations are the appropriate set of coordinate transformations for classical mechanics and hence Newton's laws of motion take the same form under these transformations, i.e. they satisfy the principle of invariance.[33] However, the Galilean transformations are not valid for the electromagnetic wave equation and Maxwell's equations since these equations do not transform invariantly under these coordinate transformations. In brief, while the laws of classical mechanics are invariant under the Galilean transformations and hence they are valid laws of physics, electromagnetic equations (i.e. wave and Maxwell's equations) are not and hence they are not valid laws of physics under the Galilean transformations.

Exercises

1. In standard setting where v represents the 1D velocity of O' frame relative to O frame along the common x-x' axis, what is the significance of v being positive, negative or zero?
2. Give an example of other conditions that are usually assumed implicitly in the state of standard setting.

1.9.1 Space and Time Transformations

For two inertial observers O and O' who are in a state of standard setting according to the above description, the Galilean transformations of space coordinates and time from O frame to O' frame are given by:

$$x' = x - vt \tag{28}$$
$$y' = y \tag{29}$$
$$z' = z \tag{30}$$
$$t' = t \tag{31}$$

---

[32] Hence v is positive if O' moves in the positive x-direction relative to O, and negative if it moves in the negative x-direction relative to O. The condition $v = 0$ represents the trivial case of the two systems being identical or they transform by a static translation along the common x-x' axis. So, based on what have been explained before (see § 1.5), v represents velocity, i.e. speed and direction, although it looks like a scalar.

[33] As we will see, all the laws of classical mechanics which are based on Newton's laws of motion are also invariant under these transformations because they are just variant forms of Newton's laws.

1.9.1 Space and Time Transformations

These transformations represent how the measurements of the spatial coordinates and time of an event are related in the two reference frames of O and O'. We can similarly obtain the opposite transformations from O' frame to O frame. This can be done by simple algebraic manipulation to the above transformations from O to O'.[34] So, for the Galilean transformation of x-x' coordinates which corresponds to Eq. 28 we have:

$$x' = x - vt \tag{32}$$
$$x' + vt = x \tag{33}$$
$$x = x' + vt' \tag{34}$$

where Eq. 31 is used in the last step. This can also be done more simply by exchanging the primed and unprimed symbols in the above equations with the reversal of the sign of relative velocity (i.e. changing v to $-v$) because if O' is moving to the right/left with speed $|v|$ relative to O, then O is moving to the left/right with the same speed $|v|$ relative to O'. Accordingly, we have:

$$x = x' + vt' \tag{35}$$
$$y = y' \tag{36}$$
$$z = z' \tag{37}$$
$$t = t' \tag{38}$$

Solved Problems
1. O and O' are two inertial frames in a state of standard setting with $v = 12$. At $t = 5$, O observes a car at position $\mathbf{r} = (3, -9.4, -0.3)$. What are the space coordinates and time of the car in frame O' at that time according to the Galilean transformations?
Answer: Because the two frames are in a state of standard setting, the two frames agree on everything except the x coordinate, that is:

$$(x', y', z', t') = (x - vt, y, z, t) = (3 - 12 \times 5, -9.4, -0.3, 5) = (-57, -9.4, -0.3, 5)$$

Exercises
1. How are the Galilean transformations of the primed variables obtained from the Galilean transformations of the unprimed variables and vice versa? Justify your answer.
2. Write down the Galilean transformations for space coordinates and time from O' frame to O frame where these frames are in a state of standard setting.
3. Two frames of reference, O and O', are in a state of standard setting where O' is moving along the common x-x' axis with velocity $v = 9$ relative to O (i.e. O' is moving in the positive x direction with speed 9). A ball which is at rest in frame O' is seen to be at position $(2.5, -3, 7.8)$ in this frame at time $t' = 0$. (a) Find the space coordinates and time of this ball in frame O at $t' = 10$ according to classical mechanics. (b) Repeat the question assuming that $v = -9$ (i.e. O' is moving in the negative x direction with speed 9).

---
[34] In fact, this manipulation is needed only for the x transformation.

4. An inertial frame of reference O' is seen in another inertial frame of reference O to have its origin at $\mathbf{r} = (3.9, -12.3, 6.1)$ at $t = 0$, where at $t = 0$ we have $t' = 10$ with t and t' being those of O and O' respectively. Put these frames in a state of standard setting assuming that the corresponding coordinate axes in these frames are parallel and have the same orientation. Also assume that the other conditions for standard setting (e.g. uniform motion along the x-x' axis only) are satisfied.
5. Two inertial frames, O and O', are in a state of standard setting where at $t = 3$ the origin of O' is seen by O to be at $x = 20$ and at $t' = 15$ the origin of O is seen by O' to be at $x' = -44$. Find the Galilean transformations of space coordinates and time between these frames.
6. A vehicle is traveling along a straight line with a constant velocity $v = 8$. An on-board siren sends a signal every 10 seconds. What is the distance traveled on the ground by the vehicle during that time interval as measured by an on-board observer and by an on-ground observer according to classical mechanics?
7. O' and O are two inertial observers in a state of standard setting with $v = 45$. According to O, the distance between two events that occur at $t = 10$ is $d = 450$. What is the distance according to O' in classical mechanics?

1.9.2 Velocity Transformations

For the above two inertial observers, O and O', the Galilean transformations of the velocity components of a physical object in the x, y and z directions from O frame to O' frame are given by:

$$u'_x = u_x - v \tag{39}$$
$$u'_y = u_y \tag{40}$$
$$u'_z = u_z \tag{41}$$

where (u_x, u_y, u_z) and (u'_x, u'_y, u'_z) are the velocity components of the object as observed by O and O' respectively. These transformations represent how the measurements of the velocity components of an object are related in O and O' frames. We note that the subscripts for the primed symbols should also be primed, but for simplicity we put the prime on the symbol only with an understanding that the prime belongs to the whole symbol. In fact, it is more appropriate to use symbols like $(u_x)'$ but this will introduce unnecessary complication in the notation.

The above velocity transformations can be obtained from the corresponding coordinate transformations (i.e. Eqs. 28-30) by taking the first time derivative of these coordinate transformations noting that v is constant and $t' = t$. For example, Eq. 39 can be obtained from Eq. 28 as follows:

$$\frac{dx'}{dt'} = \frac{d}{dt'}(x - vt) \tag{42}$$
$$\frac{dx'}{dt'} = \frac{d}{dt}(x - vt) \tag{43}$$

1.9.2 Velocity Transformations

$$\frac{dx'}{dt'} = \frac{dx}{dt} - \frac{d(vt)}{dt} \tag{44}$$

$$\frac{dx'}{dt'} = \frac{dx}{dt} - v \tag{45}$$

$$u'_x = u_x - v \tag{46}$$

where Eq. 31 is used in the second step.

Again, we can obtain the opposite transformations from O' frame to O frame by simple algebraic manipulation of Eqs. 39-41 or by exchanging the primed and unprimed symbols in these equations with reversing the sign of relative velocity, that is:

$$u_x = u'_x + v \tag{47}$$

$$u_y = u'_y \tag{48}$$

$$u_z = u'_z \tag{49}$$

It should be obvious that v is not regarded as an unprimed symbol since its "primed" and "unprimed" status is indicated by its sign.

We note that since the mass is an intrinsic property in classical mechanics and hence it is frame independent, then the momentum and kinetic energy transformations will be subject to similar rules to those of velocity transformations because momentum ($p = mv$) and kinetic energy ($E_k = \frac{1}{2}mv^2$) are dependent on velocity and mass. Accordingly, the momentum and energy problems can be easily solved with the help of the velocity transformations plus other principles of classical physics like the conservation of momentum and energy, as we will see in the exercises. We should also remark that because the subject of this book is Lorentz mechanics, we discuss the laws and rules of classical mechanics, including those related to the Galilean transformations, rather briefly and as much as needed for the discussion of Lorentz mechanics. The reader is referred to standard textbooks on classical mechanics for any missing details about this subject.

Solved Problems

1. A particle is seen in frame O to have a position given by $\mathbf{r}(t) = (6t, e^t, t^3)$. Find the Galilean velocity of the particle as a function of time in frame O' which is in a state of standard setting with frame O where O' is moving with respect to O in the x-x' direction with velocity $v = 10$. Also, find the velocity of the particle in frame O' at $t' = 5$.

    Answer: The velocity of the particle in frame O is given by:

    $$\mathbf{u} = \frac{d\mathbf{r}}{dt} = \frac{d}{dt}(6t, e^t, t^3) = (6, e^t, 3t^2)$$

    Hence, its velocity in frame O' according to the Galilean transformations is given by:

    $$\mathbf{u}' = \mathbf{u} - \mathbf{v} = (6, e^t, 3t^2) - (10, 0, 0) = (-4, e^t, 3t^2) = (-4, e^{t'}, 3t'^2)$$

    At $t' = 5$ the particle velocity in frame O' is:

    $$\mathbf{u}'(t' = 5) = (-4, e^5, 3 \times 5^2) \simeq (-4, 148.41, 75)$$

Exercises

1. How are the Galilean transformations of velocity obtained from the Galilean transformations of space coordinates and time?
2. Using the Galilean velocity transformations, show that if the momentum is conserved in an inertial frame then it is conserved in all inertial frames where you use a simple case of collision between two massive objects as a prototype in your demonstration. Repeat the question with the kinetic energy assuming this time a perfectly elastic collision.
3. A head-on collision between two massive objects, A and B, is observed by two inertial observers O and O' who are in a state of standard setting with relative velocity $v = 1$ where the mass of the objects are $M_A = 2$ and $M_B = 3$. If according to O the initial velocities of A and B are $u_{A1} = 9$ and $u_{B1} = -2$ and the final velocity of A is $u_{A2} = 4$, what are the initial and final velocities of A and B according to O' assuming that all motions are taking place in one dimension?
4. A body A of mass $m_A = 5$ is seen by an inertial observer O to move with velocity $u_{A1} = 10$ along the x axis. Following an inelastic collision with an identical massive body B which is at rest in O frame, the two bodies coalesce and continue to move along the x axis. (a) What is the total momentum of this two-body system before and after collision according to O? (b) What is the velocity of the coalesced body after collision according to O? (c) What is the total momentum of the two-body system before and after collision according to another inertial observer O' who is in a state of standard setting with O where the relative velocity between O and O' is $v = 5$?
5. Repeat the previous exercise with the kinetic energy (instead of momentum) assuming this time the collision is perfectly elastic.
6. Using the Galilean velocity transformations, show that the momentum transforms between inertial frames by a constant additive difference.

1.9.3 Velocity Composition

According to the Galilean transformations of velocity, as stated in § 1.9.2, the composition of velocities is additive, that is: if u_{x21} is the velocity of O_2 relative to O_1 and u_{x32} is the velocity of O_3 relative to O_2 then the velocity u_{x31} of O_3 relative to O_1 is given by:

$$u_{x31} = u_{x32} + u_{x21} \tag{50}$$

where all these velocities are along the same orientation which can be considered the x orientation as indicated by the subscript x. In fact, this equation is no more than another form of Eq. 47 if we note that: $u_{x31} \equiv u_x$, $u_{x32} \equiv u'_x$ and $u_{x21} \equiv v$. For example, if O_1 represents a person standing on the street, O_2 represents a moving bus with velocity v relative to the street which is the rest frame of O_1, and O_3 is a person walking inside the bus along the x orientation which is the orientation of the bus movement, then u_{x31} is the velocity of the walking person as seen by the standing person, u_{x32} is the velocity of the walking person as seen from the rest frame of the bus, and u_{x21} is the velocity of the bus as seen by the standing person.

1.9.4 Acceleration and Force Transformations

The above formula can be easily remembered if we notice that the middle "2" on the right hand side is connecting the two terms and hence it drops out from the left hand side. We note that all these 1D velocities represent signed quantities since they can be in the positive or negative x direction and this should be considered when solving numerical questions. The order of the subscripts is also important in determining the sign and hence we have $u_{xba} = -u_{xab}$ (e.g. $u_{x21} = -u_{x12}$). We also note that Eq. 50 indicates that the Galilean velocity composition follows the rule of vector addition of the involved velocities.

Solved Problems

1. A commuter on a bus is walking backward with speed 1 as the bus moves forward with velocity 5. What is the velocity of the commuter as measured by a standing observer?
   Answer: If we label the standing observer, the bus and the commuter with "1", "2" and "3" and consider the bus moving in the positive x direction, then according to the velocity composition formula of classical mechanics we have:

   $$u_{x31} = u_{x32} + u_{x21} = -1 + 5 = 4$$

   i.e. the velocity of the commuter as measured by the standing observer is 4.

Exercises

1. A predator is chasing a prey along a straight path where the velocity of the predator and prey are 13.5 and 12. What is the velocity of each relative to the other?
2. A radioactive nucleus ejects two beta particles in opposite directions where the speed of each one of these particles in the rest frame of the nucleus is $0.6c$. What is the speed of each one of these particles in the frame of the other particle according to classical mechanics?
3. Two inertial frames of reference, O and O', are in a state of standard setting with $v = 25$. An object is seen in O' to have a velocity component in the negative x direction of magnitude $u = 6.8$. What is the velocity of this object in O according to classical mechanics?
4. What are the y and z versions of the velocity composition formula according to classical mechanics?

1.9.4 Acceleration and Force Transformations

For the above described inertial observers who are in a state of standard setting with a uniform relative motion, the measured accelerations of a given object in the two frames are identical. Considering the above velocity transformations, and noting that acceleration is the time derivative of velocity, v is constant and $t' = t$, we have:

$$a'_x = a_x \tag{51}$$
$$a'_y = a_y \tag{52}$$
$$a'_z = a_z \tag{53}$$

where (a_x, a_y, a_z) and (a'_x, a'_y, a'_z) are the acceleration components of the object as observed by O and O' respectively. These transformations represent how the measurements of the

1.9.4 Acceleration and Force Transformations

acceleration components of a physical object are related in O and O' frames. As we see, the acceleration is the same in both frames, that is all inertial observers will agree in their acceleration measurements of any given object. Considering the definition of force in inertial frames and the constancy of mass and its invariance across all inertial frames, this is no more than the invariance of Newton's second law in all inertial frames.

Following the previous statements, because in classical mechanics the mass is an invariant constant and hence the force is proportional to the acceleration, the Galilean transformations of force are similar to the Galilean transformations of acceleration, that is:

$$f'_x = f_x \tag{54}$$
$$f'_y = f_y \tag{55}$$
$$f'_z = f_z \tag{56}$$

where (f_x, f_y, f_z) and (f'_x, f'_y, f'_z) are the force components as measured by O and O' respectively. The last equations can be simply obtained by multiplying the transformation equations of acceleration by the mass m, that is:

$$f'_x = ma'_x = ma_x = f_x \tag{57}$$
$$f'_y = ma'_y = ma_y = f_y \tag{58}$$
$$f'_z = ma'_z = ma_z = f_z \tag{59}$$

Solved Problems

1. Using the definition of acceleration with the Galilean transformations of time and velocity, obtain the Galilean transformations of acceleration.
   Answer: We have:
   $$a'_x = \frac{du'_x}{dt'} = \frac{du'_x}{dt} = \frac{d}{dt}(u_x - v) = \frac{du_x}{dt} - 0 = a_x$$
   $$a'_y = \frac{du'_y}{dt'} = \frac{du_y}{dt} = a_y$$
   $$a'_z = \frac{du'_z}{dt'} = \frac{du_z}{dt} = a_z$$

2. How is the mass transformed between inertial frames according to classical mechanics?
   Answer: In classical mechanics, the mass of a material object is an intrinsic property and hence it belongs to the object in itself regardless of any frame of reference or observer and hence the mass is the same for all frames and observers. Accordingly, the mass transforms between two inertial observers, O and O', as:
   $$m = m'$$
   and hence the same symbol m is usually used to represent the mass in any frame.

Exercises

1.10 Newton's Laws of Motion

1. What is the meaning and significance of the Galilean transformations of acceleration?
2. How are the Galilean transformations of acceleration obtained from the Galilean transformations of space coordinates and time?
3. How are the Galilean transformations of force obtained?
4. An object of mass $m = 0.5$ is seen in an inertial frame O' at $t' = 1.3$ to be at rest in position $\mathbf{r'} = (3.5, 4.6, 1.3)$. At that time a force $f'_x = 6$ in the x direction is applied and the object is accelerated for 3 time units. Using classical mechanics, find the position, velocity, force and acceleration of the object at time $t = 10$ in another inertial frame O which is in a state of standard setting with frame O' where O' moves with velocity $v = 5$ relative to O. Also, find the applied force and the acceleration of the object in frame O during the application of force.
5. A girl on a train throws a stone upwards with initial speed $u_0 = 3$. If the train is moving with velocity $v = 8$ along a straight railway, what is the position of the stone, as a function of time during the stone flight, as seen by the girl and as seen by a bystander on the platform who is opposite to the girl at the instant of throwing the stone?
6. Hooke's law for an ideal mass-spring system is given by:

$$ma_x = -k(x - x_0)$$

where m is mass, a_x is acceleration, k is spring constant, x is mass position and x_0 is mass equilibrium position (for more details about Hooke's law, the reader is referred to general physics textbooks). Using the Galilean transformations, show that this law takes the same form in all inertial frames.
7. Show that Newton's second law in its form: $f = ma$ is form invariant under the Galilean transformations.
8. Show that the more general form of Newton's second law: $f = \frac{dp}{dt}$ is also form invariant under the Galilean transformations.

1.10 Newton's Laws of Motion

To have a better understanding of the physical view of the world prior to the emergence of Lorentz mechanics, which will be discussed in the next chapter, we should investigate Newton's three laws of motion which are the core of classical mechanics. As stated already, these laws apply only in inertial frames.[35] These laws reflect the philosophical view, as well as the scientific and mathematical formalism, of the world at the time of Newton.[36]

---

[35] Or we should rather say: these laws characterize these frames since the procedural definition of inertial frames is based on the applicability of these laws. However, a more fundamental definition, which is based on the existence of absolute frame, can also be made, i.e. inertial frames are those frames which are at rest or in uniform translational motion with respect to the absolute frame. In fact, characterizing inertial frames by Newton's laws followed by concluding that these laws apply in inertial frames is circular. Therefore, we should have a more fundamental definition of inertial frames which is based on their state of motion relative to an absolute frame.

[36] Newtonian mechanics is also based on another pillar which is Newton's law of gravity. However, this is out of the scope of the present investigation to Lorentz mechanics.

1.10 Newton's Laws of Motion

They are the result of the persistent effort of many philosophers and scientists over many centuries and the experience of mankind since the dawn of civilization. Although these laws are attributed to Newton, they represent a collective effort whose credit should be attributed to many scholars. In fact, some of these laws predate Newton, e.g. the law of inertia which was known to Galileo and even before Galileo. Moreover, large parts of Newton's legacy, as embedded mostly in his Principia, were common knowledge at that time.[37] In reality, Newton's laws of mechanics in their relation to Newton are like Maxwell's equations of electromagnetism in their relation to Maxwell (see 12.1).

We outline Newton's three laws of motion in the following points:

1. The law of inertia which states that in the absence of a net external force, a massive object will continue in its state of rest or uniform translational motion in space, that is:

$$\sum \mathbf{f} = \mathbf{0} \quad \Leftrightarrow \quad \Delta \mathbf{u} = \mathbf{0} \tag{60}$$

   where $\sum \mathbf{f}$ is the net external force, $\mathbf{u}$ is the velocity of the object and Δ stands for change.

2. The second law which is about the relation between the force exerted on a massive object and the momentum of the object. This law states that the force is proportional to the time derivative of momentum, that is:

$$\mathbf{f} \propto \frac{d\mathbf{p}}{dt} \tag{61}$$

   where $\mathbf{f}$ is force, $\mathbf{p}$ is momentum and t is time. With a proper choice of units, the above expression becomes an equality, i.e. $\mathbf{f} = \frac{d\mathbf{p}}{dt}$. Assuming the constancy of mass, the last equation takes the following more common form:

$$\mathbf{f} = m\mathbf{a} \tag{62}$$

   where m is the mass of the object and $\mathbf{a}$ is its acceleration.

3. The law of action-reaction which states that when an object A exerts a force on another object B then B will exert a force on A which is equal in magnitude and opposite in direction, that is:

$$\mathbf{f}_{BA} = -\mathbf{f}_{AB} \tag{63}$$

   where $\mathbf{f}_{BA}$ is the force exerted by B on A and $\mathbf{f}_{AB}$ is the force exerted by A on B.

As indicated above, these three laws which form the core of classical mechanics are based on the dominant philosophical view at that time about space and time where space and time are seen as two independent, absolute and passive entities in which the physical objects exist and the events take place (refer to § 2.1 for more details). As we saw and will see (refer to § 1.9 and § 2.2), Newton's laws of motion are consistent with the Galilean relativity principle and hence the laws of classical mechanics are invariant under

---

[37] In this context we should mention the contribution of Robert Hooke, for example, among others to the law of gravity.

the Galilean transformations.[38] In other words, Newton's laws of mechanics equally apply in all inertial frames and hence they are form-invariant across these frames under the Galilean transformations (refer for example to § 1.9.4 and § 12.3). We will also see that these laws are still valid in Lorentz mechanics with some modifications in the definition of certain quantities and concepts (refer for example to § 4.3.1 and § 5.3.1).

Solved Problems

1. Discuss the issue of form-invariance with respect to the subsidiary laws of motion in classical mechanics (i.e. laws other than Newton's three laws which are the main laws of motion in classical mechanics).
   Answer: Because all the subsidiary laws of motion in classical mechanics (e.g. Hooke's law) are based on Newton's three laws of motion directly or indirectly, then all the laws of classical mechanics equally apply in all inertial frames and hence they are form invariant under the Galilean transformations. This means that by proving that Newton's three laws are form invariant (refer to § 12.3), we actually prove without extra effort that all the laws of motion in classical mechanics are also form invariant because they are no more than applications of Newton's three laws and hence they are disguised forms of Newton's laws.

2. What is the interpretation of Newton's first law?
   Answer: It means that in the absence of a net external force, an object will continue in its state of rest or uniform translational motion where this state can be understood to be ultimately with respect to the absolute frame.

3. Discuss the causes for the mass of a physical object to be variable and not constant in classical mechanics and in Lorentz mechanics.
   Answer: Regarding classical mechanics, the cause for the mass to be variable is an exchange with the environment where the object loses mass to its surrounding or gains mass from its surrounding. An example is a rocket that continuously loses mass to its surrounding due to the ejection of gases and debris. Another example is a rolling snow ball that grows in size by gaining mass from its surrounding or a drop of rain that gains mass from gathering dust in its path (it may also lose mass due to evaporation for example).
   Regarding Lorentz mechanics, there are three causes for the variability of mass:
   • An exchange with the environment, as in classical mechanics.
   • A conversion between mass and energy according to the mass-energy equivalence relation (refer to § 4.3.2, § 5.3.2 and § 8.2).
   • Change of speed if we follow the old convention in Lorentz mechanics which considers mass a function of its speed according to the formula $m = \gamma m_0$ where m is mass and m_0 is the rest or proper mass.
   These issues will be clarified further in the future.

---

[38] In fact, the Galilean transformations (which are physical in nature as they make an essential part of classical physics) are based on the same philosophical and epistemological classical view which Newton's laws are based upon. Hence, the compatibility between the Galilean transformations and Newton's laws is not accidental.

Exercises

1. Name Newton's three laws of motion.
2. State Newton's first law descriptively and mathematically in a different form to the form given in the text.
3. State Newton's second law descriptively and mathematically.
4. According to classical mechanics, what is the condition for writing Newton's second law in the form $\mathbf{f} = m\mathbf{a}$, i.e. force equals mass times acceleration?
5. Discuss the issue of value-invariance of mass in classical mechanics and Lorentz mechanics.
6. Discuss the issue of the constancy of mass in classical mechanics and Lorentz mechanics with respect to speed.
7. What is the relation between Newton's first law and Newton's second law?
8. Is there any situation, according to classical mechanics, where the above form of Newton's second law (i.e. $\mathbf{f} = m\mathbf{a}$) does not apply?
9. An object is seen to move along a straight line with a time dependent velocity given by $u = 3e^{-0.2t}$ while losing mass according to the relation $m = 6 - 0.01t$ where t is time. Find the force acting on the object at time $t = 5.3$ according to classical mechanics.
10. State Newton's third law in simple words giving an example. What is the significance of this law?
11. Discuss the following quote: "The weakness of the principle of inertia lies in this, that it involves an argument in a circle: a mass moves without acceleration if it is sufficiently far from other bodies; we know that it is sufficiently far from other bodies only by the fact that it moves without acceleration".
12. Show by a simple qualitative non-rigorous classical argument that if Newton's laws are valid in a given frame of reference, then they should be valid in all frames of reference which are in a state of rest or uniform translational motion with respect to the given frame.
13. Show by a simple qualitative non-rigorous classical argument that Newton's three laws of motion are form invariant under the Galilean transformations.
14. As stated in the text, Newton's laws of motion are supposed to be valid in Lorentz mechanics with some modifications in the definition of certain quantities and concepts. Now, someone may ask: what is the meaning of Newton's first law in the absence of absolute frame if we accept the view of special relativity that denies the existence of this frame?

1.11 Thought Experiments

Although anyone can design and analyze thought experiments to draw legitimate conclusions from their logical and scientific implications, these thought experiments can be dangerous and hence they should be done carefully like real experiments. The reason is that the designer of these experiments can be easily dragged into his subconscious illusions and stereotypes and hence make fatal mistakes. This danger will be gravely amplified by the possible existence of a subconscious presumption that these thought experiments are real experiments, or at least they are very like real experiments, and hence they have the power and authority of experimental evidence if not more. This fact is vividly seen in the literature of Lorentz mechanics where many thought experiments are designed and used to establish and prove scientific and logical arguments as if they are real world experiments. In the forthcoming chapters of this book we will meet some examples of these designed thought experiments and analyze their potential traps. In fact, many of these dubious thought experiments are raised by some scientists, let alone the general public of science, to the rank of unquestionable science.

The essential difference between real world experiments and thought experiments is that real world experiments are governed by the rules of physical world while thought experiments are governed by the rules of our thoughts, stereotypes and illusions. So, real world experiments cannot "make mistakes" and hence they are "correct" in principle, although it is still possible to make mistakes in their preparation and analysis by the experimenter, while thought experiments can be wrong in principle because the whole thinking process as well as the presumed premises can be wrong. In brief, the sources of error in thought experiments are more fundamental and numerous than the sources of error in real world experiments and that is why they are more dangerous in this sense. So in brief, without real experimental evidence to support the claimed results of thought experiments, thought experiments should be classified as science fiction. Hence, the main function of thought experiments should be restricted to demonstrating and illustrating well established scientific facts rather than establishing and proving unsupported claims.

Exercises
1. Discuss the main sources of error and traps in thought experiments.
2. Give an example of thought experiments that are widely used in special relativity.
3. Assess and criticize the use of thought experiments (or "thought methods" to be more general) in modern physics.

1.12 Requirements for Scientific Theories and Facts

For any theory to qualify to be a scientific theory, it should satisfy the following main conditions:
1. It should be completely consistent with the basic logic and the fundamental epistemological structure of mankind, i.e. it is sensible. Hence, propositions like "existence does not exist" or "apple is very intelligent" cannot represent scientific theories although they may have other values in philosophy or art or other forms of human intellectual activities.
2. It is fundamentally compliant with the principles of reality and truth (refer to § 1.6) so that it can claim to represent an honest reflection of the unique reality.
3. It is about the natural world, i.e. it records a phenomenon that occurs in the physical world and hence it is observable and inspectable by experimental methods and techniques. Therefore, a proposition like "angels live in heaven" cannot be a scientific theory

1.13 Speed 56

because it is not about the physical world and hence it is not verifiable by experimental means although it possesses values and verifiable contents (and even verified contents) in religion.

Having met the above conditions, a scientific theory will be raised in value to become a scientific fact if it is sufficiently verified by direct or indirect observational or experimental evidence to the limit that the probability of being false will become virtually zero. We note that other conditions may also be added to the requirements of scientific theory, such as being able to provide precise predictions and hence the scientific theory should have quantitative aspects, but these conditions are irrelevant to our investigation to Lorentz mechanics.

Exercises

1. Give examples of scientific theories and scientific facts.
2. In what "quantitative" sense scientific fact is a "fact"?

1.13 Speed

In this section, we briefly investigate the concept of speed and its quantification considering different forms of propagating physical phenomena and relative motion. The speed can be attributed to a projectile, like a particle, or to a wave, like sound. Moreover, the propagation can occur either in a medium or not. So, we have four main possibilities which will be investigated in the following subsections according to classical physics although we will also discuss some issues from the perspective of modern physics.

Before that, let, for a pedagogical purpose, differentiate between two types of speed: "characteristic speed" that belongs to the propagating phenomenon itself and hence in a sense it can be regarded as an intrinsic property to this phenomenon, and "observed speed" which belongs to the observer and hence it can be regarded as an extrinsic property. Now, if the propagating phenomenon does not require a medium for its propagation then the characteristic speed may be defined as the speed of the propagating phenomenon in the rest frame of the source. On the other hand, if the propagating phenomenon does require a medium for its propagation then the characteristic speed can be defined as the speed of the propagating phenomenon in the rest frame of the medium where in this definition we assume that the characteristic speed depends on the properties of the medium but not on the source of propagating phenomenon such as its speed relative to the medium.

1.13.1 Speed of Projectile

An obvious example for the speed of projectile is the speed of a massive object such as a particle of sand. Normally, the propagation (or preferably the motion) of a projectile does not require a medium and hence the observed speed of a projectile should be dependent on the relative motion between its source (or emitter) and its observer. Accordingly, if the characteristic speed of a projectile is s_c then its observed speed s_o should be:[39]

$$s_o = s_c \pm v \tag{64}$$

where v is the speed of the observer in the rest frame of the source and the $\pm$ sign depends on whether the observer is approaching or receding with respect to the source. In fact, if we treat s_o, s_c and v as vectors we can write the last formula less ambiguously as:

$$s_o = s_c - v \tag{65}$$

We should remark that although the propagation of a projectile does not require a medium, a medium may exist in the space of propagation, e.g. a particle moving in air or water. However, this is not a propagation medium since the propagation of the projectile does not need the medium. Yes, this medium may introduce a resistance force that can affect the speed of propagation but this is not within the scope of our investigation and hence we do not examine this issue. In fact, because our primary interest in Lorentz mechanics is the propagation of light in free space, the issue of the existence of a resisting medium is irrelevant even if the ether hypothesis is accepted and the light is considered corpuscular in nature because there is no suggestion in the literature of the possibility that ether can be a resisting medium to the propagation of light.

1.13.2 Speed of Wave

In classical mechanics, propagation of waves requires a medium. Now, since we assumed that the characteristic speed in this case belongs to the medium, and hence it is independent of the state of the source, then as soon as the wave is emitted it will be independent of its source and hence it will propagate in the medium with its characteristic speed s_c. So, if we have an observer who is in a state of motion with respect to the medium, then according to the additive nature of speed composition in classical mechanics, the observed speed s_o of the wave should be given by:[40]

$$s_o = s_c \pm v \tag{66}$$

where v is the speed of the observer in the rest frame of the medium and the $\pm$ sign depends on whether the observer is approaching or receding with respect to the source. Again, if we treat s_o, s_c and v as vectors we can write the last formula compactly as:

$$s_o = s_c - v \tag{67}$$

As we see, according to the rules of classical mechanics, with the presumption of generally accepted assumptions, the propagation formula for both projectile and wave takes the same form. The difference is in the meaning of v and whether it is relative to the source or relative to the medium.

---

[39] We are considering here a 1D motion along the line connecting the source to the observer.
[40] Again, we are considering here a 1D motion along the line connecting the source to the observer.

1.13.2 Speed of Wave

We note that while in classical mechanics the propagation of waves requires a medium, in modern physics the propagation of waves does not necessarily require a medium. More explicitly, while some types of wave (say material waves like sound) require a medium, other types of wave (say non-material waves like light) do not require a medium. This was introduced following the rejection of the ether hypothesis and the dominance of the special relativity view. Before that, even light waves were supposed to propagate in the ether medium whose existence was hypothesized as a necessary requirement for the propagation of electromagnetic waves (see § 2.5).

Now, if we accept the possibility of a wave propagation without need for a medium then we can ask: in the absence of a medium, what is the characteristic speed of the wave and to which entity this speed belongs? As a matter of principles and logic, there are several possibilities. For example, the characteristic speed can belong to the space if we accept the existence of absolute frame. It may also belong to the source (like projectile speed) and hence the characteristic speed of a non-material wave will be relative to the rest frame of its source. There are also other possibilities including the impossibility or non-sensibility of defining a characteristic speed as such and hence the speed (whether characteristic or observed) could be universal. However, not all these possibilities seem to be considered in the analysis of the speed of light and this, again, is due to the dominance of the special relativity view where many of these possibilities are regarded not only false but they are so absurd that they do not deserve examination, inspection or even attention. Anyway, some of these issues will be considered further in the upcoming parts of the book in the context of the speed of light (refer for example to § 1.14).

Exercises

1. Outline the main possibilities about the characteristic speed of projectile and the characteristic speed of wave.
2. Outline the main possibilities about the observed speed of projectile and the observed speed of wave and their dependence on the relative motion of the observer, source and medium.
3. A buzzer is drifting in a waterway along a straight line where the water is running uniformly with velocity $\mathbf{v} = (8, 0, 0)$. There are two observers: A who is at position $\mathbf{r}_A = (-5, 0, 0)$ and B who is at position $\mathbf{r}_B = (10, 0, 0)$. At time $t = 0$ the buzzer, which is in the rest frame of the water, passes through the origin of coordinates. What is the observed speed of sound for A and B if the characteristic speed of sound in water is 1500?
4. Make an argument in support of the claim that the speed of a classical (or material) wave does not depend on the speed of its source.
5. What can we conclude from the fact that the speed of sound inside a plane that moves at high speed is the same as the speed of sound in a room at rest.

1.14 Speed of Light

First, we should distinguish between the characteristic speed of light which is the constant c and the observed speed of light which is the actually measured value by a given observer and hence in principle it can be frame dependent and not necessarily invariant constant. There are several possibilities about the speed of light, but not all these possibilities are considered in the mainstream literature of Lorentz mechanics although some of these neglected possibilities have been considered by some scholars especially in the early days of Lorentz mechanics where these issues were under serious investigation and debate with no dominance of a particular view unlike the modern days where the dominance of special relativity inhibits these investigations and contemplations. In this section, we will try to investigate some of these possibilities. In fact, to have a proper investigation, we also need to investigate the nature of light and how it propagates and the impact of this on its speed.

Light, historically, has been considered either as a corpuscular phenomenon (Newton's view) or as a wave phenomenon (Huygens view).[41] Now, if we follow the logic of classical physics then corpuscular propagation should not need any medium for propagation and hence if we treat light particles as we treat massive particles (which seems to be the view of the corpuscular interpretation of classical physics) then the observed speed of light should depend on the relative motion between its source and its observer because the characteristic speed belongs to the rest frame of its source. On the other hand, wave propagation was classically seen to require the presence of a propagation medium, like air or water for sound. However, wave propagation according to modern physics, following the emergence of Maxwell's equations and Lorentz mechanics, does not necessarily require the presence of a propagation medium. In other words, we have two types of wave in modern physics: (a) "material waves" like sound which require a medium for their propagation, and (b) "non-material waves" like light which do not require a propagation medium although they can propagate through transparent media but this is not out of necessity and dependence in their propagation on the presence of these media. Anyway, the investigation here is about the speed of light in free space and hence conventional material media are not supposed to be present.

Now, if we follow the classical view about wave propagation then the characteristic speed of light waves (i.e. c) should belong to the medium of propagation and hence the observed speed of light as measured by an observer should depend on the movement of the observer relative to the medium (refer to § 1.13.2) which may be identified with a presumed absolute frame. On the other hand, if we follow the modern view about wave propagation, then in principle we should have two main possibilities about the speed of light: (A) the dependence of this speed on the motion of the observer (like classical view) if we consider light waves as material waves, and (B) this type of dependence or another type

---

[41] We note that in modern physic (namely quantum mechanics) light is believed to be of dual nature and hence have both of these aspects. However, we do not consider this possibility here: first to avoid unnecessary complications, and second because it is irrelevant to the issue of the speed of light which is the issue of primary interest to us in this book since the dual nature of light implies having a wave characteristic and hence the issue of the necessity of the presence of a propagation medium or not will follow anyway.

of dependence or even non-dependence at all if we assume that light waves are non-material waves and hence they do not need a medium for propagation.

The diversity or "ambiguity" in (B) can be seen as a result of the "ambiguity" about non-material waves since the daily life experiences, on which classical physics is based, associate wave propagation with a medium. Hence, propagation of waves without the presence of a medium seems odd and more difficult to imagine and this opens more possibilities about the speed of non-material waves. However, the existence of non-material waves and their supposed propagation without a medium is not only seen in modern times as something that is easy to imagine, like the existence and propagation of material waves, but it is seen as the only possibility for the propagation of light. This may partly explain why the universality of the observed speed of light and being frame-invariant, which amounts (in the absence of absolute frame) to the denial of any frame-dependence unlike the other possibilities which imply some sort of frame-dependence, is generally accepted these days. Otherwise, it will be more objective to consider other possibilities for the propagation of light, e.g. considering a quasi-material medium of propagation such as ether, or considering the existence of a universal absolute frame even if the existence of a propagation medium is denied, where both these possibilities may lead (although not necessarily as we will see) to the dependence of light speed on the frame of observation.

Finally, we should remark that the issue of the speed of light and the issue of the velocity of light in the context of frame dependency are two different issues where we will see later that the analysis of the formalism of Lorentz mechanics should lead to the conclusion that although the speed of light is frame independent (in some sense which will be clarified later), the velocity of light is frame dependent. In fact, this frame dependency of the velocity of light is part of the reason for the frame independence of the speed of light (refer for example to § 9.7.1 and § 11).

Solved Problems

1. Discuss the main possibilities about the speed of light in inertial frames of reference.
   Answer: There are many possibilities about the speed of light in inertial frames which are the domain of Lorentz mechanics. However, in very general terms we can say that there are three main possibilities:
   (a) The characteristic speed of light c belongs to a privileged absolute frame (say the ether frame or the frame of absolute space) and hence the observed speed of light for any inertial observer will be frame dependent where this dependency originates from the relative motion between the observer and the privileged frame. We may call this possibility "classical wave model".
   (b) The characteristic speed of light c belongs to the rest frame of the source of light and hence the observed speed of light for any inertial observer will be frame dependent where this dependency originates from the relative motion between the observer and the source. We may call this possibility "classical projectile model".
   (c) The characteristic speed of light c is universal and hence the observed speed of light is frame independent. However, we should still have different opinions about the cause and interpretation of this universality, e.g. if the presumed universality is in a real

1.14 Speed of Light

or apparent sense and what "apparent" means. We may call this possibility "modern physics model".

Exercises
1. What is the difference between the characteristic speed of light and the observed speed of light?
2. What are the main possibilities regarding the nature of light and its propagation according to classical physics? Also, how these possibilities affect the presumed speed of light?
3. Discuss and assess the validity of the Galilean transformations with regard to the two possibilities of the last question.
4. What is the other possibility (or possibilities) that emerged in modern physics about the propagation of light and its speed?

Chapter 2
Emergence of Lorentz Mechanics

In this chapter, we investigate the historical background for the emergence and rise of Lorentz mechanics by examining the historical events and scientific developments that took place in the second half of the 19th century and the beginning of the 20th century and led to the appearance of Lorentz mechanics. We should remark that the purpose of this overview is to put the appearance and development of the theory of Lorentz mechanics in a proper scientific context and hence to improve the understanding of the theory and its roots. Therefore, this background account should not be seen as a historical report in its conventional sense since we have no intention to get involved in historical details or follow stringent historical investigative rules and methods. In fact, there are many details and controversies that must be taken into consideration if a strict historical account is to be prepared. We should also remark that because there are controversies and contradictions in the available historical records, we generally adopt the narrative that makes more sense and looks more logical within the given context.

2.1 Classical View of the World

According to the classical view of the world, as embedded in the classical (or Newtonian) mechanics, the space and time are two independent, absolute and passive entities in which the physical objects do exist and the physical events take place. In fact, the concepts of time and space in classical mechanics, as described here, are inherited from the cultural and philosophical heritage of the old civilizations and were formulated vividly in the Aristotelian philosophy which was the dominant school of thought in the Mediterranean civilizations during the medieval ages. Accordingly, the space as we know is strictly described by the Euclidean geometry and the regular flow of time can be measured in an absolute sense by any recurrent physical process such as the repetitive swing of a free pendulum.

The classical view of the world is based on the common sense and direct experiences which are gathered from everyday life. Therefore, it is not surprising that classical mechanics, whose foundations originate from this view, was hugely successful in describing the physical world at the spatial and temporal scales that are commensurate to the human perceptual experiences which correspond to comparatively large objects (much larger than the size of atom) and relatively low speeds (much slower than the speed of light). Classical mechanics is also based on other classical principles such as determinism and preciseness although these are not in competition or conflict with Lorentz mechanics and its principles but with other branches of modern physics, mainly quantum mechanics. There are also other properties of space and time which are not usually indicated or inspected in this context such as homogeneity and isotropy but these properties are usually assumed implicitly in

2.1 Classical View of the World

the scientific theories including Lorentz mechanics and hence they are not controversial to require particular attention.[42]

To be more specific and explicit about the classical view that is relevant to the investigation of Lorentz mechanics and its opposite view, we outline in the following a number of issues concerning the physical world from the viewpoint of classical mechanics:

1. The belief in the principles of physical reality and truth and hence there is a unique physical reality that the role of science is to discover, describe, predict and quantify. Accordingly, a true scientific statement is the one that honestly reflects this unique reality by being like a mirror image of this reality. In contrast, a false scientific statement does not reflect this unique physical reality. So, if there are a number of contradicting propositions, then there is at most one true proposition while all the other propositions (and possibly all) are false because the physical reality and truth are unique.
2. All physical events take place in an absolute 3D Euclidean space. This space is the same for all observers regardless of their state of rest or motion.
3. There is a 1D continuum called "time" which evenly flows and is common to all observers and hence it is absolute and universal in this sense.
4. Events can be strictly ordered in time according to all observers so if V_1 and V_2 are two physical events with V_1 preceding V_2 by a time interval Δt according to one observer, then the temporal order of V_1 and V_2, as well as the magnitude of Δt, will be the same for all other observers.
5. The space and time are totally independent of each other and hence the time of an event will not be dependent on its location in space or the location of its observer in space and vice versa.
6. The time is also independent of the motion of the observer and hence the exact time of the occurrence of a given event is the same regardless of the state of rest or motion of the observer relative to the event. Similarly, the time duration (or interval) of the occurrence of a series of events is the same regardless of the state of rest or motion of the observer.
7. Like time, space is also independent of the motion of the observer and hence the positional separation between events is the same for all observers.

We note that some of the above points may have not been stated explicitly in the literature of classical physics prior to the emergence of Lorentz mechanics but they represented an implicit general consensus at that time according to the available records. The reason may be that they have been seen so obvious that they do not need to be explicitly stated. Some may have not even been contemplated consciously due to their deep "intuitive" nature and "instinctive" roots. However, the modern literature of classical mechanics is more explicit about these issues due to the challenges posed by the view of Lorentz mechanics. We also note that there are many other features that characterize the classical view of the physical world; the above are preferentially selected because of their intimate relation to Lorentz mechanics and its definitely or tentatively opposite view about these issues or some of these issues at least.

---

[42] We note that the assumptions of homogeneity of space and time and isotropy of space may be lifted in other scientific theories like general relativity.

2.1 Classical View of the World

We may also add to the above list of issues the following issues which are of less general nature, or they are less known or less recognized, than the issues in the above list:[43]

1. The inherent physical properties of a physical object, such as its length, do not depend on its state of rest or motion. Accordingly, the length of a 1 meter stick is 1 meter whether it is at rest or it is moving.
2. Inline with the previous point, the mass of any material object is an intrinsic property of the object and hence it does not depend on its state of rest or motion. More generally, the mass of the object does not depend on its energy and hence mass and energy are two separate physical attributes.
3. There are two main types of reference frame: inertial frames where the laws of classical mechanics (i.e. Newton's three laws of motion and their subsidiary results and formulations) hold true, and non-inertial (or accelerating) frames where these laws do not hold. Inertial frames are characterized by being at rest or in uniform motion with respect to an absolute reference frame, which can be seen as the frame of absolute space, while non-inertial frames are characterized by being in a state of acceleration with respect to this absolute reference frame.
4. Light (and electromagnetic radiation in general) propagates in the space like the motion of material objects and waves and hence its observed speed is frame dependent, i.e. it depends on the state of motion of its observer whether this motion is relative to the source or relative to the propagation medium (refer to § 1.13 and § 1.14).[44]

These issues, which are also preferentially selected, are closely linked to Lorentz mechanics and its interpretations, as will be discussed in detail later on.

Solved Problems

1. What we mean by determinism and preciseness?
   Answer: Determinism may be defined as the existence of the physical reality in a totally specified form regardless of the observer, while preciseness may be defined as the ability of the observer in principle to get an exact image of the reality, i.e. completely identified truth. So in brief, determinism is about the identity of reality while preciseness is about the identity of truth.

Exercises

1. Outline the main features and issues that characterize the classical view of the world as embedded in the philosophical and scientific framework of classical mechanics.
2. Search for some famous quotations about space and time in the Aristotelian philosophy and comment briefly.
3. Comment briefly on the following quote which is attributed to Newton about space: "Absolute space, in its own nature, without regard to anything external, remains always

---

[43] Also, the previous list is mostly philosophical in nature while the following list is mostly physical.
[44] So, in principle we can imagine light as being made of material particles ejected from the source towards the observer and hence their observed projectile speed is dependent on the relative motion between the source and the observer. We may also imagine light as a material wave (like sound wave propagating through air or water) where its observed wave speed depends on the motion of the observer relative to the rest frame of the propagation medium.

similar and immovable. Relative space is some movable dimension or measure of the absolute space, which our senses determine by its position to bodies, and which is vulgarly taken for immovable space ... Absolute and relative space, are the same in figure and magnitude; but they do not remain always numerically the same ... And so, instead of absolute places and motions, we use relative ones; and that without any inconvenience in common affairs ... etc.".

4. Search for some famous quotations about time in classical mechanics and comment briefly.
5. Analyze the concept of universality of time in classical mechanics.
6. Discuss the claim that the classical concepts about space and time are intuitive and may even be instinctive and hence they may have biological roots in our mental blueprint.
7. Outline other features and issues in the classical view of the world which are not as general and recognized as the main ones but they have direct link to corresponding issues in modern physics and Lorentz mechanics in particular.
8. Name and discuss some of the other characteristics which are usually presumed (implicitly if not explicitly) about space in both classical and modern physics.
9. Name and discuss some of the implicitly or explicitly presumed characteristics of time in physics and science in general.
10. Discuss the reasons for the success and failure of classical mechanics in relation to its reliance on the direct experience of mankind and daily practices.
11. Name and discuss some of the principles of classical mechanics which are not in conflict with the framework of Lorentz mechanics (at least according to the common belief) but with the framework of quantum mechanics.

2.2 Galilean Relativity

The fact that the form of the physical laws should be independent of the uniform motion (including rest) of the reference frame is not new to modern physics. In classical mechanics, Galileo principle of relativity states that the physical laws of mechanics take the same form in all inertial frames, and hence the state of rest and uniform motion in absolute sense cannot be detected or distinguished by performing mechanical experiments in such frames. The relativity principle of Galileo may also be stated as: there is no privileged inertial frame where the laws of physics take a special (or privileged) form. Hence, there is no experiment that can be done to reveal this privileged frame by testing if the laws will have the required special form. In brief, all observers who are at rest or move uniformly in space are equivalent.

To be more clear about this issue and to have a better understanding of its significance and the possible difference between the classical relativity principle of Galileo and the modern relativity principle of special relativity (which is commonly associated with Lorentz mechanics), let analyze this principle further. The relativity principle in general can have different interpretations; the main ones are:

1. There is no absolute rest frame (which is usually identified with the absolute space) and hence the space does not distinguish between rest frames and uniformly moving

2.2 Galilean Relativity

frames since there is no absolute frame relative to which being at rest or in motion has any sensible and realistic meaning. In brief, there is no distinction between rest frames, or between uniformly moving frames, or between rest frames and uniformly moving frames because the state of being at rest or in motion in an absolute sense does not exist since being at rest or in motion is an extrinsic property which has no meaning in the absence of an external reference frame although being at rest or in motion is still sensible between the frames themselves but this is just a conventional difference with no fundamental physical distinction. This should be the logical interpretation of the relativity principle if we deny the existence of absolute space and hence we deny the existence of an absolute frame relative to which being at rest or in motion has a sensible realistic meaning. However, as we will see later, this interpretation will not only abolish the difference between rest frames and uniformly moving frames (as well as between rest frames themselves and between uniformly moving frames themselves) but also between all these frames (i.e. resting and uniformly moving) on one hand and accelerating frames on the other hand since in the absence of absolute space even acceleration has no sensible realistic meaning because acceleration is also an extrinsic property. The distinction between accelerating and non-accelerating frames should then be, or depend on, an intrinsic property of the frame which is not obvious. Consequently, the difference between inertial and non-inertial frames, which is a physically real difference with obvious physical consequences of holding and not holding the laws of mechanics, will be difficult to explain. The relative acceleration between the frames themselves in the absence of a physically-real and universal absolute frame will be a matter of convention and hence it cannot be the cause of real physical effects.

2. There is an absolute rest frame (represented by any frame which is at rest with respect to the absolute space) but this absolute frame does not distinguish between those frames which are at rest or between those which are in uniform motion or even between those which are at rest on one hand and those which are in uniform motion on the other. However, this absolute frame (which is based on the absolute space) does distinguish between all those frames on one hand and accelerating frames on the other hand. In simple words, the distinction occurs only between inertial and non-inertial frames but not between inertial frames themselves regardless of the state of inertial frames as being at rest or in motion relative to this absolute frame. As we will see, this interpretation of the relativity principle is logically consistent since it can explain not only the uniform behavior of all inertial frames, but it can also explain the fundamental physical difference between all the inertial frames on one hand and all the non-inertial frames on the other hand.

Now, it is not very clear, according to the historical records which are available to the author, if the Galileo notion is based or not based on an absolute rest space relative to which the uniform motion is defined. This is because the relativity notion of Galileo predates the formulation of the modern concept of inertial frame which is based in its definition on the validity of Newton's laws of mechanics and its philosophical infrastructure (see § 1.10 and § 2.1). However, the notion of absolute space and its existence was at the root of classical physics (or natural philosophy) even before Galileo. So, it is very likely that the

2.2 Galilean Relativity

Galileo principle is based on the existence of an absolute space where uniform motion takes place and where this uniformity is identified relative to this absolute space. Accordingly, the essence of the principle of Galilean relativity is that: no experiment can revel the privileged frames which are at rest or in motion with respect to this absolute space. This means that according to the Galilean view the space of physical events "does not care" about who is moving and who is not moving with respect to it as long as the movement is uniform and hence the space does not discriminate between non-accelerating frames and if they are moving with respect to the space or not. Yes, this absolute space does care about non-uniform movers (i.e. accelerating frames) where it treats them differently by denying them the privilege of holding the laws of classical mechanics such as the law of inertia. So, the Galilean principle of relativity should be completely consistent with the existence of absolute space and hence it is completely different in this regard from the commonly adopted relativity principle of special relativity which is based on the denial of the existence of absolute space altogether. However, as we will see, Lorentz mechanics is not necessarily in conflict with the notion of absolute space considering other interpretations (like Lorentz interpretation) which are not inline with the interpretation of special relativity.

In this context, we note that the quote "It only ever makes sense to speak of motion relative to something else" which is attributed to Galileo may also indicate a different view where he does not consider the space as an absolute physical entity. However, the possibility of this quote indicating that Galileo was believing in the relativity principle in its special relativistic sense is highly unlikely due to the dominance of the concept of absolute space and its existence at that time. Anyway, we have no particular interest in the view of Galileo himself; what is important is the general consensus at that time and the stand of classical physics as incorporated later in Newton's laws and the associated philosophical framework where there seems to be a general acceptance (whether explicit or implicit) to the notion of absolute space by most natural philosophers of that era. Hence, it is more likely that any relativity principle (whether Galilean or else) at that time is not meant to deny this absolute space, but to deny that the frames which are at rest relative to absolute space are privileged from those frames which are in uniform motion relative to absolute space (or the other way around). By priority, this should also imply the denial of the existence of a privileged frame among the resting frames themselves and the denial of the existence of a privileged frame among the uniformly moving frames themselves, i.e. denying the privilege of any inertial frame over any other inertial frame regardless of their state of rest or motion.

Now, if we add to the above view about absolute space the fact that the Galilean time is also absolute, as it is obvious from the Galilean time transformation (refer to Eq. 31) for inertial observers at least, then we see that the Galilean notion of space and time should be totally classical, i.e. absolute rest space with absolute universal uniformly-flowing time, or what we can call absolute frame. As seen earlier, this classical notion of space and time is embedded in the Newtonian mechanics and it forms part of its philosophical infrastructure. Therefore, the Galilean relativity principle does not touch the notion of absolute space and absolute time although it states that space and time do not discriminate against any non-accelerating frame, where acceleration is relative to the absolute rest space, and hence all these frames are subject to the same laws of classical mechanics. Accordingly, the claim which may be found in some special relativity texts, that Galileo abolished absolute space by his relativity principle seems unfounded especially when we notice that Newton's mechanics, which is inline with the Galilean physics and embraces its main elements, is based on the existence of absolute space.

In this context, we should remark that inline with the Galilean principle, a **Newtonian principle of relativity** may also be formulated by some authors. The essence of this principle may be stated as that: when Newton's laws apply in a reference frame O_1 then they will equally apply in any other reference frame O_2 which is in a state of rest or uniform motion relative to O_1. This principle may also be called the **Galilean invariance**. However, it is obvious that this does not represent a different relativity principle of classical mechanics, but it is just the Galilean relativity principle with a different phrasing or at least it is physically equivalent to the Galilean relativity principle.

Exercises

1. State the Galilean relativity principle in a few and simple words. State any principal assumption about this principle. What is the obvious consequence of this principle?
2. Some authors may state the essence of the Galilean relativity principle as "absolute rest cannot be defined". Discuss this.
3. Analyze the Galilean relativity principle and its interpretation and the implication of this principle on the existence of absolute space.
4. Discuss the possibility that even in Lorentz mechanics the notion of absolute space can be accepted, i.e. there may be no contradiction between the formalism of Lorentz mechanics and the existence of absolute space in contrast to the view held by the special relativists that this mechanics put the final nail in the coffin of absolute space.
5. Is there a logical requirement for the theory of special relativity to deny the existence of absolute space, i.e. can we imagine special relativity to embrace the relativity principle in its classical sense?
6. What is the Newtonian principle of relativity? What is the Galilean invariance? What is the relation between them and the principle of Galilean relativity?
7. What can you conclude from the essential difference between inertial and non-inertial frames?
8. Make a simple argument, based on the speed of light, for the case that according to classical physics which embraces the Galilean relativity principle, there is still an absolute frame (space or ether) and hence it is incorrect to claim that the relativity principle in its classical sense abolished absolute space but kept absolute time.
9. Discuss the relation between the relativity principle, the Galilean transformations and Maxwell's equations.

2.3 Maxwell's Equations and Speed of Light

In the 1860s, James Clerk Maxwell brought together the existing knowledge of electricity and magnetism by forming a set of four concurrent equations, with introducing an additional term to one of these equations, that can describe all electromagnetic phenomena (see § 12.1 and 12.3.3 for more details). As a byproduct of Maxwell's formulation of electromagnetism, the following equation emerged:[45]

$$c = \frac{1}{\sqrt{\varepsilon_0 \mu_0}} \tag{68}$$

where c is the speed of light and ε_0 and μ_0 are the absolute permittivity and permeability of free space. Now, since this equation does not refer explicitly to any particular frame of reference, as there is no parameter in the equation that refers to a frame of reference or observer, a natural question has emerged that is the speed of light in this equation belongs to which frame? This is based on the general understanding at that time that the speed of light, like the speed of projectiles and waves (refer to § 1.13 and § 1.14), is subject to the Galilean rule of velocity composition which is additive and hence the speed of light varies between reference frames depending on the state of motion of the frame (relative to the source or relative to the medium of propagation), i.e. it is frame-dependent. Accordingly, there was a clash between classical mechanics and Maxwell's equations because the Galilean transformations of classical mechanics require the speed of light to be frame-dependent while Maxwell's equations seem to suggest that this speed is not associated with any particular frame and hence it is frame-independent.[46]

At that time, there seems to be a general consensus[47] that light, as a wave phenomenon, propagates through a transparent medium called ether and hence the natural answer to the above question should be that the speed of light (as represented by c) is relative to the rest frame of ether. This frame of ether could be the same as the above mentioned rest frame of absolute space if ether is assumed to be at rest with respect to the frame of absolute space. In fact, if we accept the ether hypothesis then ε_0 and μ_0 can be attributed more easily to the ether as some of its intrinsic properties rather than to the space itself since attributing physical properties to the space may be more difficult to imagine because space may be seen by some as a hypothetical entity or as a mere container to the physical objects and events and hence it is not eligible to have real physical properties. The possibility, which was suggested later by Poincare[48] and incorporated in the theory of special relativity,

---

[45] In fact, the constant c appears (either explicitly or implicitly as a combination of ε_0 and μ_0) even in some of Maxwell's equations, and hence the problem is not restricted to this equation although this equation may be more explicit in having no apparent reference to a frame of observation.

[46] We should remark here that the problem of the speed of light which emerged from Maxwell's equations is based on several assumptions some of which may look very obvious to us, e.g. the same set of coordinate transformations should equally apply to classical mechanics and Maxwell's equations.

[47] This consensus seems to exist at least in the context of Maxwell's equations where these equations established the status of light as a wave phenomenon that follows the wave propagation equation (refer to § 12.3.3).

[48] We note that this possibility, in an apparent sense, was also considered by Lorentz in his interpretation

that the speed of light could be invariant and hence the constant c that emerged from Maxwell's equations does not belong to a particular frame, was inconsistent with the framework of classical physics and hence it was not contemplated as a solution initially so all the investigations of possible solutions to this difficulty (i.e. having a speed of light in the electromagnetic formulation of Maxwell's equations with no explicit reference to a frame or observer) focused on other possibilities.

Solved Problems

1. Can you propose a fundamental challenge to the problem that is posed by the question that the speed of light as inferred from Maxwell's equations belongs to which frame?
   Answer: This problem is essentially based on the assumption that c represents the speed of light. However, if it is interpreted as a universal constant (which is a combination of two other universal constants) that may represent the speed of light then there should be no problem. However, this interpretation may be challenged because c obviously represents the speed of light in the electromagnetic wave equation for example. It may also be claimed that this problem is based on the confusion between the characteristic speed of light and the observed speed of light, although this can be easily challenged. More discussions and clarifications about these issues will follow later.[49]

Exercises

1. Is it possible to challenge the claim that the equation $c = (\varepsilon_0 \mu_0)^{-1/2}$ does not refer to any particular frame of reference?
2. Explain how the problem about the speed of light has emerged from Maxwell's equations. Also discuss how this problem was dealt with initially at that time.
3. Make a semi-formal classical argument that since Maxwell's equations contain the constant c, then Maxwell's equations are not form invariant under the Galilean transformations.
4. Referring to the previous question, as we still have the speed c incorporated in Maxwell's equations even after the replacement of the Galilean transformations with the Lorentz transformations, how the issue of form invariance of Maxwell's equations is then solved?

---

of Lorentz mechanics. We should also note that the frame-independence of the speed of light was also considered as a possible explanation for the null result of the Michelson-Morley experiment by other scholars prior to the emergence of special relativity and hence it is not an invention or discovery of Einstein as many special relativists claim.

[49] It is noteworthy that the problem that is posed by the question that the speed of light as inferred from Maxwell's equations belongs to which frame is basically based on the assumption of a wave model for the propagation of light since light is supposed to be wave in nature according to the wave propagation equation. The situation may change if we can assume that despite the wave nature of light it may also follow a projectile (or ballistic) propagation model in some aspects, which may be consistent with the dual nature of light which is generally accepted in modern physics. If this assumption is accepted as a possibility, then addressing the above problem may require extra effort. We do not go through these details due to their limited value. However, the reader is advised to contemplate on these issues.

2.4 Maxwell's Equations and Galilean Transformations

Apart from the above problem that emerged from Maxwell's equations about the speed of light and to which frame this speed belongs (see § 2.3), there was another problem that emerged from Maxwell's equations; this time between Maxwell's equations and the Galilean transformations.[50] While the laws of mechanics were invariant under the Galilean transformations, it appeared that Maxwell's equations and their derived results, like the electromagnetic wave equation, are not invariant under these transformations (see § 12.3.2 and § 12.3.3).

In fact, there were several possibilities for the failure of Maxwell's equations to be invariant under the Galilean transformations, the main ones are:

1. Both Maxwell's equations and the Galilean transformations are right, but the Galilean transformations apply to mechanics only and hence we should look for another set of transformations for electromagnetism. This possibility was excluded because of the general conviction of the necessity of the existence of a unified physical theory that should equally apply to mechanics and electromagnetism and hence the same set of coordinate transformations should apply to the laws of mechanics and to the laws of electromagnetism.
2. Both Maxwell's equations and the Galilean transformations are wrong. This was dismissed because Maxwell's equations were well established and generally accepted. Moreover, it was unnecessarily excessive because the problem can be addressed and fixed by dismissing only one of these two sets of equations. This may be described as "economy in science" where minimally required assumptions are used to establish a theory and minimally required adjustments are employed to fix a flawed theory.
3. Maxwell's equations are wrong and the Galilean transformations are right, and hence the Galilean transformations should equally apply to the laws of electromagnetism as well as to the laws of mechanics. Hence, the right set of electromagnetic equations should transform invariantly under the Galilean transformations. However, the suggestion that Maxwell's equations could be wrong was rejected because these equations were well established and verified by strong experimental evidence. Moreover, there were other indications that the problem originates from the Galilean transformations rather than Maxwell's equations.
4. Maxwell's equations are right and the Galilean transformations are wrong, and hence the right transformations should equally apply to the laws of electromagnetism (represented by Maxwell's equations) as well as to the laws of mechanics. The success of the Galilean transformations in mechanics so far could then be explained by claiming that within the traditional domain of classical mechanics (i.e. low speed) the Galilean transformations make a good approximation to the right set of transformations. This possibility is the one that was eventually considered as the most viable and sensible one. The focus was

---

[50] In fact, these two problems are fundamentally the same as they both arise because of the assumption of the validity of the Galilean transformations where the second is related more directly to the Galilean transformations than the first. However, for clarity and pedagogical purposes, we preferred to address them separately. Also, refer to the exercises of this section and § 2.6.

then to find the right set of transformations that apply simultaneously to mechanics and electromagnetism.

Exercises

1. Explain how another problem (this time related more directly to the Galilean transformations) has emerged from Maxwell's equations.
2. Referring to the previous question, discuss in detail all the possibilities about the validity of the Galilean transformations and Maxwell's equations.
3. Why the general opinion at that time seemed to accept Maxwell's equations and question the integrity of the Galilean transformations?
4. Discuss how the two problems that emerged from Maxwell's equations (i.e. the problem of having a specific speed of light without identification of frame of reference and the problem of Maxwell's equations being non-invariant under the Galilean transformations) are essentially identical.
5. Discuss the issue that within the traditional domain of classical mechanics (i.e. low speed) the Galilean transformations of space coordinates and time are good approximation to the right (or exact) set of transformations which are the Lorentz transformations.
6. Discuss the possibility (which is similar to the first possibility in the text) that both Maxwell's equations and the Galilean transformations are right but Maxwell's equations apply to a privileged frame or ether frame.

2.5 Light as Wave Phenomenon and Luminiferous Ether

According to Maxwell's equations, all types of electromagnetic radiation behave like waves. In fact, the belief that light is a wave phenomenon predates Maxwell and the dispute about the light nature as waves or particles (wave or corpuscular nature of light) goes back to the days of Huygens and Newton and may even be before. Young, among other physicists before Maxwell, also investigated the nature of light experimentally and has shown that light demonstrates a number of wave-like phenomena such as interference, diffraction and polarization. Now, since material waves like sound propagate through material media like air and water, there was a general assumption by the classical physicists who believed in the wave nature of light that light also requires a medium which was generally dubbed as luminiferous (i.e. light bearing) ether. In fact, the contribution of Maxwell's electromagnetic theory in this respect may only be in the extension of the belief that light propagates through ether to other types of electromagnetic radiations since according to his theory light is just one form of electromagnetic radiation where all these forms share the common feature of having the characteristic speed c in free space and behaving as waves.

Now, according to classical physics, waves propagate through media and hence their observed velocity is subject to the velocity addition rule where the velocity of the observer in the rest frame of the medium of propagation should be added vectorially to the characteristic velocity of the wave in the medium as explained in § 1.13.2. So, it was natural

2.6 Michelson-Morley Experiment

to assume that this also applies to the light waves in their propagation through ether. Accordingly, an experiment (or rather a number of experiments) to verify the dependence of the speed of light on the movement of the Earth in space (or rather its movement through ether) was natural to propose in the end of the 19th century where all the elements and theoretical motivations for conducting such an experiment were gathered. Such an experiment, if succeed, will achieve several objectives: verifying the ether hypothesis, verifying the movement of the Earth through the ether, verifying the velocity addition nature of light propagation, and measuring the velocity of the Earth relative to the ether. This leads us to the next stage of investigation where we discuss in § 2.6 one of the most famous experiments in modern science whose purpose was to achieve those objectives (or at least some) through measuring the speed of the "ether wind", i.e. the relative movement of ether that is caused by the movement of the Earth in space. As we will see, the failure of this experiment to detect the presumed ether wind had a strong direct impact on the emergence of Lorentz mechanics. We should notify the reader that more details and background information about the theory of ether and its nature are given in the exercises of this section.

Exercises
1. Summarize the essence of this section about light as a wave phenomenon and the role of luminiferous ether outlining the relation of this to the concept of absolute space (or absolute frame).
2. What are the two main types of wave? To which type of these light belongs?
3. Describe in general terms the role of ether in classical physics. Also, outline the essence of the ether theory and its logical implications on the speed of light.
4. What is the meaning of "ether wind"?
5. Investigate the possibilities of the relation between ether and absolute space in classical physics.
6. Investigate the nature of ether and its role in the propagation of light according to the old electromagnetic theory.

2.6 Michelson-Morley Experiment

Based on the above-given facts and the ideas that were brewing during the second half of the 19th century, with the general acceptance of the ether theory, a number of scientists in the late 19th century and the early 20th century tried to measure the speed of the Earth through ether. The most famous (and probably the best planned at the time) of these experiments was that of Michelson and Morley which is based on using interferometric techniques to detect and measure the speed of the Earth with respect to the luminiferous medium. As a matter of history, the original experiment was conducted by Michelson around 1881 and the refined version of the experiment (which is known as the Michelson-Morley experiment) was conducted by Michelson and Morley around 1887. The experiment is based on detecting the "ether wind" which is supposed to be caused by the movement of the Earth relative to the ether. The essence of the Michelson-Morley experiment is to

2.6 Michelson-Morley Experiment

use an interferometer to detect the phase shift of interference pattern generated by the interference of two light rays traveling in two perpendicular directions at supposedly two different speeds where this speed difference is presumably caused by this movement. The technical details of the Michelson-Morley experiment are given in § 12.2 which the reader is strongly advised to read to follow the upcoming arguments.[51]

Although there were some claims by some scientists, who conducted similar experiments, of detecting the presumed phase shift in these interference patterns and hence confirming the ether proposal, most of these experiments failed to detect this effect and hence the majority of the scientific community at that time seemed to accept the conclusion of null result which is reached by Michelson and Morley as well as by other scientists.[52] The failure of these experiments to detect the natural effect associated with the existence of ether put many question marks, e.g. on the existence of the luminiferous medium, the interpretation of the speed of light as obtained from Maxwell's equations, the relation between classical mechanics and electrodynamics, and even on the foundations of classical physics in general.

Following these developments, there were several attempts to explain the failure of these experiments; the main ones may be outlined as follows:

1. Questioning the technical aspects of these experiments and if the experimental settings and techniques were appropriate and accurate enough to detect the alleged effect. Hence, this group usually supports the other experiments which claimed to have detected the movement of the Earth through ether.
2. Questioning the logic and scientific foundations of these experiments and hence casting a shadow on the ability (in principle not because of technicalities or other marginal factors) of these experiments to detect the variation of the speed of light, in its dependency on the movement of the luminiferous medium relative to the experimental device, by the employed techniques. A prominent example of this sort of attempts is the proposal of an ether drag caused by the moving equipment which makes it impossible to detect the movement through the ether. The essence of this proposal is that these experiments are relying in their conclusion on the assumption that the ether medium in the immediate neighborhood of the experimental device is in a relative motion with respect to the device. Now, if we assume that the part of the ether in the immediate neighborhood of the device is dragged (like water and air surrounding objects moving through these material media) and hence it is not in relative motion with respect to the device, then this sort of experiments will not be able to detect the alleged effect because there is no relative motion of the ether that surrounds the device. However, the ether drag hypothesis was challenged by stellar aberration where telescopes are required to

---

[51] We put this in an appendix to avoid interruption and distraction from the main objectives of this chapter.

[52] In fact, according to some records a phase shift was observed but it was much smaller than it should be according to the Michelson-Morley assumptions and analysis (see § 12.2). If so, then this may be explained by the followers of the ether theory as supporting evidence to the ether theory, and hence the failure of the phase shift to be of the anticipated magnitude may be attributed to the wrong assumptions (e.g. the presumed speed of the ether wind) and analysis of Michelson and Morley or even to technical flaws.

be tilted to observe light from astronomical objects because of the Earth motion. But the debate did not end here as there were arguments and counter arguments from both sides where the physical setting and the effect of ether drag and stellar aberration were disputed.[53]

3. The attempt to explain the failure of these experiments by suggesting a compensating physical effect that nullifies the result of any relative motion through the ether. The essence of this attempt is that the ether is real, and there is a relative direction-dependent motion between the experimental device and the ether, but if we assume that there is a real physical effect that shortens the length of the device in the direction of motion in exactly the same amount to compensate for the time difference[54] in the light travel in the two perpendicular directions then the Michelson-Morley experiment (and any similar experiment which is based on the same physical principles) must fail to detect the alleged effect. This explanation is represented by the FitzGerald-Lorentz proposal of length contraction which will be thoroughly investigated in § 2.7. As we will see, this attempt eventually prevailed and led to the emergence of Lorentz mechanics although the exact nature of this proposed length contraction, which is supposedly caused by ether, did not survive for long time due to the eventual dominance of the view of special relativity theory. Consequently, the null effect was seen as demonstration of the proposal that the speed of light is the same for all inertial observers regardless of the meaning of this and if it is real or apparent and regardless of its cause as being length contraction or something else (or length contraction plus something else).

Anyway, if we accept the results of the Michelson-Morley experiment (i.e. we do not question its technical or logical or scientific validity), then we should reach the logical conclusion that the Galilean transformations are not the right transformations (at least for electromagnetism) and hence we should search for another set of transformations as we will see in § 2.7.

Solved Problems

1. Was the Michelson-Morley experiment (and its alike) the first attempt to detect the ether?

    Answer: In fact, there were precursors to the Michelson-Morley experiment to detect the ether and its movement relative to the Earth such as the so-called telescope effect experiment which is based on the hypothesis of changing the focus of the observed astronomical objects like the Sun when observed by a telescope during different phases of the Earth orbital movement because of the supposed change of direction of the ether

---

[53] Another example of this category of attempts is questioning the assumptions and analysis of the Michelson-Morley experiment in the context of the claimed detection of a smaller phase shift than anticipated, as indicated in a previous footnote. A third example is what we will discuss later (see the exercises of this section and § 12.2) about the main limitation of the Michelson-Morley experiment and its alike in adopting a classical wave propagation model and hence they cannot rule out a projectile (or ballistic) propagation model which can easily explain the null results with no necessity for the modification of classical physics and the rejection of the Galilean transformations (at least for this reason although other reasons, such as non-invariance of Maxwell's equations, may still remain).

[54] This time difference is supposed to originate from speed variation and its dependency on direction.

wind. Also, the so-called Fresnel drift effect, which the Fizeau experiment (refer to the exercises of § 4.2.2) is based upon, was originally proposed for the detection of the ether. We note that some of these attempts are much earlier than the Michelson-Morley experiment and its alike as they precede the formulation of Maxwell's equations which mark the rehearsal to Lorentz mechanics. Moreover, these precursors are generally based on different technical (and some even theoretical) approaches to that of Michelson and Morley. In fact, most (if not all) of these precursors cannot be considered as conclusive as Michelson-Morley experiment due to theoretical or technical question marks and uncertainties and hence the main "evidence" for the null conclusion about the existence of ether should be attributed to the Michelson-Morley experiment.[55]
2. Compare the FitzGerald-Lorentz proposal of length contraction (to explain the null result of the Michelson-Morley experiment) with other proposals in this regard.
Answer: Some valid comparison points are:
• While other proposals assume that the Michelson-Morley experiment and its null result are wrong, the FitzGerald-Lorentz proposal accepts the experiment and its result.
• While other proposals are largely based on trying to salvage classical physics unscathed (or almost), the FitzGerald-Lorentz proposal started as an attempt to salvage classical physics by a minimal change (through the suggestion of a rather minor non-classical effect of length contraction without direct threat to the Galilean transformations and classical framework) but it eventually led to a radical change through the emergence of Lorentz transformations and Lorentz mechanics.

Exercises
1. Outline the main proposals that were suggested to explain the null result of Michelson-Morley experiment.
2. What is the common factor that is shared between the problem of the speed of light that emerged from Maxwell's equations and the problem that emerged from the null result of the Michelson-Morley experiment and its analysis?
3. What property should be attributed to ether if the ether drag hypothesis is to be sensible?
4. Assess and criticize the Michelson-Morley experiment and similar experiments.

2.7 FitzGerald-Lorentz Proposal and Emergence of Lorentz Transformations

As indicated earlier, there were a number of attempts by several renown scientists at that time (such as FitzGerald, Lorentz, Poincare, Voigt and Larmor) to explain the failure of the Michelson-Morley experiment to detect the Earth motion through ether. One of these attempts was made by FitzGerald and Lorentz who proposed a length contraction effect in the direction of motion that compensates for the time difference in the light journeys and hence explains the null result of the Michelson-Morley experiment. The essence of this proposal is to nullify the effect of the time difference in the light journeys along the

---
[55] As we will see, even the Michelson-Morley experiment is not as conclusive as it is supposed to be.

2.7 FitzGerald-Lorentz Proposal and Emergence of Lorentz Transformations 77

two arms of the Michelson-Morley apparatus by adjusting the effective length of the arm in the parallel direction to the motion so that the phase between the two beams becomes independent of the orientation of the device and hence there will be no phase shift in the interference pattern between the two light beams as the device is rotated (see § 12.2).

This proposal was eventually developed further by Lorentz to a set of mathematical transformations of space coordinates and time (to be known later as Lorentz transformations) that can explain this failure and make both the laws of mechanics and the laws of electromagnetism form invariant. The essence of these transformations (refer to § 4.2.1 for details) is to adjust the Galilean transformations (see § 1.9) by modifying some old terms and introducing new terms so that these transformations take into account the effect of motion on the coordinate measurements. These changes resulted in nullifying the effects of any movement through the luminiferous medium in the Michelson-Morley experiment and unified the laws of mechanics and the laws of electromagnetism to become form invariant under the new set of coordinate transformations. The new spacetime coordinate transformations and their main consequences will be investigated in detail in the upcoming parts of the book (see for example § 4.2 and § 5).

The proposal of the new set of spacetime coordinate transformations, to replace the old Galilean transformations of space coordinates and time, is the main event that marks the birth of the mechanics of Lorentz transformations as a new branch of modern physics. Many developments in the formulation and interpretation of these transformations and their physical, mathematical and logical consequences have followed in the subsequent years and decades where many scientists have contributed with original ideas and techniques to these developments. However, all these developments originate from the fundamentally novel idea of having a set of time and space coordinate transformations that differ from the intuitive Galilean transformations of classical mechanics. As we will see in § 5, all the formalism of Lorentz mechanics is based on these new spacetime coordinate transformations where all the details of Lorentz mechanics can be derived, directly or indirectly, from these coordinate transformations following simple standard mathematical and physical methods such as employing the rules of differential calculus, using the standard concepts and definitions of the physics of motion like velocity and acceleration, and employing fundamental scientific and epistemological principles like form-invariance of physical laws. Accordingly, considering the proposal of the new set of spacetime coordinate transformations as the birth of the mechanics of Lorentz transformations is fully justified. Consequently, the common belief that this mechanics is born with the publication of the Einstein 1905 paper is totally unfounded.

Exercises

1. In appendix § 12.2 about the Michelson-Morley experiment, it was shown that the times for the light in its PP_1P journey and in its PP_2P journey are different according to the logic and analysis of Michelson and Morley. Using the derived expressions for these times and assuming $L_1 = L_2$, derive the FitzGerald-Lorentz length contraction formula. Comment on the result and how this length contraction effect is supposed to nullify the effect of any time difference in this analysis.

2. Outline in a few words the essence of the FitzGerald-Lorentz proposal of length contraction as can be understood from the analysis of the previous question.
3. Investigate the problems that have been addressed by the proposal of Lorentz transformations.
4. Make a brief comparison between the Galilean transformations and the Lorentz transformations (refer to § 1.9 and § 4.2.1).
5. Can length contraction alone explain the null result of the Michelson-Morley experiment and its alike?

2.8 Poincare Suggestion and Subsequent Developments

Although the proposal of length contraction in the direction of motion by FitzGerald and Lorentz and the following development of Lorentz spacetime coordinate transformations by Lorentz did not question the existence of ether as a required medium for electromagnetic propagation, it led Poincare in 1900 to question the existence of ether and the necessity for its presence in the electromagnetic propagation theory. In fact, Poincare did not stop here but he developed a number of proposals, ideas and procedures that formed the basic elements and the framework of what will be known later as the special theory of relativity. This includes, for example, questioning the difference between time and local time and hence questioning the concept of absolute time. Despite this challenge to the ether, it kept its status as a scientific fact, or at least a working principle, by many scientists, including Lorentz himself who is one of the main founders of this mechanics, until the dominance of special relativity as will be discussed next.

Following the main contributions of FitzGerald, Lorentz and Poincare[56] in originating Lorentz mechanics and proposing its main elements, formalism, framework and interpretation, a number of other scientists in the beginning of the 20th century made further attempts to explain and elaborate this mechanics and take it to its natural conclusions. Among these attempts is that of Albert Einstein who merged the formalism of Lorentz with the interpretative ideas and framework of Poincare and added some amplifications and extensions. Although Einstein attempt (which will be known later as the special theory of relativity or special relativity) did not receive a particular attention or treatment by the scientific community when it emerged, it later dominated Lorentz mechanics and hence it became equivalent to this branch of physics so that Lorentz mechanics meant special relativity and special relativity meant Lorentz mechanics. This can be explained by a number of historical factors, the main one is the rise of Einstein to fame following the alleged confirmation of the general theory of relativity by the 1919 solar eclipse expedition to testify one of the predicted effects of this theory, i.e. light bending by the gravitational field of the Sun.

Exercises

---

[56] As well as other contributors who participated in the debates and the development of the fundamental ideas of Lorentz mechanics like Larmor and Voigt.

2.8 Poincare Suggestion and Subsequent Developments 79

1. Summarize the main historical events and scientific developments that led to the emergence and rise of Lorentz mechanics.
2. Discuss the claim by some special relativists that even Lorentz attributed the credit for Lorentz mechanics (under the label of special relativity) to Einstein in the quote that is attributed to Lorenz whose essence is that: the relativity theory belongs to Einstein.

Chapter 3
Introduction to Lorentz Mechanics

This chapter introduces a number of concepts and techniques that are commonly used in the conceptualization and formulation of Lorentz mechanics either in this book or in the general literature of this subject. The chapter also includes general discussions about important issues like the different possibilities about the speed of light, the restrictions on the applicability of Lorentz mechanics and its relation to other branches of mechanics.

3.1 Lorentz Mechanics versus Other Mechanics

In § 1.1 we briefly discussed the relationship between Lorentz mechanics and other branches of mechanics which compete with Lorentz mechanics in its field of application or have some link with it. In this section, we extend our investigation about these relations for the purpose of completeness.

The principal physical factor that distinguishes **classical mechanics** from Lorentz mechanics is speed. While classical mechanics applies to objects that move at normal speeds (i.e. $v \ll c$), Lorentz mechanics applies to objects that move at any physical speed including those comparable to the speed of light (i.e. $v \sim c$). Accordingly, classical mechanics can be seen as a special case or as an approximation to Lorentz mechanics that applies in the limit of low speeds. As indicated before, there is no exact speed limit that separates the region where classical mechanics can be used as a good approximation and the region where Lorentz mechanics must be employed. Although $v = 0.1c$ may be set by some as a practical upper limit for the applicability of classical mechanics as a valid approximation, this should be treated with caution because the validity of any approximation depends on a number of factors such as the required accuracy and allowed error and the nature of the problem and hence there is no such a universally-acceptable limit.

Quantum mechanics is the mechanics that was originally developed to deal with atomic and subatomic phenomena. On the other hand, Lorenz mechanics was originally developed to rectify the defects of classical mechanics (or rather classical physics to include electromagnetism later) which basically deals with phenomena that are commensurate to human perception. However, although Lorentz mechanics was originally conceptualized and formulated using classical (or non-quantum) concepts, there is no restriction in principle on its applicability with regard to size. Accordingly, Lorentz mechanics was eventually extended to include quantum phenomena in what is called "relativistic quantum mechanics". So in brief, quantum mechanics meets with Lorentz mechanics in the investigation of small objects that move at high speeds. However, both Lorentz mechanics and quantum mechanics have their own distinct field of applicability where only one of these disciplines is necessarily required. So, Lorentz mechanics is specifically needed to deal with large objects that move at high speeds, while quantum mechanics is specifically needed to deal with small objects that move at low speeds.

The relation between Lorentz mechanics (usually in the form of special relativity) and **general relativity** and whether the former is a special case of the latter or not is a controversial issue, and hence while some scholars and authors try to make special relativity a special case for general relativity, others claim that these are totally different theories. General relativistic mechanics is concerned with gravity and its influence on the behavior of physical objects such as matter and radiation where the formalism is based on a geometric approach by considering the curvature of the spacetime which is supposedly caused by the presence of matter and energy. So, in principle general relativity and special relativity are two independent theories where one deals with the mechanics in inertial frames[57] while the other deals with the physics of gravity. However, because of the presumed equivalence between gravity and acceleration, where the equivalence principle is a pillar of general relativity, general relativity is also concerned with accelerating (i.e. non-inertial) frames. Accordingly, some may consider special relativity as a special case for general relativity because inertial frames are a special case for accelerating frames, i.e. inertial frame is an "accelerating frame" with zero acceleration. However, the two theories are based on different sets of postulates, axioms, and assumptions as well as two different physical systems (i.e. inertial systems and gravitational systems). Furthermore, the proposed equivalence between gravity and acceleration is not unconditional and hence some of the consequences of the equivalence principle are arguable. Anyway, the important thing is that the formalism of special relativity cannot be obtained directly from the formalism of general relativity, i.e. Lorentz spacetime coordinate transformations and their derived consequences and formulations which consist the main body of Lorentz mechanics cannot be obtained from the field equations of general relativity. Accordingly, it seems more appropriate to regard these as two different theories. This seems more obvious when we compare this with classical mechanics as a special case for Lorentz mechanics, and hence we obtain the main formalism of classical mechanics from the formalism of Lorentz mechanics (see § 4.5), where this sense of "special case" does not equally apply to special relativity in its relation to general relativity.

Exercises
1. Discuss the relation between classical mechanics on one hand and Lorentz mechanics and quantum mechanics on the other hand.
2. Discuss the difference between the relation of Lorentz mechanics to classical mechanics and the relation of Lorentz mechanics to quantum mechanics.
3. Discuss the relation between the special theory of relativity (as a representative of Lorentz mechanics) and the general theory of relativity.
4. Discuss the following statement about the equivalence principle: "According to the equivalence principle of general relativity, there is no experiment conducted in a small confined space that can distinguish between a uniform gravitational field and an equivalent uniform acceleration".

---
[57] We would rather say: Lorentz mechanics deals with how space and time should be transformed between inertial frames.

3.2 Restrictions and Conditions on Lorentz Mechanics

Like any physical theory (and even non-physical theory), Lorentz mechanics applies under certain restrictions. Moreover, its effects that distinguish it from classical mechanics are only tangible under certain conditions. In the following points we outline the main restrictions and conditions that should be imposed on the validity and applicability of Lorentz mechanics and the limits on its distinguishing consequences:

1. Since Lorentz mechanics is restricted to inertial frames, all the statements related to Lorentz mechanics are based on assuming inertial frames of observation although in many cases this condition is not stated explicitly. Hence, Lorentz mechanics is not the right type of physical theory for accelerating frames.

2. Based on the commonly accepted view of general relativity, which is based on a conditional equivalence between gravity and acceleration, the validity of Lorentz mechanics theory is also restricted to frames that are void of gravity, i.e. frames which are not affected by gravitational fields due to the presence of aggregates of matter and energy. However, this is based on the view of relativistic mechanics (whether special or general) where the physical forces are interpreted in terms of geometric factors representing the spacetime manifold with this manifold being flat in special relativity and curved in general relativity. A related (and controversial) issue is the claim by some that special relativity can be (formally or logically) obtained as a special or limiting case to general relativity representing the absence of acceleration and aggregates of matter and energy. However, all these details are not within the scope of the present book and may be addressed in the future. In general, any force field (whether gravitational or not) can be considered within the framework of Lorentz mechanics as long as it does not affect the inertial nature of the frame of observation.[58] In this context, we note that the inertiality of a frame is basically determined by its state of rest or motion with respect to an absolute frame and hence in principle it should not be affected by any external field that does not affect its state of rest or motion.[59] All other effects and characterizations of inertial frames, like holding the laws of classical mechanics, are consequences of this fundamental characterization. We also note that the characterization of space as being flat or curved are just mathematical models and artifacts and hence they should not be considered as physical facts in any scientific theory unless they have tangible physical effects. Mathematical consistency, beauty and aesthetic factors do not guarantee the reality and correctness of any physical theory or give it any legitimacy in the absence of tangible experimental evidence in support of these factors.[60]

---

[58] This is not in conflict with what we indicated earlier that we are not dealing with gravity in our investigation of Lorentz mechanics.

[59] This could be based on the claim that the geometry of the spacetime is not affected by the force fields due to the presence of matter and energy and hence the flat space of Lorentz mechanics is still valid for formulating gravitational theories. Although this is obviously in conflict with general relativity, it grants legitimacy to any theory that can formulate the physics of gravity without resort to the curvature of spacetime.

[60] To be more explicit and clear about this important issue, we can say: it may be the case that special relativity is not applicable in the presence of gravitational fields according to the general relativistic

3.2 Restrictions and Conditions on Lorentz Mechanics 83

3. Considering the mathematical definition of the speed ratio β and the Lorentz factor γ (see Eq. 5 and Figure 2), which are generally present in the formulation of Lorentz mechanics and they give it most of its distinctive features, all the non-classical effects that are predicted by Lorentz mechanics are significant and detectable only at high speeds, i.e. speeds that are comparable to the characteristic speed of light ($v \sim c$) or at least they make a significant fraction of the speed of light. This is because at low speeds, the speed ratio becomes very close to zero and the Lorentz factor becomes very close to unity and hence the laws and rules of classical mechanics apply very well in such cases since the Lorentzian formulation will reduce practically to the classical formulation. A quick comparison between the Lorentzian formulae and their classical counterparts will reveal this fact, i.e. the formulation of Lorentz mechanics will reduce practically to the formulation of classical mechanics when the speed v is low. For example, the formulae of Lorentz mechanics for transforming the spatial coordinates (Eqs. 90-92), time (Eq. 93) and velocity (Eqs. 99-101) will reduce to their Galilean counterparts of classical mechanics (i.e. Eqs. 28-30, Eq. 31, and Eqs. 39-41) when $v \ll c$.[61] Accordingly, classical mechanics becomes a good approximation at the limit of $v \ll c$ as stated above. The reader is referred to § 4.5 for more details.

A matter that is related to the restrictions on Lorentz mechanics is the characteristic speed of light c as a speed that is restricted to light and as an ultimate physical speed for any physical entity. As indicated earlier (see § 1.4 and § 1.5) and will be discussed further in the future (see § 3.8.1), according to the contemporary formalism of Lorentz mechanics, all physical speeds within the framework of this mechanics should be subject to the condition that $v < c$ with the equality (i.e. $v = c$) being allowed only for light. This is because the Lorentz γ factor has a singularity when $v = c$ and hence it becomes infinite. Moreover, this factor becomes imaginary when $v > c$. However, this does not rule out the possibility of speeds reaching or exceeding the speed of light c outside the domain of validity of Lorentz mechanics which is inertial frames. So, it is quite possible in principle to have physical speeds that reach or exceed c within a different theory that is related to another domain of validity although this is categorically denied by the followers of the relativity theories. Our assertion, however, should not be understood to mean that we wish to confirm that such speeds (i.e. $v \geq c$) are possible in reality with no further investigation, but it means that we cannot automatically deny the possibility of reaching such speeds because of the speed limit that is imposed by Lorentz mechanics as a requirement for the finity and reality of Lorentz factor.

In fact, it is possible in principle that this condition (i.e. $v < c$) can be violated even within the domain and framework of Lorentz mechanics itself if the current formulation of

---

formulation (and even interpretation) of gravity because the special and general relativistic formulations are incompatible. But this does not mean that the gravity is not compatible with inertial frames under different formulation of the physics of inertial frames and/or the physics of gravity. In other words, the above non-applicability can be caused by the particular formalism and interpretation of the physics of inertial frames and the physics of gravity in the relativity theories.

[61] For the velocity transformations, we need another condition as will be discussed later. This is also the case with the transformation of time.

this mechanics is extended or modified. For example, Lorentz mechanics may be shown in the future to be a special case for a more general mechanical theory (whether for inertial frames or more general) as it was shown in the past that classical mechanics (which once was believed to be the ultimate mechanics) is a special case for other mechanical theories like Lorentz mechanics. What is important to notice and what we want to emphasize is that we should diversify our options to consider possibilities other than those allowed by contemporary theories especially when there are many question marks about these theories. This means that we should dismiss all these dogmatic beliefs that such speeds are impossible in principle because of the requirement of special relativity or even general relativity. Science cannot make real progress under the banner of such slogans and the rule of assertive authoritarianism. Accordingly, c as a restricted and ultimate physical speed should not be treated as a universal and absolute fact, as it is currently treated in modern physics, and hence we should always consider the possibility of physical speeds reaching and even exceeding the characteristic speed of light c although this may be shown in the future to be physically impossible based on real physical evidence not on assertive claims or postulates of a theory or even as a mathematical requirement for the formalism of a potentially restricted theory.

Exercises
1. What we mean by "speed" when we say: classical mechanics is a good approximation at low speeds? Also what we mean by "low"?
2. Explain how the reduction of the Lorentzian formulation to the classical formulation at low speeds can be used as a validation test for the derived formulation of Lorentz mechanics.
3. Analyze the expression that is used by some authors that classical mechanics is a good approximation to Lorentz mechanics (labeled as special relativity) when c tends to infinity.
4. Assess the "sacred" rule of modern physics that c is a speed restricted to light and it is the ultimate speed for any physical object.

3.3 Space Coordination

Since mechanics is the science of describing motion in space, a coordinate system is needed to locate and identify any point in space.[62] This identification should be unique and thorough so that each point in the space has a unique identification that distinguishes it from all other points in the space. Accordingly, length measurement equipment, like measuring sticks, as well as a method for identifying directions and orientations in space, are required. Unique and thorough identification of all points of the space can only be achieved if we assign to each point a number of mutually-independent spatial coordinates that is equal to the number of dimensions of the space.

Space coordination will not be thorough if the number of coordinates is less than the

---
[62] We mean by "coordination" assigning a coordinate system to space, although this use of the word may not be conventional. Some may use the term "coordinization" for this purpose.

3.4 Time Measurement and Synchronization 85

number of space dimensions, e.g. two coordinates can only identify a surface in a 3D space. Also, space coordination will not be unique if the number of independent coordinates is more than the number of space dimensions, e.g. three coordinates will result in multiple identification of each point in a 2D space since the third coordinate is a free parameter that does not correspond to any dimension in the space. The latter case can also be ruled out because of employing redundant surplus coordinates or even because of being meaningless when there is no extra dimension in the physical setting to be represented by the surplus coordinates. The coordinates should be mutually independent to represent the mutually independent dimensions of the space. If the coordinates are not independent then they can only represent a sub-space of the given space. For example, if we represent the xy plane by $(x, 2x)$ coordinates then this type of coordination will only identify a line in the plane.

There are many types of coordinate systems that can be used in the coordination of the space of Lorentz mechanics. However, because Lorentz mechanics is based on an underlying flat space with a flat metric, a rectangular Cartesian system is usually used to coordinate the space and formulate the science. The advantage of using this type of system is that, as well as being sufficient for achieving the required purpose, it is intuitive and hence it is easy to comprehend, manipulate and visualize. Moreover, it is simple in its use and application and hence it usually leads to more simple mathematical formulations. As we saw earlier (refer to § 1.5), this type of coordinate system is commonly used in coordinating the space of inertial frames which are in a state of standard setting.

Exercises
1. Why space coordination is necessary in mechanics?
2. List the main requirements of a coordinate system. What is the purpose of these requirements?
3. What will happen if the requirements of the previous question are violated?
4. What are the advantages of using rectangular Cartesian coordinate systems in space coordination?
5. What is the justification of the common use of rectangular Cartesian coordinate systems in the setting and formulation of Lorentz mechanics?

3.4 Time Measurement and Synchronization

Since mechanics is the science of describing motion in space according to its timeline, a time measurement procedure is required. A regular repetitive physical process, such as the swing of a free pendulum, is normally used to measure time according to its supposedly regular flow.[63] Since any reference frame requires time measurement which should be universal for all points of the space that underlies the frame, a procedure to synchronize the clocks in all positions of the frame is required so that all the clocks in the frame read the same time at any instant. Hence, the essence of the required synchronization procedure is to measure the time at different points of the space consistently by a given observer so

---
[63] As discussed elsewhere (see § 9.7), this is based on using discrete units in measuring the flow of time.

3.4 Time Measurement and Synchronization

that all the clocks in the frame at all points of the space will read the same time according to that observer. We note that this requires the notion of a universal time for a particular observer, i.e. there is a meaningful sense of "same time at every point in the space" (refer to the exercises for more details).

There are two main synchronization methods:

1. Gathering all the clocks in one location where they are synchronized to read the same time. The clocks are then moved to their positions in the space where they are wanted to be. This method can be justified only if we assume that moving the clocks (which requires an accelerated motion as well as potentially uniform motion) will not affect their time-measuring mechanism, that is the clocks will keep running uniformly at their normal rate despite the motion and acceleration. Although this assumption is justified in principle in classical mechanics, it may not be justified in Lorentz mechanics where movement can cause changing the count rate of the time measurement mechanism due to the change in the "count rate" (or flow) of time itself according to the time dilation effect at least as a possible cause even before inferring and formulating this effect as a physical fact.

   Accordingly, an argument may be made that this time synchronization procedure should be excluded as a possible synchronization procedure in Lorentz mechanics as long as time dilation effect (or indeed any time-changing effect to be more general) stands as a possibility (let alone being a physical reality) within the framework of Lorentz mechanics. Yes, in classical mechanics time is a frame independent absolute entity with a universally regular flow and hence all we need to ensure is that the time counting mechanism is not affected accidentally by the movement of the clock (so that the counting mechanism keeps in pace with the flow of real time) which is rather easy to assume and even implement. So, according to this argument this time synchronization procedure belongs to classical mechanics but not to Lorentz mechanics. However, we will see in the exercises that for this argument to be fully established, it needs another assumption that is the time counting mechanism necessarily follows the flow of real time of the frame of the counting mechanism, i.e. we cannot find a clock that runs with a fixed counting rate regardless of its state of rest and motion since all types of clock necessarily follow in their count rate the flow of frame time.[64] Although this is a generally accepted assumption, especially in the special relativity circles where there is very little distinction (if there is any) between time counting mechanism and time flow in the frame of the mechanism, it is not a well established fact because there is no logical inconsistency in assuming such a counting mechanism. Anyway, if we assume (let be the impossible) that we can find a counting mechanism that is independent of the flow of time in the frame, so that it keeps in pace with a master clock which is at rest in the synchronized frame rather than with the time of the frame of the moving clock, then this synchronization procedure should also be applicable in Lorentz mechanics not only in classical mechanics. As we will see, we should distinguish in the time counting mechanisms (or

---

[64] The essence of this is the impossibility of creating a "faulty" clock that can be made to follow a counting rate that is not in pace with the flow of time of its frame, i.e. we cannot make a deliberately defective clock that reads wrong time according to a certain pattern. This impossibility is not obvious.

3.4 Time Measurement and Synchronization 87

clocks) between fundamental physical processes such as atomic transitions which may necessarily follow in their counting rate the flow rate of real time of their frame and the synthetic or composite counting mechanisms, like ordinary clocks, which do not necessarily follow the flow rate of real time of their frame. We believe that the latter, at least, can be used in this synchronization method even within the framework of Lorentz mechanics where the flow rate of real time is supposedly frame dependent. We should also distinguish between the physical processes that are used in the definition, calibration and measurement of time and hence although some fundamental processes will necessarily be affected by the motion, other fundamental processes may not be affected, and this depends on what we call the "defining concept". For example, if we use the speed of light as the defining concept then the fundamental physical processes that are based on this speed and used in the measurement of time will necessarily be affected, but other fundamental processes may not be affected since they may not be correlated to the defining concept or depend on it (see the exercises).

2. The second method, which belongs to Poincare and is commonly used in the literature of special relativity, uses the speed of light to synchronize the clocks with no necessity for moving the clocks after synchronization. According to this method, the clocks are set in a stand-still state where each clock is placed in its final position in the space. The clock at the origin of coordinates,[65] which we may call the master clock, is then set to read zero time (00:00) while each one of the other clocks is set to read the time required for a light signal sent from the origin of coordinates to reach that point. If we assume the constancy of the speed of light at all positions (i.e. homogeneity) and in all directions (i.e. isotropy) then the offset (or advance) time of each clock should be $t_o = d/s$ where t_o is the offset time of the particular clock, d is the distance of that clock from the origin of coordinates, and s is the speed of light which is assumed to be a universal constant in that frame. A light signal is then sent from the origin of coordinates in all directions with the start of the master clock at the origin. As soon as any one of the other clocks receives the light signal, it starts running and because of the time-offset of that clock, its reading will be identical to the reading of the clock at the origin. Hence, eventually all these clocks in this 3D array will read the same time as seen by the observer in that frame and in fact by any observer in another frame. This synchronization method is based on the constancy of the speed of light in that frame regardless of direction, location or time, i.e. it is based on the homogeneity and isotropy of the space and the homogeneity of time with respect to the speed of light. It is also based on a number of other generally-accepted assumptions, e.g. the time-measuring mechanism is not affected by the clock location (due for example to different ambient physical conditions like temperature and external force fields), and the rate of time measurement is identical in all clocks as they use identical or similar time-counting mechanisms. In fact, these general assumptions also apply to the first method and indeed to any synchronization method.

In this regard we note that the Poincare synchronization procedure, as described in the

---

[65] This reference or master clock is not required to be at the origin of coordinates although the procedure will be more tidy and intuitive if this is the case.

second point above, does not require for its validity the second postulate of special relativity about the invariance of the observed speed of light across all inertial frames and hence if the above assumptions of homogeneity and isotropy are accepted, this synchronization method will be valid regardless of accepting or rejecting special relativity and its postulates. However, the procedure requires the constancy of the speed of light in any location and in any direction in the given inertial frame and hence if we assume that the value of s is a frame-dependent parameter, we can still use light for the above synchronization procedure as long as it is assumed constant in that frame.[66] Accordingly, there is no particular significance in using light in this synchronization procedure and hence any type of transmitted signal (such as sound waves) can be used in this synchronization procedure once we assume it satisfies the above requirements, i.e. its speed is independent of location, direction and time and the synchronization belongs to an inertial frame. However, we should remark that it is difficult to accept the assumption of the independence of direction if we accept a velocity transformation rule that depends on direction, like the Galilean velocity transformation,[67] and hence we need to assume the constancy of the speed of light with respect to direction if the above-described procedure is to be valid although this speed may still be frame-dependent. Nonetheless, the procedure will still be valid, even if the speed is dependent on direction, as long as we can find and implement a correction mechanism that accounts for the effect of this direction-dependency. In fact, such a mechanism is easy to find and implement as long as the nature of the direction-dependency is well known and can be easily quantified.

Exercises
1. Try to make a deep analysis and criticism to the concept of synchronization.
2. Discuss the validity of the first method of time synchronization in classical mechanics and in Lorentz mechanics.
3. Make a more fundamental distinction between the physical processes whose time rate should be intrinsically affected by the motion and those which should not.
4. Apart from the universality of the speed of the employed signal within the given frame, there is another fundamental condition for the validity of the Poincare synchronization procedure. What is this?
5. Discuss if the invariance of the speed of light across all inertial frames (i.e. the second postulate of special relativity) is needed in the Poincare time synchronization procedure.
6. Assess a Poincare-like time synchronization procedure that is based on the use of sound, instead of light, where the source of sound is in the rest frame of the medium of propagation, e.g. air or water.
7. Is the Poincare method of synchronization proprietary to Lorentz mechanics or it can also be used in classical mechanics?
8. Discuss the validity of the Poincare time synchronization procedure according to clas-

---

[66] In brief, as long as s is constant across any given frame it is fine, even if s is not invariant across different frames.
[67] This reservation should mainly apply to a classical wave propagation model but not to a projectile propagation model.

3.5 Calibration of Space and Time Measurement 89

sical mechanics.
9. A clock that is at $\mathbf{r}_c = (0.1a, -1.3a, 8.4a)$ where $a = 3 \times 10^8$ is to be synchronized by the Poincare procedure. If the master clock is at $\mathbf{r}_m = (-0.35a, 3.4a, 0.4a)$ what is the offset time of the clock assuming that the speed of light in that frame is c.
10. What sort of symmetry the offset time of clocks in the Poincare procedure should have? Also, where is the location of the center of symmetry?

3.5 Calibration of Space and Time Measurement

As indicated earlier and will be seen in more details later, the whole Lorentz mechanics, as a theory for spacetime coordinate transformations, is based on adopting the characteristic speed of light as an ultimate standard for calibrating space and time measurements. Now, since the standard for space and time measurements inevitably enters in the definition of space and time, the spacetime in any frame will be dependent on this adopted standard of calibration. Hence, a completely consistent theory, which is different from Lorentz mechanics and based on different transformations, can in principle be obtained by adopting a different standard for the calibration of space and time measurements. However, this does not necessarily mean that one of the two theories should be wrong or less accurate (e.g. like classical mechanics as an approximation to Lorentz mechanics) because these two theories can describe the unique reality from different perspectives by using different concepts and formulations. Yes, even if the two theories are correct and accurate, one of them may be theoretically or practically superior and may have more potential for extension and generalization. This should motivate the search for an alternative theory to the theory of Lorentz mechanics even if we believe that Lorentz mechanics is a correct and accurate theory.

3.6 Reference Frame

As seen earlier, a reference frame is a combination of a coordinate system with a time assignment mechanism so that all the points of the spacetime can be uniquely and unambiguously identified by a set of spatial and temporal coordinates and hence all the events in the spacetime can be monitored and recorded. In the following subsections we investigate the construction of reference frames and their various types.

3.6.1 Construction of Reference Frame

The following procedure can be used to build a reference frame:
1. We start by introducing a 3D coordinate system to the space. This coordinate system is usually chosen to be a rectangular Cartesian system although this is not a required condition.
2. We then introduce to the space a 3D array of clocks positioned equidistantly in each one of the three coordinate directions.

3. The clocks are then synchronized by following one of the synchronization methods; the main synchronization methods are described in § 3.4. Accordingly, all the clocks in the 3D array will read the same time.

On constructing a reference frame for a given spacetime manifold, an observer in that frame or in another frame can assign spacetime coordinates to any event that occurs in the event space by recording the spatial coordinates of the array point nearest to the event and reading the clock at that point. Any degree of accuracy of these records and readings can be achieved by employing a sufficiently resolved densely-packed array to fulfill the required accuracy. As discussed before, building a reference frame by the above procedure applies only to inertial frames where the laws of physics on which the reference frame is based and built are known to apply.

Exercises
1. Describe in detail how to construct a simple and tidy frame of reference using the Poincare synchronization procedure.
2. Discuss and assess the frame construction procedure that you described in the previous exercise.
3. Explain how an observer can assign spacetime coordinates to events that occur in a given frame.

3.6.2 Inertial and non-Inertial Frames

Inertial frame is a frame of reference in which Newton's first law (i.e. the law of inertia) holds true (see § 1.10) and this explains why it is known as "inertial". The validity of Newton's first law is a sufficient and necessary condition for a frame to be inertial. However, the definition of inertial frames may be based on the validity of all three Newton's laws of motion (as we did earlier) since the validity of the first law logically implies the validity of the other two laws, or at least this is the case according to our observations of the world as formulated in classical mechanics and hence this implication is empirically established. In fact, characterizing inertial frames by Newton's first law only is based on the sufficiency of this condition; otherwise it is legitimate to characterize and define inertial frames by holding the laws of classical mechanics (not only Newton's first law or Newton's three laws of motion) since these laws, which are generally based on Newton's laws of motion, also hold in these frames. In brief, the validity of Newton's first law qualifies the frame to be inertial and since the validity of Newton's laws of motion, and even the laws of motion of classical mechanics in general, is a characteristic feature to all inertial frames, these frames can also be characterized by the validity of these laws. However, this extension may reduce the definition to be redundant or circular if the definition of inertial frame is supposed to be used to identify the frames in which these laws hold true, i.e. inertiality implies validity of these laws. Accordingly, a more fundamental definition[68] of inertial frames that is based on the existence of an absolute frame (i.e. inertial frames are those frames which are in a state of rest or uniform motion with respect to the absolute frame)

---

[68] We usually label the previous definition (i.e. the one based on Newton's laws) as procedural.

3.6.2 Inertial and non-Inertial Frames

may be adopted instead (refer to the following paragraph and exercises). As we will see, any reference frame which is at rest or in a state of uniform motion relative to an inertial frame is also inertial, while a reference frame which is in a state of accelerated motion with respect to an inertial frame is non-inertial.[69]

We note first that the above procedural definition of inertial frames may be considered as a definition by the symptoms rather than by the cause. The more fundamental characteristic of inertial frames, which any profound definition should be based upon, is the state of motion of the frame as being non-accelerating, i.e. its kinematic state. This state of non-acceleration should be the cause for the frame to hold Newton's laws as well as other laws of mechanics, as we will see later. However, defining inertial frames as non-accelerating frames may be seen as a verbal rather than a technical definition (i.e. it is only explanatory and lacks real technical content) since acceleration requires another frame to which the acceleration is referred. But this criticism is legitimate only if we deny the existence of an absolute frame to which all states of rest, uniform motion and accelerated motion are referred. So, if we accept the existence of an absolute frame then this definition will be full of technical content since it defines the state of the frame with reference to this real physical entity, i.e. absolute frame. We should also note that the procedural definition of inertial frames may be stated differently by different authors (e.g. frames in which no acceleration is observed in the absence of external forces), but the essence of most of these definitions is the same that is Newton's laws of motion, and the law of inertia in particular, hold true in these frames.

In the following, we list a number of properties that characterize inertial and non-inertial frames; most of these properties are equivalent and some are even identical but they are phrased differently:

1. In inertial frames, a massive object will be seen in a state of rest or uniform motion in the absence of external forces.
2. In non-inertial frames, a massive object will be seen accelerating in the absence of external forces.
3. Any frame that is in a state of rest or uniform motion with respect to an inertial frame is also inertial.
4. Any frame that is accelerating with respect to an inertial frame is non-inertial.
5. A frame that is in a state of rest or uniform motion with respect to a non-inertial frame is also non-inertial.
6. A frame that is accelerating relative to a non-inertial frame can be inertial or non-inertial.
7. The three Newton's laws of mechanics hold true in inertial frames but not in non-inertial frames. As indicated above, this also applies to all the laws of classical mechanics which are based on Newton's laws since these are no more than disguised forms of Newton's laws. This should also extend to all the laws of physics which are known to apply in

---

[69] These statements should equally apply to the fundamental and procedural definitions. However, although this may be conventional with respect to the fundamental definition, it is physical (or "real") with respect to the procedural definition since it has observable physical consequences.

3.6.2 Inertial and non-Inertial Frames 92

these frames although this may not be valid as a characterizing definition.[70]

8. All inertial frames are equally valid for formulating the laws of physics. Accordingly, the laws of physics should be invariant under time and space transformations between inertial frames. This should apply both to classical physics where the Galilean transformations are applicable and to Lorentz mechanics where the Lorentz transformations are applicable. As we saw earlier (refer to § 2), the failure of this criterion in classical physics (i.e. the failure of the classical transformations to transform the laws of electromagnetism invariantly) played an important role in the emergence of Lorentz mechanics.[71]

We should remark that some frames may be approximated as inertial frames although they are not really inertial. For example, the Earth may be treated as an inertial frame in some cases although it is non-inertial frame due to its accelerated motion around the Sun and within the Milky Way Galaxy as well as its spin around its spherical axis.[72] In this context, we note that such approximations depend on the case and context and hence these approximations may be valid in one case or context but not in another case or context. In other words, it may be legitimate to treat a given non-inertial frame as inertial in certain circumstances but it is illegitimate to treat this non-inertial frame as inertial in other circumstances. The difference may be attributed to the scale of the considered phenomenon and whether it is large or small (i.e. has significant spatial and temporal extension or not where this may be linked to the curvature of spacetime which is supposed to be approximately-flat at small scales);[73] however this is only one factor and hence other factors should also be considered in these approximations. In brief, the important issue that determines the validity or invalidity of these approximations is whether the

---

[70] In fact, this extension is the essence of the extended classical invariance principle (and even the relativistic principle subsequently) which originates from the Galilean relativity (that is apparently restricted to mechanics) and which was developed during (and perhaps even before) the investigations (by Lorentz and others) about the necessity of the invariance of Maxwell's equations under the right set of transformations of space and time which led to the development of Lorentz transformations (to replace the Galilean transformations) and the emergence of Lorentz mechanics. Also see the next point.

[71] From this (rather twisted) presentation, it should be obvious that the extended invariance principle predates the emergence of special relativity. In fact, there was no sufficient justification for the emergence of Lorentz transformations and Lorentz mechanics without the existence of this extended version of the invariance principle in classical physics (i.e. prior to the emergence of special relativity). This should clarify an important issue about the credit for developing these fundamental ideas and principles.

[72] We assume that the resultant of all these diverse forms of motion (particularly the rotational) will violate the state of inertiality without need for detailed consideration of these forms and any particular external frame. This is based on the assumption that it is very unlikely that the rest frame of the Earth is the frame of absolute rest or it is in uniform motion with it. However, the treatment of the Earth as an inertial frame is justified when the effects of acceleration are marginal. This should also be consistent with what will be claimed later (see § 11.2) that the Earth may be a good approximation to the absolute frame.

[73] As discussed earlier, the geometric approach of physical theories like general relativity characterizes the spacetime of inertial frames as flat in contrast to the spacetime of non-inertial frames (including their gravitational equivalents) which is characterized as curved.

acceleration is negligible (considering the context and objective of the problem) or not and hence such approximations are acceptable if the effects of acceleration are within the allowed error margin of the problem; otherwise they are not.

Exercises
1. Define inertial frame in a memorable way.
2. Compare between the procedural definition (i.e. the one based on Newton's laws) and the fundamental definition (i.e. the one based on the state of motion relative to absolute frame) of inertial frame.
3. Characterize inertial and non-inertial frames in terms of their velocity and acceleration in space.
4. Assess and criticize the use of concepts like velocity and acceleration in the definition of inertial and non-inertial frames where these concepts may require a "master reference frame" to which these concepts are referred.
5. Show that a frame which is in a state of acceleration relative to a non-inertial frame can be inertial or non-inertial and hence its inertiality cannot be easily identified from simple observations. Demonstrate this graphically.
6. Use a graphical argument, similar to the one in the previous exercise, to show that the state of inertiality of any frame can be easily identified from simple observations from any given inertial frame.
7. Show, using a simple non-rigorous argument, that in the definition of inertial frames the condition of holding Newton's first law and the condition of holding all three Newton's laws are equivalent.
8. A camera is installed on a rotating platform, and another camera is installed on the ground. Assess how the two cameras will record the events that occur in their surroundings.

3.6.3 Difference between Inertial and non-Inertial Frames

The difference between inertial and non-inertial frames is very important not only for mechanics but for the whole of physics, and therefore we feel this point, which has been examined rather briefly in the last subsections, requires further clarification. So, we ask again: what is the difference between inertial and non-inertial frames, i.e. why in some frames Newton's laws hold true while in other frames they do not? The simple answer is that because inertial frames are at rest or uniformly moving while non-inertial frames are accelerating. But then another question will emerge that is being at rest or uniformly moving or accelerating are extrinsic properties and hence with respect to what frame these attributes are held? If we believe in the existence of a universal rest frame, such as the frame of absolute space if we believe in the existence of absolute space, then the answer is simple that is inertial frames are those frames which are in a state of rest or uniform motion with respect to this universal absolute frame, while non-inertial frames are those frames which are in a state of acceleration relative to this universal absolute frame.

The difference between the inertial and non-inertial frames, by holding and not holding Newton's laws of motion, can then be easily justified because the absolute space can privilege the inertial frames by the distinction of holding these laws and deprive the other frames from this privilege. The essence of Newton's laws will then be that: absolute space is the absolute and unique authority for defining force and acceleration where force means a physical agent that produces acceleration with respect to this absolute space. Consequently, the first law about inertia and the second law about force will be no more than definitions of this intrinsic property of the space because to have any sensible meaning to being absolute and unique authority for defining force and acceleration we should have no acceleration in the absence of force (which is the first law) and we should have acceleration in the presence of force (which is the second law). In fact, even the third law can be explained in this way if we add to the above the condition that within this absolute space forces exist only in couples, and this could be another intrinsic property of the absolute space. Further discussion about this issue will be given later (see for example § 9.4).

Exercises
1. Discuss the necessity for the existence of an absolute frame for the reality of the physical difference between uniformly moving frames and accelerating frames as being inertial and non-inertial.
2. Discuss the suggestion that what discriminates between inertial and non-inertial frames is not the existence of absolute space (or rather absolute frame) but the actual large scale distribution of matter and energy in our space that produces these different effects and makes some frames behave as inertial while others as non-inertial.

3.7 Causal Relations

A causal relation between two events, V_1 and V_2, in the spacetime means that V_1 is the cause of V_2 or V_1 is caused by V_2. It is generally assumed that for any event to be a cause for another event, there should be some sort of communication between the two events by a signal that should be sent from the cause event to the caused event so that the cause can influence the caused and avoid the presumably physically-impossible "action at a distance". This will put a limit on the causal relations between events in the spacetime, that is any causal relation in this space is restricted by the speed of the communication signal and hence no causal relation can exist between two events if the time of the occurrence of the caused event is earlier on the timeline than the time for the signal to travel from the cause event and reach the location of the caused event. For example, if V_1 is supposed to cause V_2 by sending a signal from V_1 at $t = 0$ and this signal requires 10 seconds to reach the location of V_2, then if V_2 is actually occurring at $t = 5\,\mathrm{s}$ then V_2 cannot be caused by V_1 since V_2 occurs before the signal of V_1 can arrive to the location of V_2. As indicated above, this may be seen as a form of action at a distance which is supposed to be physically impossible.

We should remark that the generality (at least) of the requirement of a communication signal for establishing causal relations may be questioned in the case of co-positional or identical events (see § 1.3) unless it is claimed that such causal relations are impossible which is not obvious. The problem with co-positional is having zero distance and the problem with identical is having zero time as well and hence in both cases the communication between the events by a signal of a finite speed is difficult to imagine. This may be more obvious in the case of co-positional where the signal should travel in time only and hence whatever the presumed speed of the signal (assuming it is finite, i.e. $0 < v \leq c$ or $0 < v < \infty$) it is difficult to imagine. In fact, if we do not impose a lower limit (i.e. being greater than zero) on the speed of the alleged communication signal then we may argue that there should be no reason to impose an upper limit (i.e. not exceeding c according to special relativity) because it is not an ordinary physical signal. The unusual nature of this alleged causality communication signal is its speed variation since it should be able to take any value over a wide range even if it is restricted by an upper limit according to special relativity.[74] There are many issues and question marks about this alleged signal but we cannot go through these details due to limited space.

Exercises
1. Assess Newton's law of gravitation and Coulomb's law of electrostatic force in the light of the concept of action at a distance.
2. Based on the nature of causal relations, what is the relation between two simultaneous events?

3.8 Speed of Light

As discussed previously, there are several possibilities for the speed of light in free space as observed from an inertial frame, the main ones are:
1. The speed of light takes the characteristic value c only in a privileged absolute frame, and therefore the observed speed of light is variable for all observers that move with respect to the absolute frame where the speed v of the observer with respect to the absolute frame will be added vectorially to the value c, i.e. $c \pm v$.[75] In fact, this possibility treats light like a material wave according to the classical view (refer to § 1.13.2). If a medium, like ether, for the propagation of light is assumed and this medium is supposed to be at rest in the absolute frame then the characteristic speed c belongs to the frame of ether which can also be identified with the frame of absolute space.
2. The speed of light is like the speed of projectile (refer to § 1.13.1) and hence its characteristic value c is with respect to the frame of its source. Therefore, the observed speed of light is frame dependent since it depends on the motion of the observer relative to the source, i.e. the observed speed will be given by $c \pm v$ where v is the relative speed between the observer and the source. This possibility does not require the existence of an absolute frame although it is not fundamentally in conflict with such a frame, i.e. it is logically consistent with the existence of absolute frame although a propagation medium such as ether is not needed.

---

[74] Anyway, if the causality communication signal is not required in some cases (e.g. spatially identical events) then this will put a question mark on its requirement in other cases. Moreover, if this signal can have variable speed, then this will put a question mark on the upper limit restriction, i.e. $v \leq c$.
[75] As indicated before, we are considering in this context a 1D motion along the line connecting the source to the observer.

3. The characteristic speed of light c is a universal constant for all inertial observers and hence the motion of the observer and the motion of the source relative to any frame do not affect this universal speed of light. This is the traditional view of special relativity theory which is commonly based on the denial of the existence of any absolute frame (whether space or ether) or at least irrelevance of such a frame as far as Lorentz mechanics is concerned. More details about this issue will come later (see for example § 9.3.2 and § 11).

We note that according to the first and second possibilities the characteristic speed of light c should be neither restricted to light nor ultimate (although this may be disputed), and therefore we could have speeds of physical objects other than light that reach or exceed c.[76] This is consistent with the framework of classical mechanics as represented by the Galilean velocity composition which is additive in nature. In contrast, according to the third view (following the special relativity interpretation) the characteristic speed of light c is both restricted to light and ultimate and hence light (and all electromagnetic radiations and indeed all massless physical objects) propagate through free space only with this characteristic speed c, while everything else cannot reach or exceed this speed.

Exercises

1. Discuss and assess the main possibilities about the speed of light in inertial frames of reference.

3.8.1 Speed of Light as Restricted and Ultimate Speed

In this subsection, we discuss the commonly held view that no physical speed can reach or exceed the characteristic speed of light c. This is one of the claims that associate the special relativity interpretation of Lorentz mechanics. Accordingly, several claims of speeds reaching or exceeding c both in experimental work and in astronomical observations have been rejected based on the premise that the characteristic speed of light c is restricted to light so no physical object other than light[77] can move with this speed; moreover this speed is the ultimate physical speed for any physical object and hence no physical object including light can exceed this speed.[78]

Now, it may be claimed that the impossibility of reaching or exceeding c is embedded in the formalism of Lorentz mechanics regardless of the postulates and consequences of special relativity, because the Lorentz factor γ has a singularity when $v = c$, while this factor becomes imaginary when the characteristic speed of light c is exceeded, i.e. $v > c$.[79] However, this may be challenged by the following:

---

[76] It should be obvious that the observed speed of light can also be lower or higher than c.
[77] In this context, "light" should be extended not only to all types of electromagnetic radiation but to all massless objects as well.
[78] We should refer to a number of apparently superluminal phenomena that have been interpreted in such a way to avoid the possibility of exceeding c.
[79] The Lorentz velocity transformations may also be used to establish the claim that the limit on the speed is imposed by the formalism of Lorentz mechanics (refer to § 4.2.2).

3.8.1 Speed of Light as Restricted and Ultimate Speed

1. It is possible that the Lorentz factor γ as given by Eq. 5 is just an approximation and hence the exact expression of γ may contain terms of higher orders in β. Although, these terms are negligible in practice, they can avoid some of the consequences of reaching or exceeding the speed of light, i.e. singularity of the Lorentz γ factor when $v = c$ and becoming imaginary when $v > c$.

2. It is also possible that the whole Lorentz mechanics can be an approximation to another mechanics which is valid for objects reaching or exceeding the characteristic speed of light and hence Lorentz mechanics is just an approximation to that mechanics, like classical mechanics as an approximation to Lorentz mechanics. Accordingly, the current formulation of Lorentz mechanics is not final and another more comprehensive and general theory could emerge where both classical and Lorentz mechanics are valid approximations to this more general theory. If this is the case, then classical mechanics will be a good approximation to that general mechanics at low speeds (i.e. $v \ll c$) while Lorentz mechanics will be a good approximation to that general mechanics at speeds comparable to the characteristic speed of light (i.e. $v \sim c$). The general mechanics will then apply to phenomena where speeds can reach or even exceed the characteristic speed of light. Accordingly, the impossibility of reaching or exceeding the speed of light may be valid in the framework of Lorentz mechanics in its current formulation, and in the framework of special relativity in particular, but it does not necessarily represent a fundamental law as it is claimed by special relativists. So, although the current formulation of Lorentz mechanics is not valid for speeds reaching or exceeding c, this possibility is not physically impossible within another theory which is more comprehensive and precise than Lorentz mechanics.

3. Even if we accept that c is restricted to light and it is the ultimate speed within the domain of validity of Lorentz mechanics which is inertial frames, we cannot, on the ground of Lorentz mechanics, rule out the possibility of reaching or exceeding c outside this domain, i.e. non-inertial frames.

4. Even if we accept that c is characteristic to light and it is the ultimate speed for light, we cannot rule out the possibility of exceeding (or falling below) this speed by a physical entity other than light because what enters in the formalism of Lorentz mechanics is the speed of light c and hence the speed restrictions that are imposed by this formalism should be related to light. In other words, we need an independent evidence to extend the restrictions imposed by the formalism of Lorentz mechanics to all types of (massless) physical objects other than light. Further clarifications about this issue will be given in the future.

We should remark in this context that reaching or exceeding the speed of light (although at lower value than c) in material media such as water is an established fact according to modern physics and this could equally apply to free space even if this possibility is invalid according to the current formalism of Lorentz mechanics. So, despite the commonly held opinion that c is restricted to light and it is ultimate speed, and hence no physical object other than light can reach this speed and no physical object including light can exceed this speed, this should not be considered as an established and universal fact and hence it should be possible in principle to reach and exceed this speed even if this possibility is

3.8.1 Speed of Light as Restricted and Ultimate Speed

ruled out by the current formalism of Lorentz mechanics.

Some special relativists may try to prove the claim that c is at least an ultimate speed by the second postulate of special relativity because the constancy of the speed of light in all inertial frames, according to this postulate, means that c is an ultimate speed because it cannot be exceeded. We note first that the second postulate of special relativity is a postulate and not an established fact; moreover it is not part of the formalism of Lorentz mechanics if we follow the method of special relativity by deriving the Lorentz transformations from the two postulates. We also note that the experimental evidence in support of Lorentz mechanics does not imply the validity of this postulate because the formalism of Lorentz mechanics can be established by another framework and interpretation and even without any framework or interpretation. As discussed earlier and will be investigated further, special relativity with its postulates and principles is only an interpretation of the formalism of Lorentz mechanics and hence the rejection or collapse of special relativity and its postulates and principles do not necessitate the rejection and collapse of the formalism of Lorentz mechanics. Similarly, the evidence in support of the formalism is not an evidence in support of special relativity and its postulates. Moreover, even if the second postulate of special relativity is accepted as a basis for the formalism of Lorentz mechanics this does not rule out the possibility of exceeding c in non-inertial frames or within a more comprehensive theory, as explained earlier. We should also repeat our claim that the formalism of Lorentz mechanics can be established with no resort to any postulate other than the Lorentz spacetime coordinate transformations themselves supported by experimental evidence. Adding to all these arguments, there are a number of question marks about the validity of the logic of the derivation of Lorentz spacetime coordinate transformations from the postulates of special relativity. As we will see in § 12.4, most of the available derivations are not sufficiently rigorous.

It may also be claimed that exceeding the speed of light means traveling back in time which is inconsistent with causality. However, these claims and arguments are either based on certain interpretations and assumptions which are not well established, or based on false logic and hence the propositions do not lead to the conclusions. Hence, these claims and arguments are not worth further investigation. We should remark that there are a number of phenomena and observations, especially in astronomical studies, that were linked to superluminal speeds. All these were allegedly explained and justified by special relativists within the framework of special relativity with no violation of the second postulate of this theory, i.e. according to special relativists these phenomena appear to be superluminal but they are not actually superluminal. However, at least some of these explanations and justifications can be challenged.

Also, there are several claims by respected scientists of violations of the constancy of the speed of light where this speed is claimed to be frame dependent in contradiction with the second postulate of special relativity. In fact, some of these claims may even be a challenge to the current formalism of Lorentz mechanics with regard to the velocity composition rule (which is not supposed to be Galilean) if the frames of observation are really inertial which may be difficult to establish. All these claims, however, were rejected by special relativists with no examination or consideration. The main basis for the rejection of these claims is their conflict with special relativity which allegedly is a firmly established theory.

We should remark that the issue of c being restricted to light and ultimate physical speed concerns the speed of light in the direction of propagation and hence it does not include lateral speed where the source moves in a perpendicular direction to the direction of propagation causing a lateral movement of the projection of the light beam. An example of this is a lighthouse with a rotating light source where the projection of the light beam on a distant mountain, for instance, moves laterally. In such cases, the lateral speed (which may also be called the sweep speed) of light can exceed c. This issue will be investigated further in the exercises of this section and in § 10.

Exercises

1. Summarize the stand of special relativity about the limits and restrictions on the speed of physical objects.
2. Why we insist on the claim that reaching and exceeding the characteristic speed of light c is possible in principle if it is denied by the formalism of Lorentz mechanics?
3. What are the main restrictions that should be imposed on the claim that no massive object can reach or exceed the speed of light?
4. Explain how the lateral speed of light can exceed the constant c.
5. It has been shown in the literature of special relativity that the speed restrictions are required for preserving causal relations between events. Can this be regarded as a proof for these restrictions?[80]

3.8.2 Measuring the Speed of Light

We should distinguish between the one-way measurement of the speed of light and the two-way measurement of this speed where the first case is based on measuring the speed in a single direction while the second case is based on measuring the average speed over two opposite directions, i.e. over the forward-backward journey. In fact, most (if not all) of the methods of measuring the speed of light are two-way measurements. However, there are claims in the recent research literature of developing and implementing methods of one-way measurements. But these claims have been criticized and dismissed and hence they are difficult to accept. It is largely believed that the one-way measurement of the speed of light is impossible.[81]

We note that the two-way measurement methods of the speed of light are supposed to require the assumption of isotropy if these measurements are used to infer the actual speed of light and not the average speed over the two specific directions. Moreover, it may be claimed that these methods require the second postulate of special relativity for their validity because if the speed of light is frame dependent (i.e. subject to the Galilean

---

[80] It is claimed in the literature of special relativity that the desire to preserve causal relations is the basis for c being regarded as an ultimate speed.

[81] There are many issues about the one-way and two-way speed of light that deserve to be investigated and discussed. However, these issues are not of primary interest to us as they are not within the objectives of this book and therefore we do not go through these details. The interested reader should consult the literature.

velocity composition rule) then the effect of frame dependency will be masked since the average speed will smooth out any direction dependent difference between the speeds in the two opposite directions. However, we will see in the exercises that any potential direction dependency of the speed can still be inferred from the dependence of the average speed on the orientation by rotating the measuring apparatus and observing any variation in speed, similar to the working principles of Michelson-Morley experiment which are based on the validity of the Galilean transformations[82] (refer to § 2.6 and § 12.2) although this can be done by using a single beam instead of two interfering beams as in the Michelson-Morley experiment.[83]

Exercises

1. Assess the methods of one-way and two-way measurement of the speed of light considering the issue of possible dependency of this speed on direction.
2. A scientist is trying to measure the "ether wind" that is caused by the Earth movement in space. He uses a device that measures the time of flight of light in different orientations by rotating the device and recording the time as a function of orientation. The device simply measures the time of the two-way trip of light over a length $L = 15$ meter in its forward-backward journey. The scientist observed a maximum two-way trip time of 1.0001×10^{-7} second in the east-west orientation and a minimum two-way trip time of 1.00005×10^{-7} second in the north-south orientation. Assume the validity of the Galilean velocity composition rule and use the Michelson-Morley analysis method (refer to § 2.6 and § 12.2) which is based on classical mechanics to infer the speed of the ether wind and its orientation according to this method.

3.9 Spacetime in Lorentz Mechanics

The concept of spacetime which permeates Lorentz mechanics is a generalization of the concept of ordinary "spatial" space where a temporal dimension is added to the ordinary space to produce a spacetime "space" or manifold. While space and time in classical mechanics are generally considered to be two separate and independent entities, space and time in Lorentz mechanics form parts of a single entity which is commonly known as spacetime. Spacetime in Lorentz mechanics is a flat manifold and hence it is Euclidean in nature although it may not be considered Euclidean technically (refer to § 6.3 for more details).

As indicated earlier and will be investigated in detail later, inertial observers who are in relative motion have different measurements of length and time. Now, because spatial distance and time intervals are not universal but they depend on the observer, we should

---

[82] We should also note that the Michelson-Morley experiment is based on a classical wave propagation model where the source and observer are in relative rest with each other and in relative notion with the medium.

[83] If we accept the working principles of Michelson-Morley experiment then we can calculate v (as done in the exercises) and hence we can also find the observed speed of light in each direction. Regarding the characteristic speed of light, it is c which is a given constant whose value facilitates the calculation of v and the observed speed in each direction.

3.9 Spacetime in Lorentz Mechanics

try to find an invariant parameter that corresponds to the infinitesimal line element (or arc length) ds in the ordinary "spatial" space.[84] This invariant is the spacetime interval which is given in its infinitesimal form by:

$$(d\sigma)^2 = \left(dx^0\right)^2 - \left(dx^1\right)^2 - \left(dx^2\right)^2 - \left(dx^3\right)^2 \qquad (69)$$

This $d\sigma$ which is the infinitesimal "line element" or "arc length" of spacetime is universal and hence all inertial observers will have the same measurement of this spacetime parameter, i.e. all inertial observers will agree on the value of $d\sigma$ between two given events that occur in the spacetime manifold under the Lorentz spacetime coordinate transformations (refer to § 3.9.4). As seen earlier, the temporal coordinate in the spacetime is ct and not t and hence all the coordinates in the spacetime have the physical dimension of length, i.e. the above mathematical expression for spacetime interval is dimensionally consistent.

In fact, this invariance of spacetime interval may be seen as the origin of the idea of "spacetime" because in the ordinary 3D spatial manifold, the invariant ds defines the metric of the underlying "spatial" space, and hence when we see that the invariant in the manifold of Lorentz mechanics is this mix of space and time (represented by $d\sigma$ which defines its metric) that possesses this invariant property in this 4D manifold then the underlying "space" of Lorentz mechanics is not the space or the time but the "spacetime". In brief, the role of ds in the space of classical mechanics is taken by $d\sigma$ in the "space" of Lorentz mechanics, and hence the role of ordinary space of classical mechanics is taken by "spacetime" in Lorentz mechanics. In this merge of space and time, the two separate manifolds of classical mechanics (i.e. the 3D spatial manifold and the 1D temporal manifold) are unified in a single 4D manifold and that is why space and time are said to be merged or mingled or mixed in Lorentz mechanics.

We will see (refer to § 3.9.4) that while ds and dt are separately invariant in classical mechanics under the Galilean transformations, they are not invariant in Lorentz mechanics under the Lorentz transformations; instead what is invariant is their mix as represented by $d\sigma$. This is an indication not only to the strong bond between space and time in Lorentz mechanics but also to the nature of the underlying spaces in the two mechanics. As indicated earlier and will be seen in more details later (see § 3.2, § 3.3 and § 6.3), the spacetime of Lorentz mechanics is a flat manifold, and hence in this regard it is similar to the time and space manifolds of classical mechanics.

In this context, we should remark that the concept of spacetime in its formal and mathematical form was originated by Minkowski who developed the idea of spacetime diagrams, spacetime invariant and 4D vectors although the qualitative idea can be traced back to Poincare and may be even before.[85] Later, this concept was found useful not only in the formulation of Lorentz mechanics, but it was found useful in the formulation of other physical theories as well. Eventually, the concepts of spacetime and spacetime interval were incorporated in Lorentz mechanics and other branches of physics and nowadays the concepts and techniques of spacetime have wide spread use in physics.

---

[84] We note that in ordinary space we have: $(ds)^2 = \left(dx^1\right)^2 + \left(dx^2\right)^2 + \left(dx^3\right)^2$.

[85] We are following claims that we found in the literature; some of these claims at least may require further verification.

Exercises

1. Why we are forced to travel in time and only in forward direction but not in space? Does this break the symmetry (or equivalence) of the spacetime coordinates and hence space and time are still not completely merged in a spacetime manifold (which should usually have equivalent dimensions) even in Lorentz mechanics?
2. Discuss the issue of invariance of ds, dt and $d\sigma$ in classical mechanics and in Lorentz mechanics and the implication of this on the underlying spaces of these mechanics.

3.9.1 Spacetime Diagram and World Line

Spacetime diagram is an abstract device that is commonly used in Lorentz mechanics to graphically represent objects and events that exist and evolve in spacetime. So, the purpose of spacetime diagram is the graphical representation of physical occurrences in the 4D manifold of spacetime. The diagram represents the event space from the perspective of a particular inertial observer and hence it belongs to that observer. According to the available historical records, spacetime diagram is invented by Minkowski in 1908 and hence it is commonly known as Minkowski diagram. A physical event can be represented on the spacetime diagram by a point, while a line or a curve that connects two points in the spacetime diagram is described as world line and it represents the development of a particular series of events or the existence of an object in the spacetime. As we will see, a 2D spacetime diagram having a single spatial coordinate plus the temporal coordinate is commonly used (refer to Figure 4 for an example). This 2D diagram may represent one-dimensional motions (based for example on a state of standard setting) and may even originate from a cross section of a 3D light cone diagram of a given event (refer to § 3.9.2).[86]

In fact, spacetime diagram is no more than a generalization of the diagram that represents a spatial coordinate system by adding a temporal coordinate. However, as indicated earlier the convention is to represent the temporal coordinate of spacetime by ct rather than t so that all the coordinates of spacetime diagram have the physical dimension of length. Accordingly, spacetime diagram is a diagram representing the "coordinate system" of spacetime. Because it is not possible to make a diagram (whether 2D or 3D) that realistically represents a system with four dimensions, the number of spatial coordinates represented on the diagram are usually reduced to two or even one. Hence, a spacetime diagram is commonly simplified by plotting it as the temporal coordinate against a single spatial coordinate, e.g. ct versus x as seen in Figure 4. This type of spacetime diagram represents motions in one dimension which is inline with the previously-described standard setting that is commonly employed to simplify the formulation of coordinate transformations between two frames as well as other formulations of Lorentz mechanics (refer to § 1.5).

---

[86] We note that the 2D diagram is more appropriate for the spacetime representation of occurrences and motions related to a state of standard setting (refer to the solved problems) since the y and z coordinates are usually not needed in this representation.

3.9.1 Spacetime Diagram and World Line

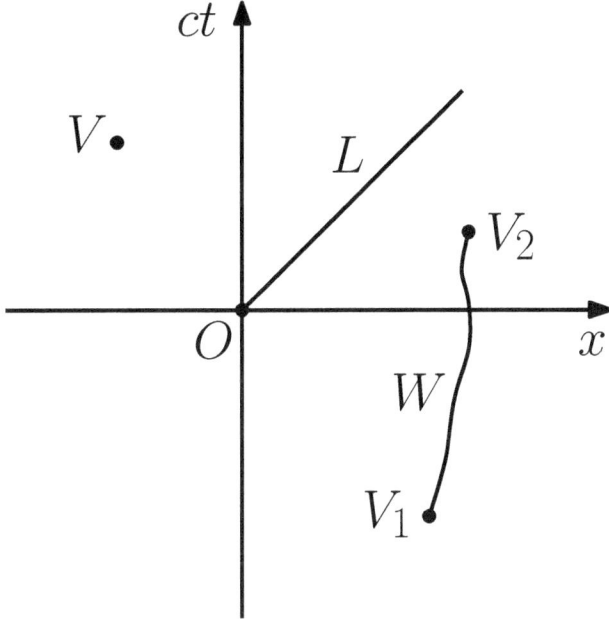

Figure 4: A spacetime diagram with a single spatial dimension where V symbolizes an event in the spacetime, W represents a world line connecting two events V_1 and V_2, and L is the world line of a light ray which is bisecting the angle between the x and ct axes and hence it is the line $ct = x$.

To illustrate the idea of spacetime diagram and clarify it further, we plot in Figure 5 a simplified 2D spacetime diagram with a number of straight lines representing the world lines of a number of uniformly moving objects in the spacetime where we note the following:

1. The segment L_1 is the world line of an object O_1 that is at rest in the x dimension, so while its temporal coordinate is changing its x spatial coordinate remains the same during this time interval.
2. The segments L_2 and L_3 represent the world lines of objects, O_2 and O_3, that are moving uniformly at speeds lower than the speed of light. However, O_2 is slower than O_3 since the speed of O_2 (which is proportional to the slope $\frac{dx}{dct}$ of L_2) is lower. We note that we define the slope in this context unconventionally as the horizontal change over the vertical change (instead of vertical over horizontal which is inline with the convention) for the purpose of representing the speed (or in fact a speed factor which is proportional to speed) rather than its reciprocal. This is due to the exchange of the temporal and spatial axes in this diagram (i.e. vertical and horizontal) from their conventional position (i.e. horizontal and vertical).
3. The segment L_4 (which is the line $ct = x$ or $x^0 = x^1$) is the world line of an object O_4 that is moving at the speed of light and hence its slope is equal to 1. If we assume that nothing but light can move at the characteristic speed of light c, then this segment can only represent the world line of a light signal.
4. The segment L_5 is the world line of an object O_5 that moves at a speed higher than

3.9.1 Spacetime Diagram and World Line

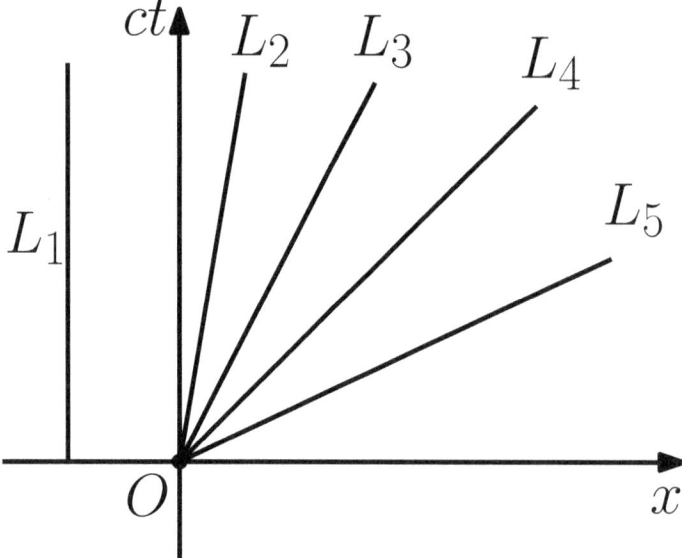

Figure 5: Graphical illustration of the world lines of a number of objects on a 2D spacetime diagram.

the characteristic speed of light c. Now, if we assume that c is the ultimate speed and hence no physical object can exceed this speed, then the world line represented by L_5 cannot represent the movement of any physical object.

Similarly, we plot in Figure 6 a 2D spacetime diagram with the world line C_1 of a uniformly accelerating (i.e. having constant acceleration) object O_1 and the world line C_2 of a non-uniformly accelerating object O_2. We note that the slope $\frac{dx}{dct}$ (which is proportional to the instantaneous speed) at any point on these world lines should not reach or exceed 1 if we assume that the speed c is restricted to light and it is the ultimate speed for any physical movement.

As outlined earlier, a continuous line or curve that connects two points on a spacetime diagram is called world line. World line represents the continuous existence of a physical object in the spacetime (or its movement in this space) or represents a continuous series of correlated events. World line may also be used as a label for the actual path in the spacetime rather than a curve on the diagram of spacetime.

Solved Problems

1. Plot a simple 2D spacetime diagram on which you represent two inertial frames in a state of standard setting with $v > 0$ as their origins coincide at $t = t' = 0$.

    Answer: Let have two frames, O and O', where O is the primary frame that is represented as usual (e.g. as in Figure 4) and hence its coordinate axes are orthogonal. Now, we need to draw the axes of O' (i.e. ct' and x') on this diagram and hence we need to use the Lorentz coordinate transformations to find how these coordinate axes will look in frame O.

    The ct' axis is given by the condition $x' = 0$, and hence we use the x' coordinate

3.9.1 Spacetime Diagram and World Line

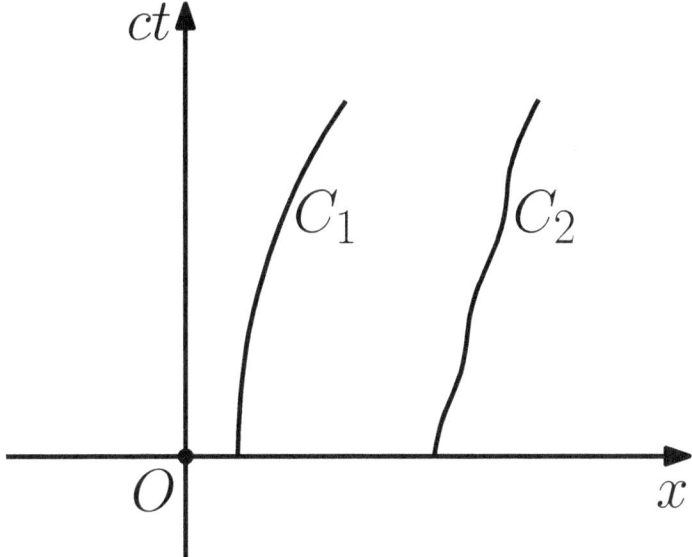

Figure 6: A 2D spacetime diagram on which the world line C_1 of a moving object with a uniform acceleration, and the world line C_2 of a moving object with a non-uniform acceleration, are represented.

transformation to find its equation in frame O, that is (refer to Eq. 90):

$$\begin{aligned} x' &= \gamma(x - vt) \\ 0 &= \gamma(x - vt) \\ x - vt &= 0 \\ vt &= x \\ ct &= \frac{c}{v}x = \beta^{-1}x \end{aligned}$$

where the third step is justified by $\gamma \neq 0$.

Similarly, the x' axis is given by the condition $ct' = 0$, and hence we use the ct' coordinate transformation to find its equation in frame O, that is (refer to Eq. 94):

$$\begin{aligned} ct' &= \gamma\left(ct - \frac{vx}{c}\right) \\ 0 &= \gamma\left(ct - \frac{vx}{c}\right) \\ ct - \frac{vx}{c} &= 0 \\ ct &= \frac{v}{c}x = \beta x \end{aligned}$$

Accordingly, the diagram should look like Figure 7.

2. Referring to the previous question, how the x' and ct' axes vary as the relative speed v varies?

   Answer: As the relative speed v increases, the x' and ct' axes will converge toward

3.9.1 Spacetime Diagram and World Line

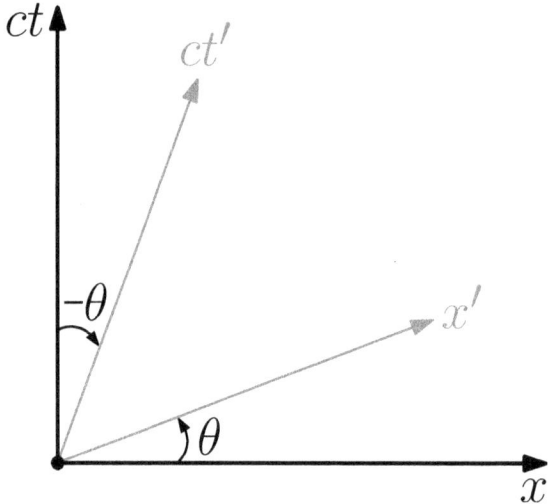

Figure 7: A 2D spacetime diagram representing two inertial frames in a state of standard setting at $t = t' = 0$.

the line $ct = x$ which represents the world line of a light ray for both frames due to the invariance of the speed of light (refer to § 4.2.2). In other words, as v increases the angle θ will increase.

3. Referring to Figure 7, how the spacetime coordinates of events are obtained from a 2D spacetime diagram representing two frames?
   Answer: It is simply done by drawing parallel lines to the coordinate axes of each frame and reading the coordinate axes at the intersection points, and hence the coordinates in the two frames are easily determined and correlated. This is demonstrated in Figure 8.

4. Referring to Figure 7, what you note about the axes of O' compared to the axes of O?
   Answer: We note that the spatial x' axis is rotated relative to the spatial x axis by an angle $\theta = \arctan \beta$ while the temporal ct' axis is rotated relative to the temporal ct axis by an angle $-\theta$. This is unlike ordinary rotation of one 2D (spatial) coordinate system with respect to another 2D (spatial) coordinate system where both spatial axes are rotated by the same angle and in the same sense. Hence, "rotation" (as a transformation that keeps certain invariance properties) in the Minkowski spacetime is different from ordinary rotation in ordinary space.

Exercises

1. Define the concept of "event" in spacetime.
2. Determine if the coordinates of spacetime, as represented by its "coordinate system", have the same physical dimension or not.
3. Plot a simple 2D spacetime diagram on which you draw the world lines of the following objects: (a) An object that moves in time but not in space. (b) An object that moves in space but not in time. (c) An object that moves in space and in time. (d) An object that moves neither in space nor in time. Comment on these objects.

3.9.2 Light Cone

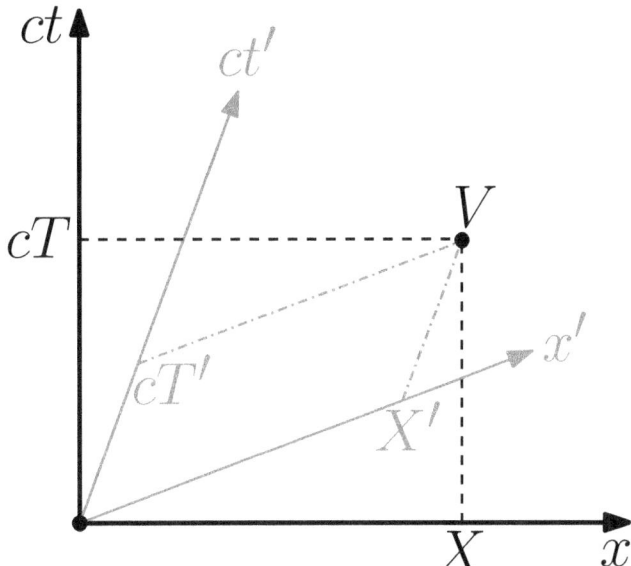

Figure 8: The coordinates of an event V on a 2D spacetime diagram representing two inertial frames where (cT, X) and (cT', X') are the coordinates in O and O' respectively.

4. Based on your answer to the previous question, determine if moving at the speed of light means stopping in time as claimed by some special relativists.
5. Briefly define world line.
6. Describe the world line of a light signal.
7. To which concept in the normal "spatial" space the world line concept corresponds?
8. What is the world line of a free particle in the Minkowski spacetime?
9. Compare the world line of a free massive particle with the world line of a light ray.
10. Compare the concepts of "null geodesic" and "geodesic" in the Minkowski spacetime.
11. What is the world line of a massive particle moving under the influence of an external force?

3.9.2 Light Cone

The light cone, which is also called the null cone,[87] is an abstract graphical device used to represent and demonstrate the state of events in the spacetime as seen from the perspective of a given event. The light cone in the spacetime of Lorentz mechanics is a double cone whose apex is positioned at the origin of spacetime coordinates[88] where the concerned event is located. The cone itself (i.e. its surface) is represented mathematically by the following equation which is a statement of a vanishing spacetime interval (see § 3.9.3) in the event space:[89]

---

[87] In fact, "null cone" is usually used to label the cone in the diagram rather than the diagram itself.
[88] Being at the origin of spacetime coordinates is not necessary although this is usually done.
[89] If the light cone is not centered at the origin of coordinates then the x^μ should be replaced by Δx^μ.

3.9.2 Light Cone

$$(x^0)^2 - (x^1)^2 - (x^2)^2 - (x^3)^2 = 0 \tag{70}$$

where the frame O in which this equation applies is inertial. Now, since the spacetime interval is invariant across all inertial frames, then in any other inertial frame O' the light cone equation should take the same form, that is:

$$(x'^0)^2 - (x'^1)^2 - (x'^2)^2 - (x'^3)^2 = 0 \tag{71}$$

where in this equation we are assuming that O and O' share the origin of coordinates at the time of the event.

Due to the difficulty or impossibility of representing the 4D manifold of spacetime graphically on a 2D surface (e.g. on paper) and even on a 3D model, the light cone diagram normally represents events taking place in a plane representing two spatial dimensions in the event space. Accordingly, the spacetime diagram on which the light cone is plotted consists of three spacetime coordinates: the temporal coordinate plus two spatial coordinates. To outline the characteristic features of the light cone diagram, we plot in Figure 9 a 3D spacetime diagram (i.e. one temporal and two spatial dimensions) with a schematic plot of a light cone that belongs to a given event V. From Figure 9, we note the following observations:

1. This diagram represents the light cone of a given event V which is at the apex of the light cone where this apex is located at the origin of spacetime coordinates as shown in Figure 9. As indicated earlier, being at the origin is not a requirement. Therefore, each point in the spacetime has its own light cone and hence more than one light cone can be drawn on a single spacetime diagram where each light cone is attached to a particular event of interest.

2. The cone is a geometrically regular cone where the temporal axis x^0 is its axis of symmetry while the plane of the two spatial axes x^1 and x^2, which passes through the apex of the cone, is perpendicular to the temporal axis. Accordingly, any plane that is parallel to the $x^1 x^2$ plane will cut the cone in a circular cross section.

3. The $x^1 x^2$ plane (i.e. $x^0 = 0$) represents the present, the part below this plane represents the past and the part above this plane represents the future. All these are relative to the time of event V which is supposed to occur at $t = 0$.

4. Assuming that c is the ultimate speed and causal relations can only be established through communicating signals, no causal relation can occur between event V and any other event V_1 unless V_1 is inside the cone or on its surface. Hence, no causal relation (i.e. V influences or is influenced by) can be between V and any event in the "spacelike" zone that surrounds the cone.

5. The zone inside the lower and upper parts of the cone are "timelike". Events in these parts can have a causal relation with V where events inside the lower part (i.e. backward in time) can influence V (i.e. they cause V) but cannot be influenced by V while events inside the upper part (i.e. forward in time) can be influenced by V (i.e. they are caused by V) but cannot influence V.

6. The surface of the cone is "lightlike". Events there can have causal relation with V *iff* they can communicate with V at the speed of light. Hence, if we assume that c is

3.9.2 Light Cone

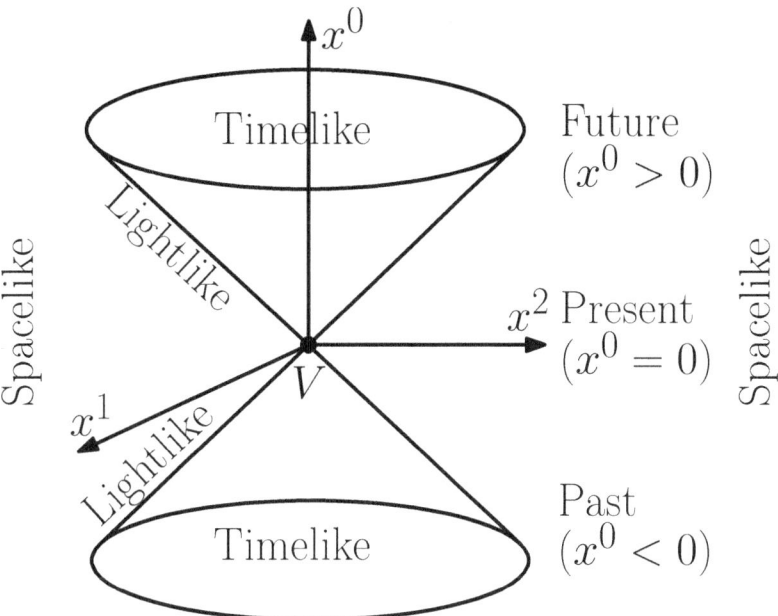

Figure 9: The projection of light cone (as seen by an observer located at the apex) onto the hyperplane $x^3 = 0$ where x^1 and x^2 are the other two spatial coordinates while x^0 is the temporal coordinate. The causal relation to the event of interest V (which is located at the apex of the cone) is not considered in the labels past, present and future.

restricted to light, then any influence between V and events on the surface of the light cone should be communicated only by light signals. As before, events on the lower part of the cone can cause V while events on the upper part of the cone can be caused by V.

7. As a matter of terminology, communications between V and events inside the light cone are conducted by timelike trips, communications between V and events on the light cone are conducted by lightlike trips, and communications between V and events outside the light cone are conducted by spacelike trips. It is obvious that the possibility of lightlike and spacelike trips is subject to the Lorentzian speed restrictions and hence the communication between events by these types of trip are impossible when the speed required for communication is physically impossible.

We note that many authors label the inside of the lower part of the cone as past, label the inside of the upper part of the cone as future and label the apex point of the cone as present while all other parts (i.e. outside the cone) are labeled as "elsewhere". This is essentially based on considering the potential causal relation of the event V to other events as being in its past or future or present or causally inaccessible, while we used these terms in the above description and Figure 9 to classify the parts of the time coordinate relative to the time of V regardless of any potential causal relation. We also note that the light cone diagram may be simplified by including the temporal coordinate and only one spatial coordinate and hence it is represented by a 2D plot representing a vertical cross section of the 3D plot or the trace of the light cone on a single spatial dimension, as seen

3.9.2 Light Cone

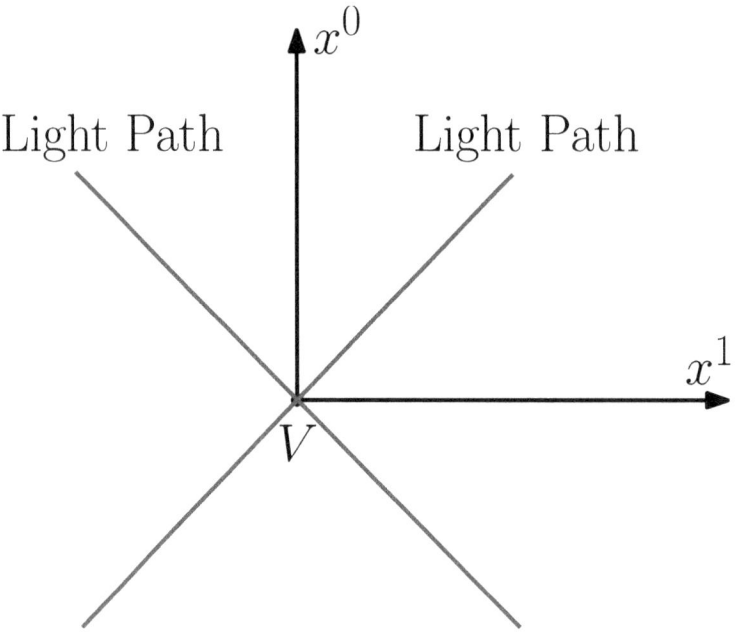

Figure 10: A 2D light cone diagram.

in Figure 10. In fact, this is no more than a 2D spacetime diagram with the intersection of the light cone with the $x^0 x^1$ plane. We should also note that the surface of the light cone is made of the null geodesics of the light signal that originates from the apex of the cone in the 4D event space, as discussed earlier. In fact, this applies only to the upper part of the cone if we consider the temporal dependency; the lower part then represents reflection of the upper part in the $x^1 x^2$ plane which may be seen as an inversion in time.[90]

Solved Problems

1. Discuss possible causal relations between an event V and other events in its elsewhere region.

   Answer: As seen in § 3.7, there is a limit on the causal relations between events in the spacetime, that is any causal relation in this space is restricted by the speed of the communication signal and hence no causal relation can exist between two events if the time of the occurrence of the caused event is earlier on the timeline than the time for the signal to travel from the location of the cause event and reach the location of the caused event. Now, if we assume that c is the ultimate speed (at least within the framework of Lorentz mechanics) then in the spacetime of Lorentz mechanics, no causal relation can exist between V and any event in the elsewhere region of the light cone diagram of V and hence this region is causally inaccessible to V, i.e. no event in this region can cause V or be caused by V.

2. What is the relation between spacetime and light cone?

---

[90] It may also be interpreted as representing incoming light rays converging on the origin prior to $t = 0$, but this may not be physically viable; moreover it does not apply if we assume that the light signal originates from the apex.

Answer: Light cone introduces a partition in the spacetime from the perspective of a given observer at a given instant of time and in a given position in space. So, the spacetime is divided by the light cone according to the causal relation between the event of interest V (which is at the apex of the light cone and hence it represents the view of the given observer) and other events that exist in the spacetime into the following four main parts:
• The part of the spacetime where the events there can influence V but cannot be influenced by V. This part may be called past of V and it is represented by the lower part of the cone (i.e. surface of cone and its inside).
• The part of the spacetime where the events there are taking place and can, in principle, influence V and be influenced by V. This part may be called present of V and it is represented by the apex of the cone.
• The part of the spacetime where the events there can be influenced by V but cannot influence V. This part may be called future of V and it is represented by the upper part of the cone (i.e. surface of cone and its inside).
• The part of the spacetime where the events there can neither influence V nor be influenced by V. This part may be called elsewhere of V and it is represented by the outside of the cone.

We note that this causal divide is based first on the assumption that the characteristic speed of light c is the ultimate speed which is inline with the current formulation of Lorentz mechanics but it may not be true in general, as discussed before. The divide is also based on the premise that causal relations can only be established by communicating signals. A third presumption is that the caused cannot precede the cause which may originate from the impossibility of traveling back in time due to the unique direction of time flow. In fact, the third (and possibly even the second) may be considered as part of the definition of causal relation. As we see, some of these issues are scientific but other issues are philosophical and epistemological.

Exercises
1. Make a simple sketch of a 3D (i.e. one temporal and two spatial) light cone diagram where the causal relations with the event of interest are considered in the labels: past, present and future.
2. What are the main characteristics of the light cone diagram and the relations between the events in the event space that are represented by the light cone diagram?
3. Outline the two main types of light cone diagram with regard to the number of dimensions.
4. Discuss if the use of light cone diagram is dependent on accepting special relativity and its second postulate.
5. What a cross section of the light cone made by a plane perpendicular to the x^0 axis at a given value of x^0 represents?
6. Put appropriate restrictions (according to the Lorentzian speed restrictions) on the world line of (A) a light ray passing through a given point in spacetime and (B) a massive object passing through such a point.

3.9.3 Spacetime Interval

The spacetime interval between two events, $V_1(\mathbf{x}_1)$ and $V_2(\mathbf{x}_2)$, is a quantity that belongs to the spacetime and is defined in its finite form $\Delta\sigma$ by the following expression:

$$(\Delta\sigma)^2 = (\Delta x^0)^2 - (\Delta x^1)^2 - (\Delta x^2)^2 - (\Delta x^3)^2 \tag{72}$$

where $\Delta x^\mu = x_2^\mu - x_1^\mu$ ($\mu = 0, 1, 2, 3$). The spacetime interval may also be defined in an infinitesimal form by replacing Δ with d in the above expression, that is:

$$(d\sigma)^2 = (dx^0)^2 - (dx^1)^2 - (dx^2)^2 - (dx^3)^2 \tag{73}$$

We note that although the square (i.e. the quadratic form) of the spacetime interval is real, the interval can be imaginary, i.e. when the square is negative. We also note that the magnitude (which is real) of the spacetime interval may be called spacetime distance or spacetime length. Moreover, unlike its counterpart in ordinary space, this magnitude can be zero even when the two events have different spacetime coordinates. In brief, neither the quadratic form or the interval or its magnitude is positive definite.

There are three types of spacetime interval that connects two events in the event space (refer to Figure 9):[91]
1. Spacelike interval when the square of the interval between the events is negative. The label "spacelike" is based on the dominance of *space* over time.
2. Timelike interval when the square of the interval between the events is positive. The label "timelike" is based on the dominance of *time* over space.
3. Lightlike interval (also known as null interval) when the square of the interval between the events is zero. The label "lightlike" is based on the equality of distance and time which is a property of the light trajectory in the event space.

The spacetime interval is invariant across all inertial frames under the Lorentz transformations of spacetime coordinates but it is not invariant under the Galilean transformations (refer to § 3.9.4). Because of this invariance, the type of the spacetime interval (i.e. being spacelike, timelike or lightlike) is also invariant under the Lorentz transformations but not under the Galilean transformations.

We remark that the spacetime interval may be represented by temporal coordinate minus spatial coordinates, as in Eq. 72, or the other way around, that is:[92]

$$(\Delta\sigma)^2 = (\Delta x^1)^2 + (\Delta x^2)^2 + (\Delta x^3)^2 - (\Delta x^0)^2 \tag{74}$$

It is obvious that the above classification of the spacetime interval (i.e. spacelike, timelike or lightlike) is based on the first form, and hence according to the second form the

---

[91] We note that in the following classification of the interval type, it should be more appropriate to use Δ (which represents a finite change) instead of d (which represents infinitesimal change) in the expression of spacetime interval since the intervals that correspond to these types usually represent straight line segments of finite length in the event space.

[92] Or rather: $(\Delta\sigma)^2 = (\Delta x^1)^2 + (\Delta x^2)^2 + (\Delta x^3)^2 - (\Delta x^4)^2$.

3.9.3 Spacetime Interval

spacelike and timelike (in their relation to the sign of the square of the interval) should be interchanged since the dominance of time or space in its relation to sign will be reversed.[93]

Solved Problems

1. What is the physical dimension of spacetime interval?
   Answer: Spacetime interval ($\Delta\sigma$ or $d\sigma$) has the physical dimension of length.
2. What is the characteristic feature of the null geodesics?
   Answer: The null geodesics, which are the light paths in the spacetime of Lorentz mechanics, are characterized by having zero spacetime interval, that is: $\Delta\sigma = 0$. Accordingly, they are associated with lightlike intervals. This also implies that light follows straight trajectories in the spacetime of Lorentz mechanics since these trajectories are geodesics in this flat space.
3. What is the significance of the invariance of the type of spacetime interval across frames? Try to link this to light cone.
   Answer: This invariance means that although different observers do not agree in general on the spacetime coordinates of events and their spatial and temporal separation, they should agree on which event is inside the light cone of a given event, or outside the cone or on the cone and hence causal relations stay invariant across different frames.[94] This makes it impossible to have different causal relations across frames which is inconsistent with the rules of physical reality since these relations are real intrinsic physical attributes and hence they should not be frame dependent.
   From practical point of view, this invariance (in conjunction with light cone diagram) is very useful in facilitating the determination of causal relations between events by graphic (and graphic-like) techniques.
4. Discuss the issue of positive definiteness of the spacetime interval, its magnitude and its quadratic form.
   Answer: None of these is positive definite. In more details:
   - Spacetime interval $\Delta\sigma$: this can be positive, zero or imaginary.
   - Magnitude of spacetime interval $|\Delta\sigma|$: this can be positive or zero.
   - Quadratic form of spacetime interval $(\Delta\sigma)^2$: this can be positive, zero or negative.

Exercises

1. Define, descriptively and mathematically, the concept of spacetime interval using its infinitesimal form.

---

[93] To avoid possible ambiguity and confusion, it is more appropriate to classify the types of spacetime interval according to the relative size of the temporal and spatial separations instead of the sign of interval. Hence: $(\Delta x^0)^2 < \sum (\Delta x^i)^2$ for spacelike, $(\Delta x^0)^2 > \sum (\Delta x^i)^2$ for timelike and $(\Delta x^0)^2 = \sum (\Delta x^i)^2$ for lightlike.

[94] In fact, this partially accounts for the preservation of causal relationships since it eliminates some causes of the violation of this preservation, e.g. we still need to identify which is the cause and which is the caused although this may be determined from the temporal order of the events. Also, to be more specific we should restrict these relationships to those involving the event of concern (i.e. the event to which the light cone belongs which is at the apex of the cone) although this can be easily generalized since each event has its own light cone.

2. What differences you note between the space interval ds and the spacetime interval $d\sigma$? Also, what are the common features between ds and $d\sigma$?
3. Is the spacetime of Lorentz mechanics flat or curved? Explain why.
4. Discuss if the invariance of the spacetime interval across all inertial frames under the Lorentz transformations is based on accepting special relativity and its second postulate.
5. Write the expression of the lightlike interval in a way that makes it more sensible.
6. List and describe the types of spacetime interval.
7. Compare the terms: timelike, lightlike and spacelike with the terms: past, present, future and elsewhere.
8. Calculate the spacetime intervals between the following pairs of physical events: (a) V_1 and V_2, (b) V_1 and V_3 and (c) V_2 and V_3 where $V_1\,(13,-2,6,11)$, $V_2\,(10,9,2,-1)$ and $V_3\,(0,1,2,-1)$ and the parenthesized numbers represent (x^0, x^1, x^2, x^3). Also, determine the type of each interval.
9. What is the type of the spacetime interval between two simultaneous events and between two co-positional events?
10. Two planets, A and B, are separated by a distance $d = 9 \times 10^{10}$ kilometers. Two teams of scientists are holding two conferences: the conference of the first team is on planet A which starts at 8:00 o'clock and lasts for 60 minutes, and the conference of the second team is on planet B which starts at 9:30 o'clock and lasts for 30 minutes. Is it possible for a member of the first team to fully attend both conferences? What about partial attendance?
11. Assuming that causal relations are subject to the speed restrictions, correlate the type of spacetime interval to the causal relation that potentially exist between the two events linked by the interval.
12. Analyze the significance of the sign of the square of spacetime interval $(\Delta\sigma)^2$ between two events and its invariance across inertial frames. Also, try to demonstrate the analysis by using the light cone of an event.

3.9.4 Invariance of Spacetime Interval

Following the style of ordinary spaces where the space interval (or infinitesimal line element) ds is an invariant of the space and hence it is the same for all coordinate systems, the spacetime interval $d\sigma$ is an invariant (under Lorentz transformations) of the spacetime and hence all inertial observers[95] will measure the same spacetime interval between any two events although they may disagree about their spacetime coordinates and their spatial and temporal separations.[96] Now, since they measure the same spacetime interval they should also agree on the type of interval.[97] In the following, we show how the spacetime interval is invariant under the Lorentz transformations of spacetime coordinates where

---

[95] Inertial observers represent inertial frames and hence they correspond to coordinate systems.

[96] We note that for this reason, Lorentz transformations may be seen as rotation in the 4D spacetime manifold since they leave this interval invariant like rotation in ordinary space which leaves the spatial interval invariant.

[97] This partially implies the agreement on causal relationships (or rather on the nature of potential causal relationships).

3.9.4 Invariance of Spacetime Interval

we use in this demonstration the Lorentz transformations as given by the upcoming Eqs. 95-98[98] plus standard algebraic techniques as well as the expression of spacetime interval. For the purpose of notational simplicity, we use ct for the temporal separation (instead of Δx^0) and x, y, z for the spatial separations (instead of $\Delta x^1, \Delta x^2, \Delta x^3$) in the expression of spacetime interval.

$$c^2 t^2 - x^2 - y^2 - z^2 \tag{75}$$

$$= c^2 \gamma^2 \left(t' + \frac{vx'}{c^2} \right)^2 - \gamma^2 (x' + vt')^2 - y'^2 - z'^2$$

$$= c^2 \gamma^2 \left(t'^2 + 2t' \frac{vx'}{c^2} + \frac{v^2 x'^2}{c^4} \right) - \gamma^2 \left(x'^2 + 2x' vt' + v^2 t'^2 \right) - y'^2 - z'^2$$

$$= c^2 \gamma^2 t'^2 + 2c^2 \gamma^2 t' \frac{vx'}{c^2} + c^2 \gamma^2 \frac{v^2 x'^2}{c^4} - \gamma^2 x'^2 - 2\gamma^2 x' vt' - \gamma^2 v^2 t'^2 - y'^2 - z'^2$$

$$= c^2 \gamma^2 t'^2 + 2\gamma^2 t' vx' + \gamma^2 \frac{v^2 x'^2}{c^2} - \gamma^2 x'^2 - 2\gamma^2 x' vt' - \gamma^2 v^2 t'^2 - y'^2 - z'^2$$

$$= c^2 \gamma^2 t'^2 + \gamma^2 \frac{v^2 x'^2}{c^2} - \gamma^2 x'^2 - \gamma^2 v^2 t'^2 - y'^2 - z'^2$$

$$= \left(c^2 \gamma^2 t'^2 - \gamma^2 v^2 t'^2 \right) + \left(\gamma^2 \frac{v^2 x'^2}{c^2} - \gamma^2 x'^2 \right) - y'^2 - z'^2$$

$$= \gamma^2 c^2 t'^2 \left(1 - \frac{v^2}{c^2} \right) - \gamma^2 x'^2 \left(1 - \frac{v^2}{c^2} \right) - y'^2 - z'^2$$

$$= \gamma^2 c^2 t'^2 \frac{1}{\gamma^2} - \gamma^2 x'^2 \frac{1}{\gamma^2} - y'^2 - z'^2$$

$$= c^2 t'^2 - x'^2 - y'^2 - z'^2$$

that is:
$$c^2 t^2 - x^2 - y^2 - z^2 = c^2 t'^2 - x'^2 - y'^2 - z'^2 \tag{76}$$

and hence the spacetime interval is invariant under the Lorentz transformations of spacetime coordinates. In the following problems and exercises we show that spacetime interval is not invariant under the Galilean transformations of space coordinates and time, although ds and dt are independently invariant under the Galilean transformations. We also show that ds and dt are not invariant under the Lorentz transformations.

Solved Problems

1. Show that the spacetime interval is not invariant under the Galilean transformations of space coordinates and time.
   Answer: Using the spacetime interval expression and the Galilean transformations, all of which are given in the text (see § 1.9.1 and § 3.9.3), we have:
   $$c^2 t^2 - x^2 - y^2 - z^2 = c^2 t'^2 - (x' + vt')^2 - y'^2 - z'^2$$

---

[98] These transformations are obviously based on standard setting, but this does not damage the generality of the result.

3.9.4 Invariance of Spacetime Interval

$$= c^2 t'^2 - x'^2 - 2x'vt' - v^2 t'^2 - y'^2 - z'^2$$
$$= c^2 t'^2 - x'^2 - y'^2 - z'^2 - \left(2x'vt' + v^2 t'^2\right)$$

and hence the spacetime interval is invariant under the Galilean transformations only if $v = 0$ which is just a special case and hence it does not represent a general invariance relation. In fact, the case $v = 0$ means that the two frames are essentially the same from the viewpoint of Lorentz mechanics due to the absence of relative motion between the two frames where on this relative motion the formulation of Lorentz mechanics with its features that distinguish it from classical mechanics (in particular the Lorentz γ factor) is based.

2. Compare the invariance of spacetime interval $d\sigma$ in Lorentz mechanics with its counterpart in classical mechanics.

   Answer: From the previous and upcoming discussions, we can see that the "space-time" of classical mechanics has two separate invariances: ds for the space and dt for the time. This is because in classical mechanics space and time are absolute, universal and independent entities and hence each one of these is "preserved" independently under the "space-time" coordinate transformations (i.e. the Galilean) from one inertial frame to another inertial frame. However, in Lorentz mechanics we have only one invariant quantity which is the particular combination of these two as represented by the spacetime interval. This highlights the issue of the strong bond between space and time in Lorentz mechanics and the merge of these into a single spacetime entity which does not exist in classical mechanics. This also indicates the importance of length contraction and time dilation effects in Lorentz mechanics and how they interact and vary in such a way to keep the spacetime interval invariant despite the violation of the invariance of ds and dt due to the action of these effects. These issues will be discussed in detail in the future.

Exercises

1. What is the essential condition for the invariance of spacetime interval between frames of reference?
2. What it means to say that the infinitesimal line element of a space is invariant? How this applies to the spacetime interval $d\sigma$?
3. What is the significance of the invariance of spacetime interval in Lorentz mechanics?
4. Use the fact that the spacetime interval is invariant to analyze the movement through space and time.
5. Show that ds and dt are separately invariant under the Galilean transformations.
6. Show that ds and dt are not invariant under Lorentz transformations.
7. Compare and discuss in general terms the issue of invariance in classical mechanics and in Lorentz mechanics.

Chapter 4
Formalism of Lorentz Mechanics

In this chapter, we present the formalism of Lorentz mechanics while in the next chapter we will present the derivation of the main parts of this formalism. The reason for this separation is to have a clearer picture to the formalism first followed by a deeper inspection and verification later on. This will help the readers to understand the formalism without the confusion of the more complex mathematical arguments and derivations. Also, some readers may not be interested in the formal derivations and hence they can skip the derivation chapter with no loss of continuity or integrity. The nature of the exercises in these two chapters is also different where in the formalism chapter the focus is on the numerical side while in the derivation chapter the focus is on the analytical and symbolic side. We hope the separation according to this plan will help the readers in their effort to understand the subject and fulfill the objectives of this book.

4.1 Physical Quantities

In this section, we present the main physical quantities that form the main building blocks of the formalism of Lorentz mechanics and compare them with their counterparts in classical mechanics. As we will see, the basic definitions of most physical quantities are the same in classical mechanics and Lorentz mechanics, although some of these definitions may differ in their symbolic expression or in their quantitative values due to the difference between classical mechanics and Lorentz mechanics in the definition of some involved concepts or in the transformation rules that apply between the frames of observation.

4.1.1 Length

While in classical mechanics the length of a physical object is an intrinsic property, and hence it is independent of the frame of observation, in Lorentz mechanics it is frame dependent. As we will see (refer for example to § 7.2), according to Lorentz mechanics the length L of an object having a proper length L_0 where this object is in a state of uniform motion with speed v relative to an inertial observer is measured by this observer to be:[99]

$$L = \frac{L_0}{\gamma} \qquad (77)$$

where γ is a function of v. It is noteworthy that the length in this relation corresponds to the dimension of the object in the direction of the relative motion between the object and the frame of observation and hence the dimensions of the object in the perpendicular directions to the direction of motion will not be affected, i.e. they keep their proper values.

---
[99] This is simply the length contraction formula.

4.1.1 Length

Solved Problems

1. A missile whose proper length is 10 is seen from an inertial frame to move at a constant speed v and have a length of 9. Find v.
   Answer: We have:
   $$\begin{aligned} L_0 &= \gamma L \\ 10 &= \frac{9}{\sqrt{1-\beta^2}} \\ \sqrt{1-\beta^2} &= 0.9 \\ 1-\beta^2 &= 0.81 \\ \beta^2 &= 0.19 \\ v &= c\sqrt{0.19} \end{aligned}$$

2. A rod whose proper length is 5 is seen by an inertial observer to move in the x direction along the orientation of its length with speed $v = 0.35c$. What is the time required for this rod to pass a given point on the x axis in the frame of the observer?
   Answer: The improper length of the rod is given by:
   $$L = \frac{L_0}{\gamma} = 5 \times \sqrt{1 - 0.35^2} = 5\sqrt{0.8775}$$
   Hence, the time required for the rod to pass a given point on the x axis is given by:
   $$\Delta t = \frac{L}{v} = \frac{5\sqrt{0.8775}}{0.35c} \simeq 4.4607 \times 10^{-8}\,\text{s}$$

Exercises

1. What is the physical origin of the relation between the proper and improper values of the length of a physical object according to Lorentz mechanics? What is the restriction on this relation?
2. A moving object is observed to be contracted in the direction of motion by 40%. Find its relative speed.
3. A square whose proper area is 1 is seen by an inertial observer to have an area of 0.85. Assuming that the square is moving uniformly with two of its sides oriented along the direction of motion, what is its speed relative to the observer? Repeat the question assuming this time that one diagonal of the square is oriented along the direction of motion.
4. O and O' are two inertial frames in a state of standard setting. A stick which is in the xy plane of frame O makes an angle with the x axis of $\pi/4$ in frame O and an angle of $\pi/3$ in frame O'. (a) What is the relative speed v between the two frames? (b) If the proper length of the stick is 1 in frame O what is its length in frame O'?

4.1.2 Time Interval

5. A cube of sides $s = 1$ m in its rest frame is seen by an inertial observer O to be moving along one of its sides with speed $v = 0.6c$. What are the dimensions of this cube as measured by O? What is its volume?
6. Repeat the previous question but this time the cube is seen to be moving along one of its face diagonals.
7. The radius of a circle is measured in its rest frame to be 1. The area of this circle is measured to be 2 by a moving observer. What is the speed of the moving observer relative to the rest frame of the circle?

4.1.2 Time Interval

While in classical mechanics the time interval between two events is independent of the frame of observation and hence it is a universal invariant quantity, in Lorentz mechanics it depends on the frame of observation. As we will see (refer for example to § 7.3), in Lorentz mechanics the time interval Δt as measured by an inertial observer from a moving frame is related to its proper value Δt_0 in its rest frame by the formula:[100]

$$\Delta t = \gamma \Delta t_0 \tag{78}$$

Solved Problems

1. Two inertial observers are in a state of relative uniform motion with speed $v = 0.35c$. The proper time interval between two events is measured as 1 s in the frame of one observer. What is the size of this time interval as measured by the other observer?
   Answer: Here, we have $\Delta t_0 = 1$ and hence:

   $$\Delta t = \gamma \Delta t_0 = \frac{1}{\sqrt{1 - 0.35^2}} \simeq 1.0675 \, \text{s}$$

2. An inertial spaceship, which is moving with a constant velocity of $0.85c$ relative to planet A, passes this planet at 1:00 o'clock where the clock of the spaceship and the clock of planet A were synchronized at that instant. The spaceship then passed planet B at 1:15 o'clock according to the clock of the spaceship. Assuming that the time on the two planets, which are in a state of relative rest, is synchronized and the time dilation effect takes place in the frame of the spaceship:
   (a) What was the time on planet B when the spaceship passed planet B?
   (b) What is the distance between the two planets according to an observer on planet A and according to the pilot of the spaceship?
   (c) Comment on the results.
   Answer:
   (a) If we label the time of the spaceship with t_s and the time of the planets with t_p then according to the time dilation effect we have:

   $$\Delta t_p = \gamma \Delta t_s = \frac{15}{\sqrt{1 - 0.85^2}} \simeq 28.4747 \, \text{min}$$

---

[100] This is simply the time dilation formula.

4.1.2 Time Interval

i.e. the time on the planets when the spaceship passed planet B was about 1:28:28.

(b) The distance d_p between the two planets according to an observer on planet A is:

$$d_p = v\Delta t_p \simeq 0.85c \times 28.4747 \times 60 \simeq 4.3566 \times 10^{11}\,\mathrm{m}$$

The distance d_s between the two planets according to the pilot of the spaceship is:

$$d_s = v\Delta t_s = 0.85c \times 15 \times 60 \simeq 2.2950 \times 10^{11}\,\mathrm{m}$$

(c) The results of this exercise show how time dilation and length contraction produce effects on time and length which are equal in magnitude (although they are seen as opposite in sense from different proper-improper perspectives) because we have:

$$\frac{\Delta t_p}{\Delta t_s} = \gamma = \frac{d_p}{d_s}$$

This highlights an important issue that characterizes the whole of Lorentz mechanics, that is the temporal and spatial coordinates of the spacetime of Lorentz mechanics are subject to similar effects under the influence of motion. In fact, these effects (i.e. time dilation with respect to time and length contraction with respect to space) are essentially identical. The results also explain why the relative speed is the same in both frames, that is (refer to the previous equations):

$$v = \frac{d_p}{\Delta t_p} = \frac{d_s}{\Delta t_s}$$

because both time and distance contract (or shrink) by the same factor in the frame of spaceship, or alternatively both time and distance dilate (or expand) by the same factor in the frame of the planets. In fact, this also highlights another issue which is discussed in § 5.1.5 that is time dilation and length contraction effects are labeled as such according to different proper-improper perspectives from which these effects are observed; otherwise both effects can be seen to act in the same sense, i.e. both can be seen as dilation in time and length, or as contraction in time and length.

3. Assuming that the speed of light is the same in all inertial frames, if the pilot in the previous exercise sends a light signal when he arrives to planet B, when this signal will arrive to planet A according to the spaceship time and according to the planets time? Comment on your result.

Answer:
For the spaceship, the time required for the signal to travel from B to A is:[101]

$$\Delta t_s = \frac{d_s}{c} = \frac{0.85c \times 15}{c} = 12.75\,\mathrm{min}$$

---

[101] We note that c in the following two equations can be in any units (e.g. m/min) since it will be canceled.

4.1.2 Time Interval

and hence the time of signal arrival to A according to the spaceship time is 1:27:45. For the planets, the time required for the signal to travel from B to A is:

$$\Delta t_p = \frac{d_p}{c} \simeq \frac{0.85c \times 28.4747}{c} \simeq 24.2035 \, \text{min}$$

and hence the time of signal arrival to A according to the planets time is about 1:52:41.
Comment: from the answer we see that:

$$c = \frac{d_s}{\Delta t_s} = \frac{(d_p/\gamma)}{(\Delta t_p/\gamma)} = \frac{d_p}{\Delta t_p}$$

This means that the assumption of the universality of the speed of light in all inertial frames should be interpreted in apparent sense due to the contraction of both length and time or the dilation of both length and time by the same factor in each one of these frames. This issue will be discussed further in the future (also see § 1.6). The universality of c should also be linked to the identicality of the relative speed v between the two frames, as seen in the previous question.

4. Obtain time dilation and length contraction formulae as special cases of Lorentz space-time coordinate transformations (refer to § 4.2.1) and comment on the results.
Answer: The Lorentz time transformation for the time interval between two events, A and B, is given by:

$$t_{BA} = \gamma \left(t'_B + \frac{v x'_B}{c^2} \right) - \gamma \left(t'_A + \frac{v x'_A}{c^2} \right) = \gamma \left[(t'_B - t'_A) + \frac{v}{c^2} (x'_B - x'_A) \right]$$

where t_{BA} stands for the difference between t_B and t_A. If the two events are co-positional in O' (i.e. $x'_B = x'_A$) then $x'_B - x'_A = 0$ and we have:

$$t_{BA} = \gamma (t'_B - t'_A) = \gamma t'_{BA}$$

which is the time dilation formula.
Similarly, the Lorentz space transformation for the space interval (or separation) between two events, A and B, along the x orientation is given by:

$$x'_{BA} = \gamma (x_B - v t_B) - \gamma (x_A - v t_A) = \gamma \left[(x_B - x_A) - v (t_B - t_A) \right]$$

where x'_{BA} stands for the difference between x'_B and x'_A.[102] If the two events are simultaneous in O (i.e. $t_B = t_A$) then $t_B - t_A = 0$ and we have:

$$x'_{BA} = \gamma (x_B - x_A) = \gamma x_{BA}$$

which is the length contraction formula.
Comment: the obtained results mean that time dilation formula applies when the two

---

[102] We consider the x transformation because according to the state of standard setting x is the dimension that is affected by the motion and hence it is subject to length contraction.

4.1.2 Time Interval

events are co-positional in the transformed frame while length contraction formula applies when the two events are simultaneous in the transformed frame;[103] otherwise the full formulae of Lorentz spacetime coordinate transformations in which spatial and temporal coordinates are considered in both space and time coordinate transformations should be used. Another important thing to note is that, as indicated earlier and will be explained further later on, time dilation and length contraction effects are labeled as such according to different proper-improper perspectives; otherwise both can be labeled as dilation or as contraction of the temporal and spatial coordinates. So, if we unify our perspectives we should have: $t_{BA} = \gamma t'_{BA}$ and $x_{BA} = \gamma x'_{BA}$, or $t'_{BA} = \gamma t_{BA}$ and $x'_{BA} = \gamma x_{BA}$.

Exercises
1. What is the physical origin of the relation between the proper and improper values of the time interval between two events according to Lorentz mechanics?
2. Describe in a few words the relation between the proper and improper values of length and time interval according to Lorentz mechanics.
3. Inertial observers O and O' are in a state of relative motion with speed $v = 0.8c$. The time interval between two events which are co-positional in the frame of O' is measured by O' to be $\Delta t' = 3$. What is the time interval Δt as measured by O?
4. A cosmonaut is planning to travel at a constant velocity to a planet that is 1 Ly (i.e. light year) away from the Earth in 5 years of his time (i.e. he ages five years during this journey). What is the required speed for this journey? Solve this problem once from the frame of the Earth and once from the frame of the cosmonaut to check if the two solutions are consistent and comment on the results. Assume in your answer that the Earth frame is inertial and it is in a state of rest relative to the distant planet. Also assume that the time dilation effect takes place in the frame of the cosmonaut.
5. An astronaut fired a rocket from the back of his inertial spaceship in the forward direction. In the frame of the astronaut the length of the spaceship is 100 m and the speed of the rocket is $0.5c$. What is the time required by the rocket to reach the front of the spaceship (a) according to the frame of the astronaut and (b) according to an inertial frame O in which the spaceship is moving along the direction of the rocket with a speed of $0.7c$? Comment on this question.
6. Two inertial observers are in a state of relative motion. The time interval between events as measured by one of these observers scales by a factor of 0.8 by the other observer. What is the relative speed between these observers?
7. The lifetime of an elementary particle in its rest frame is Δt. In frame O, which is moving at $v_1 = 0.4c$ relative to the particle rest frame, its lifetime is measured to be 3 μs. What is its lifetime in a frame O' which is moving at $v_2 = 0.5c$ relative to the particle rest frame?
8. Discuss the claim that time dilation effect extends to include even the "biological clock"

---
[103] Being co-positional in the transformed frame (i.e. O') is equivalent to taking the time measurements at the same location, while being simultaneous in the transformed frame (i.e. O) is equivalent to taking the coordinate measurements at the same time.

4.1.3 Mass 123

in the body of living organisms and hence traveling creatures age less.

4.1.3 Mass

The rest mass or proper mass is the mass of the object as measured by an observer who is at rest relative to the object. In the old formalism of Lorentz mechanics, mass was considered as frame dependent and hence we have a proper (or rest) mass m_0 and an improper mass m where these are related by the formula:

$$m = \gamma m_0 \qquad (79)$$

where γ is a function of the relative speed between the object and the observer. However, the modern formalism of Lorentz mechanics generally considers the mass as an invariant property of the object, and hence any presumed variation in the mass (according to the old formalism) due to the relative motion is absorbed in the expressions of the other Lorentzian quantities that involve mass such as energy and momentum. For example, in the expression for momentum, which is given by $p = \gamma m_0 u$, the variation that is introduced by the Lorenz γ factor belongs to the momentum (see § 4.1.6) according to the modern formalism and not to the mass as it may be seen by the old formalism. Now, since the mass is invariant according to the modern formalism, the subscript 0 in the symbol m_0 can be dropped and hence the mass of the object is symbolized by m and it belongs to all frames. Accordingly, the momentum in the above example will be expressed as $p = \gamma m u$.

Hence, if we follow the old formalism then the mass in any frame according to Lorentz mechanics is given by the above relation (i.e. Eq. 79) which links the proper and improper values, while if we follow the modern formalism then the mass in Lorentz mechanics is the same as the mass in classical mechanics, i.e. it is an intrinsic frame-independent quantity that does not depend on the state of motion of the massive object and hence we have $m = m_0$ which represents the invariance of mass across frames.[104] In fact, both the old and the modern conventions about mass have certain advantages. In the present book, we generally follow the modern convention and hence we symbolize the mass (rest and non-rest) with m and cease to use m_0 outside the present subsection except in some exceptional circumstances where the use of m_0 is required or advantageous.

Solved Problems

1. Let assume that we follow the old formalism of Lorentz mechanics about the variability of mass with speed. What is the mass of a neutron traveling at a speed $u = 0.2c$? What is the percentage change from its rest value?
    Answer: The mass of a moving object is its rest mass times the Lorentz γ factor, that is:
    $$m = \gamma m_0 = \gamma m_n \simeq \frac{1.6749 \times 10^{-27}}{\sqrt{1 - 0.2^2}} \simeq 1.7095 \times 10^{-27} \, \text{kg}$$

---

[104] However, as we will see the rest mass in Lorentz mechanics, unlike classical mechanics, can include all forms of non-kinetic energy according to the mass-energy equivalence relation.

The relative change is given by:

$$\Delta m_r = \frac{m - m_0}{m} = \frac{\gamma m_0 - m_0}{\gamma m_0} = \frac{\gamma - 1}{\gamma} \simeq 0.0202$$

i.e. the mass is increased by about 2%.

2. Let assume that we follow the old formalism of Lorentz mechanics about the variability of mass with speed. What is the speed required to increase the mass of an object to twice its rest value?
Answer: We have:

$$\begin{aligned} m &= 2m_0 \\ \gamma m_0 &= 2m_0 \\ \gamma &= 2 \\ \sqrt{1 - \beta^2} &= 0.5 \\ \beta^2 &= 0.75 \\ u &= c\sqrt{0.75} \simeq 0.8660c \end{aligned}$$

Exercises
1. Discuss the issue of mass as an intrinsic or extrinsic property according to classical mechanics and according to Lorentz mechanics.
2. What is the non-rest mass of an electron whose speed is $u = 0.4c$ according to classical mechanics and according to Lorentz mechanics?
3. Let assume that we follow the old formalism of Lorentz mechanics about the variability of mass with speed. A massive object is observed to have an improper mass m_1 at speed $u_1 = 0.05c$ and to have an improper mass $m_2 = 1.25m_1$ at speed u_2. What is u_2?

4.1.4 Velocity

The basic definition of velocity, as the first derivative of the spatial coordinates with respect to the temporal coordinate, is the same in Lorentz mechanics as in classical mechanics, that is:

$$\mathbf{u} = \frac{d\mathbf{r}}{dt} \qquad (80)$$

However, due to the different coordinate transformation rules in the two mechanics, the mathematical formulation of velocity is different in the two mechanics when the physical situation involves more than one observer and hence the velocity as transformed between different observers is the subject of interest. This will be discussed in detail later in this chapter (refer to § 4.2.2 and § 4.2.3). We note that, like in classical mechanics, many velocity problems in Lorentz mechanics can be solved using the Lorentzian form of the mathematical expression of physical quantities that contain velocity in their definition such as momentum. This also applies to speed problems where quantities involving speed in their definition, such as kinetic energy, are used. Some examples of this approach will be given in the solved problems and exercises of this subsection.

Solved Problems

1. An observer in an inertial frame observed that the trajectory of an object as a function of time is given by:
$$\mathbf{r}(t) = \left(3t^2, e^t, \sin t\right)$$
What is the velocity of the object according to classical mechanics and according to Lorentz mechanics?
Answer: The basic definition of velocity in classical mechanics and Lorentz mechanics is the same, and hence in both mechanics the velocity is defined as the first time derivative of position, that is:
$$\mathbf{u} = \frac{d\mathbf{r}}{dt} = \left(6t, e^t, \cos t\right)$$

2. Given that in Lorentz mechanics the rest energy is given by $E_0 = mc^2$ and the total energy is given by $E_t = \gamma mc^2$, what is the speed of a massive object whose total energy is 10% higher than its rest energy?
Answer: We have:
$$\frac{E_t}{E_0} = 1.1$$
$$\frac{\gamma mc^2}{mc^2} = 1.1$$
$$\gamma = 1.1$$
$$1 - \beta^2 = 1/1.1^2$$
$$\beta \simeq 0.4166$$
$$u \simeq 0.4166c$$

Exercises

1. What is the definition of velocity in classical mechanics and in Lorentz mechanics? What is the reason for the difference between the two mechanics in the formulation of velocity?
2. Given the fact that in classical mechanics the 1D linear momentum of an object of mass m and velocity u is given by $p = mu$ while in Lorentz mechanics it is given by $p = \gamma mu$, what is the velocity of a massive object whose mass is $m = 1$ and whose momentum is $p = 10^6$ according to these mechanics? What is the percentage difference between the velocities in these mechanics? Comment on the results.

4.1.5 Acceleration

Like velocity, the basic definition of acceleration, as the second derivative of the spatial coordinates with respect to the temporal coordinate, is the same in Lorentz mechanics as in classical mechanics. However, due to the different coordinate transformation rules in the two mechanics, the mathematical formulation of acceleration is different in the two mechanics when the physical situation involves more than one observer and hence the acceleration as observed by different observers is the subject of concern. This will be discussed in detail later in this chapter (refer to § 4.2.4).

Solved Problems

1. Find the acceleration, according to classical and Lorentz mechanics, of an object that was seen by an inertial observer to follow the following trajectory:

$$\mathbf{r} = \left(3, 5\cos 3t, 6t^4\right)$$

Answer: The basic definition of acceleration is the same in classical and Lorentz mechanics. Hence, in both mechanics we have:

$$\mathbf{u} = \frac{d\mathbf{r}}{dt} = \left(0, -15\sin 3t, 24t^3\right)$$

$$\mathbf{a} = \frac{d\mathbf{u}}{dt} = \left(0, -45\cos 3t, 72t^2\right)$$

Exercises

1. Discuss the concept and mathematical formulation of acceleration in classical mechanics and in Lorentz mechanics.

4.1.6 Momentum

According to Lorentz mechanics, the linear momentum of an object of mass m and velocity u, as measured by an inertial observer, is given by:

$$p = \gamma m u = \frac{mu}{\sqrt{1-(u^2/c^2)}} \tag{81}$$

The reason for using u in the expression of γ here, instead of the common symbol v, is to make it distinct from the relative speed between two inertial frames because here we are considering a single frame from which an observer O is measuring the momentum of the object. However, in formulae like this we may also use v for convenience where the meaning is obvious in the given context. We should also remark that the above equation is based on a 1D motion along a single coordinate axis (say the x axis) and hence it looks like a scalar although the momentum as a 1D vector should be distinguished by a direction which is indicated by a sign (i.e. plus in the positive direction and minus in the negative direction) as well as a magnitude.

From the above expression, it can be seen that the Lorentzian momentum will be reduced to its classical form (i.e. $p = mu$) when the magnitude of u is low compared to the characteristic speed of light (i.e. $|u| \ll c$) since in this case $\gamma \simeq 1$. As indicated before, in the formula $p = \gamma m u$ we can consider γm as the improper mass and hence the formula resembles its classical counterpart. We can also consider γ as part of the whole formula and hence it is a correction factor to the classical formula of momentum. As discussed earlier, the former view may be held by the old formalism of Lorentz mechanics while the latter view is commonly held by the modern formalism of Lorentz mechanics.

Solved Problems

1. What is the classical and Lorentzian momentum of an object whose mass is 0.5 and whose velocity as seen by an inertial observer is $u = 0.1c$? Also, find the relative error between the classical and Lorentzian values.
Answer:
Classical momentum:

$$p_c = mu = 0.5 \times 0.1c = 0.05c \simeq 15000000 \, \text{kg.m/s}$$

Lorentzian momentum:

$$p_l = \gamma mu = \frac{0.5 \times 0.1c}{\sqrt{1 - 0.1^2}} = \frac{0.05c}{\sqrt{0.99}} \simeq 15075567 \, \text{kg.m/s}$$

Relative error:

$$\Delta p_r = \frac{p_l - p_c}{p_l} = \frac{\gamma mu - mu}{\gamma mu} = \frac{\gamma - 1}{\gamma} = 1 - \frac{1}{\gamma} = 1 - \sqrt{1 - 0.1^2} \simeq 0.005$$

i.e. about 0.5%.

Exercises

1. Find the momentum of a proton whose velocity is $u = 0.35c$ according to classical mechanics and according to Lorentz mechanics.
2. What is the mass of an object whose momentum is $p = 10^8$ and whose velocity is $u = 10^7$ according to classical mechanics and according to Lorentz mechanics? Also, find the percentage error in the classical value.
3. Given that the mass of neutron is $m_n \simeq 939.57 \, \text{MeV}/c^2$, find the Lorentzian momentum of a neutron whose velocity is $u = 0.55c$ in units of MeV/c.

4.1.7 Force

The definition of force in Lorentz mechanics is essentially the same as in classical mechanics, that is force is the time derivative of linear momentum where the momentum is as defined in Lorenz mechanics by Eq. 81.[105] Accordingly, we have:

$$f = \frac{dp}{dt} = \frac{d}{dt}(\gamma mu) = ma\gamma^3 \tag{82}$$

where a is the acceleration. The last equality (i.e. $f = ma\gamma^3$), which is the Lorentzian form of Newton's second law that will be derived in § 5.2.2, applies when the mass m is

---
[105] In fact, this definition of force (which is based on Newton's second law) is inherited from classical mechanics with the amendment of the definition of momentum, as will be discussed in the future.

constant, i.e. it does not vary due to exchange with its surrounding (refer to the exercises of § 1.10 and the exercises of this subsection). It should be obvious from the expression of momentum that is used in the above equation that we follow the modern formalism of Lorentz mechanics where mass is regarded as a frame independent property. We note that due to the presence of the γ factor, the Lorentzian force (unlike the Galilean force) is not proportional to the acceleration.

Solved Problems

1. The equation of motion of an object of mass m along the x axis is seen by an inertial observer to have the following form:

$$x = e^{2t}$$

   where t is time. What is the force, as a function of time, that acts on this object?
   Answer: We have:

$$u = \frac{dx}{dt} = 2e^{2t} \qquad \text{and} \qquad a = \frac{du}{dt} = 4e^{2t}$$

   Hence:

$$f = ma\gamma^3 = \frac{4me^{2t}}{[1 - (4e^{4t}/c^2)]^{3/2}}$$

Exercises

1. What is the assumption about the mass in the following Lorentzian form of Newton's second law:

$$f = ma\gamma^3$$

2. What is the force that acts on an object whose mass is $m = 1$ and whose trajectory along the x axis is described by the following equation:

$$x = 3t + 12$$

   where t is time? Comment on your answer.

3. Repeat the previous question but the equation of motion is given this time by:

$$x = \cos At$$

   where A is a constant and t is time.

4.1.8 Energy

In Lorentz mechanics we have rest energy E_0, kinetic energy E_k, and total energy E_t which is the sum of the rest and kinetic energies.[106] These three types of energy are discussed in the following paragraphs.

---
[106] The reader should notice that "total energy" in this context means rest energy plus kinetic energy. Hence, if it should really represent the total energy, all other forms of energy, like thermal energy, should be included as part of the rest energy due to the mass-energy equivalence (refer to § 4.3.2).

4.1.8 Energy

The rest energy E_0 of a massive object is the energy contained in the mass of the object due to the equivalence between mass and energy (refer to § 4.3.2 and § 5.3.2). The rest energy is given by the Poincare formula:

$$E_0 = mc^2 \qquad (83)$$

where m represents the invariant mass (i.e. $m = m_0$) according to the modern convention which we follow.

The kinetic energy of a massive object, which is the energy associated with the motion of the object, is given by:

$$E_k = mc^2 (\gamma - 1) = E_0 (\gamma - 1) \qquad (84)$$

Now, by the binomial theorem or Taylor series expansion we have (see Eq. 13):

$$\gamma = \frac{1}{\sqrt{1 - (u^2/c^2)}} = 1 + \frac{1}{2}\frac{u^2}{c^2} + \frac{3}{8}\left(\frac{u^2}{c^2}\right)^2 + \cdots \qquad (85)$$

Hence, for $u \ll c$ where the quadratic and higher order terms in u^2/c^2 are negligible we obtain the following approximation:

$$\gamma \simeq 1 + \frac{1}{2}\frac{u^2}{c^2} \qquad (86)$$

and hence

$$E_k = mc^2 (\gamma - 1) \simeq \frac{1}{2}mu^2 \qquad (87)$$

which is the classical expression for the kinetic energy that is valid at the limit of low speeds, i.e. $u \ll c$. In Figure 11 we plot the kinetic energy as a function of β for both classical mechanics and Lorentz mechanics. As we see, while the kinetic energy curve of Lorentz mechanics rises sharply as β tends to unity (i.e. $u \to c$), due to the presence of the Lorentz γ factor, the kinetic energy curve of classical mechanics continues its quadratic rise at, and beyond, $\beta = 1$. The figure also shows that the classical form approximates the Lorentzian form very well over a considerable range of low speed.

The total energy of a massive object is the sum of its rest and kinetic energies and hence it is given by:

$$E_t = E_0 + E_k = mc^2 + mc^2 (\gamma - 1) = \gamma mc^2 = \gamma E_0 = \frac{mc^2}{\sqrt{1 - (u^2/c^2)}} \qquad (88)$$

As we will see in § 4.4, the total energy of a massive object is conserved in Lorentz mechanics as in classical mechanics. Now, since the total energy includes the rest energy associated with the mass of the object, then what is conserved is the combination of mass-energy. Energy can also be converted from one form to another (e.g. kinetic and electromagnetic) where the total is also conserved during these conversion processes (refer to § 4.4 and § 5.4).

4.1.8 Energy

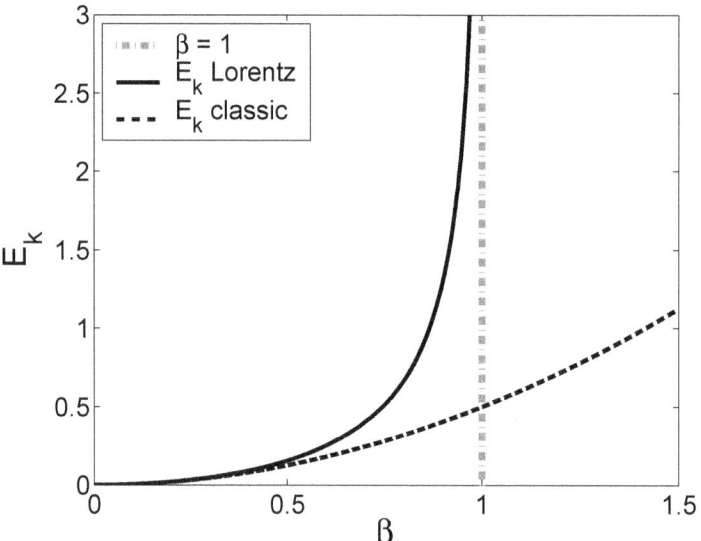

Figure 11: The kinetic energy E_k as a function of β according to classical mechanics (dashed curve) and according to Lorentz mechanics (solid curve) alongside the vertical line $\beta = 1$ (dashed-dotted line) which is an asymptote to the E_k curve of Lorentz mechanics as β approaches unity (i.e. $u \to c$) from below. The vertical axis represents E_k in units of E_0. As seen, the classical curve is very good approximation to the Lorentzian curve at low β.

From the above expression for the total energy (i.e. Eq. 88) and Figure 11, we see that as the speed of a massive object approaches the characteristic speed of light (i.e. $u \to c$) the energy approaches infinity, and hence it may be concluded that no massive object can reach the speed of light as this requires an infinite amount of energy to accelerate the object to the speed of light. Therefore, c is the speed limit for all massive objects. As discussed previously, this restriction is based on the current formalism of Lorentz mechanics and within its domain of validity. For massless objects, the above formula for the energy cannot be used due to a mathematical absurdity that is based on the fact that $m = 0$ while $E_t = \gamma m c^2 \neq 0$. So, instead of the above formula, the expression $E = pc$ is used. This will be discussed further in § 4.3.3 and § 5.3.3. Accordingly, the above conclusion about the speed restriction does not extend to massless objects (at least as implication of the above formulae for total or kinetic energy or the formulae of other quantities of massive objects like $p = \gamma m u$ for momentum) without further evidence.

Solved Problems

1. Use the Lorentzian formulae to calculate the rest energy, the kinetic energy and the total energy of a proton moving at a speed $u = 0.35c$. Also calculate the kinetic energy according to classical mechanics and compare the result with the result of Lorentz mechanics. State your answers in standard SI units.
   Answer: The mass of proton is $m_p \simeq 1.6726 \times 10^{-27}$ kg.

4.1.8 Energy

Rest energy:
$$E_0 = m_p c^2 \simeq 1.6726 \times 10^{-27} \times \left(3 \times 10^8\right)^2 \simeq 1.5054 \times 10^{-10} \text{ J}$$

Lorentzian kinetic energy:
$$E_{kl} = m_p c^2 (\gamma - 1) \simeq 1.5054 \times 10^{-10} \times \left(\frac{1}{\sqrt{1 - 0.35^2}} - 1\right) \simeq 1.0164 \times 10^{-11} \text{ J}$$

Lorentzian total energy:
$$E_t = m_p c^2 \gamma \simeq \frac{1.5054 \times 10^{-10}}{\sqrt{1 - 0.35^2}} \simeq 1.6070 \times 10^{-10} \text{ J}$$

Classical kinetic energy:
$$E_{kc} = \frac{1}{2} m_p u^2 \simeq 0.5 \times 1.6726 \times 10^{-27} \times \left(0.35 \times 3 \times 10^8\right)^2 \simeq 9.2203 \times 10^{-12} \text{ J}$$

Comparing the Lorentzian and classical kinetic energy:
$$\Delta E_r = \frac{E_{kl} - E_{kc}}{E_{kl}} \simeq 0.0929$$

So, the Lorentzian kinetic energy is about 9.29% larger than the classical kinetic energy, which is a significant difference.

2. Calculate the percentage increase in the total energy of an object moving at a speed $u = 0.3c$ as compared to its rest energy.
   Answer: The relative increase in the total energy is:
   $$\Delta E_r = \frac{E_t - E_0}{E_t} = \frac{\gamma m c^2 - m c^2}{\gamma m c^2} = \frac{\gamma - 1}{\gamma} = 1 - \frac{1}{\gamma} = 1 - \sqrt{1 - 0.3^2} \simeq 0.0461$$

   Hence, the percentage increase in the total energy of the object is about 4.61%.

3. What is the equivalent mass, m_{eq}, of a photon of wavelength $\lambda = 1$ angstrom?
   Answer: We have:
   $$E = h\nu = h\frac{c}{\lambda}$$
   where h is Planck's constant and ν is the frequency. Therefore:
   $$m_{\text{eq}} = \frac{E}{c^2} = \frac{h}{\lambda c} \simeq \frac{6.62607 \times 10^{-34}}{10^{-10} \times 3 \times 10^8} \simeq 2.2087 \times 10^{-32} \text{ kg}$$

4. Find the kinetic energy of an electron in units of E_0 (i.e. $m_e c^2$ where m_e is the mass of electron) that moves at a speed of $0.3c$ (a) according to classical mechanics and (b) according to Lorentz mechanics. Also find the relative difference between the two cases.
   Answer:
   (a) For classical mechanics we have:
   $$E_{kc} = \frac{1}{2} m_e u^2 = 0.5 \times m_e \times 0.09 c^2 = 0.045 m_e c^2$$

4.1.8 Energy

(b) For Lorentz mechanics we have:

$$E_{kl} = m_e c^2 (\gamma - 1) = m_e c^2 \left(\frac{1}{\sqrt{1 - 0.09}} - 1 \right) \simeq 0.0483 m_e c^2$$

The relative difference ΔE_{kr} is:

$$\Delta E_{kr} = \frac{E_{kl} - E_{kc}}{E_{kl}} \simeq 0.0680$$

i.e. the classical value has a relative error of about -6.8% compared to the Lorentzian value.

5. Give an example of energy units, other than joule, that are commonly used in the applications of Lorentz mechanics with justification for the use of these units.
Answer: Since one of the application fields of Lorentz mechanics is atomic and particle physics where charged particles are usually the subject of investigation, a commonly used unit of energy is the electron volt, which is usually abbreviated as eV. However, because this unit is too small in these fields where these particles generally move at very high speeds, this unit is normally scaled up by a large factor like mega (i.e. 10^6) which is abbreviated as M, and giga (i.e. 10^9) which is abbreviated as G and hence MeV and GeV units are used as common energy units.

Exercises
1. Provide more clarification about the term "total energy".
2. Classify the rest energy, the kinetic energy and the total energy as intrinsic or extrinsic properties.
3. Discuss the impact of the Poincare mass-energy equivalence relation (see § 4.3.2 and § 5.3.2) on the use of the mass and energy units.
4. A mother particle of mass $m_m = 400 \, \text{MeV}/c^2$, which is at rest in the laboratory frame, decays into two daughter particles each of mass $m_d = 150 \, \text{MeV}/c^2$. What is the kinetic energy of each one of the daughter particles?
5. Find a general relation between the Lorentzian momentum and the Lorentzian total energy. Also, find a general relation between the Lorentzian momentum and the Lorentzian rest and kinetic energies.
6. The momentum of a proton is $p = 100 \, \text{MeV}/c$. What is its kinetic energy?
7. What is the speed, according to classical and Lorentz mechanics, of an electron whose kinetic energy is 10 MeV? Comment on the results.
8. Given that the mass of electron, proton and neutron in kilogram are: $m_e \simeq 9.109384 \times 10^{-31}$, $m_p \simeq 1.672622 \times 10^{-27}$ and $m_n \simeq 1.674929 \times 10^{-27}$, find their rest energies in joules.
9. Find the Lorentzian speed of a massive object whose kinetic energy is half its rest energy.
10. Find the Lorentzian speed of a massive object whose classical speed is equal to the characteristic speed of light c. Solve the question by equating the classical kinetic energy to the Lorentzian kinetic energy. Comment on the result.

4.1.9 Work 133

11. What is the condition that should be imposed on the speed if the maximum allowed relative error in the kinetic energy between the Lorentzian and classical formulae should be less than 1%?
12. What is the momentum of a neutron whose kinetic energy is 100 MeV?

4.1.9 Work

As in classical mechanics, the work done by a force to move a massive object from position P_1 to position P_2 is equal to the difference in the kinetic energy of the object at the two positions.[107] Hence we have:

$$W = E_{k2} - E_{k1} = mc^2 \left(\gamma_2 - 1\right) - mc^2 \left(\gamma_1 - 1\right) = mc^2 \left(\gamma_2 - \gamma_1\right) \tag{89}$$

where E_{k1} and E_{k2} are the kinetic energy at P_1 and P_2, m is the mass of the object, and γ_1 and γ_2 are the Lorentz factors at P_1 and P_2 which are functions of the speed of the object at P_1 and P_2.

Solved Problems
1. Find the amount of work required to accelerate an electron from rest to $u = 0.9c$.
   Answer: We have:

$$W = m_e c^2 \left(\gamma_2 - \gamma_1\right) \simeq 0.511 \, \text{MeV} \left(\frac{1}{\sqrt{1 - 0.9^2}} - 1\right) \simeq 0.661 \, \text{MeV}$$

Exercises
1. Find the amount of work required to accelerate an object of mass $m = 10^{-6}$ kg from $u_1 = 0.2c$ to $u_2 = 0.5c$.

4.2 Physical Transformations

As seen earlier, to explain the null result of the Michelson-Morley experiment, a set of transformations that correlate the spatial and temporal coordinates in two inertial frames have been proposed by Lorentz.[108] Like the Galilean transformations, the Lorentz transformations are a set of mathematical relations that transform the spacetime coordinates of an object in a given inertial frame O_1, to the spacetime coordinates of the object in another inertial frame O_2 which is in a state of uniform motion relative to O_1. The Lorentz transformations originated from an earlier attempt by FitzGerald and Lorentz to explain the null result of the Michelson-Morley experiment through compensation by length contraction for the presumed variable speed of light in the two arms of the Michelson-Morley

---
[107] This is subject to certain conditions (e.g. non-dissipative force with no change of potential) that can be found in general physics texts.
[108] In fact, the null result of Michelson-Morley experiment was the original motivation for the development of these transformations but there were other reasons (e.g. requirement of invariance of Maxwell's equations) for this development.

4.2 Physical Transformations

apparatus (refer to § 2.6, § 2.7 and § 12.2). A distinct feature of the Lorentz transformations is that Maxwell's equations of electromagnetism, as well as the laws of classical mechanics (with some alterations), are invariant under these transformations unlike the Galilean transformations which transform the laws of classical mechanics invariantly but not Maxwell's equations (see § 12.3).

In the next subsection, we present the main Lorentz transformations of spacetime coordinates. This will be followed by other subsections where other subsidiary transformations that are derived from the main transformations of spacetime coordinates are presented. In § 5.1, these subsidiary transformations will be derived from the main Lorentz transformations of spacetime coordinates. As in the past, we assume in the following formulations that the two frames are in a state of standard setting as described earlier (refer to § 1.5 and Figure 3). We note that, inline with our plan which we explained earlier, the presentation of Lorentz transformations in this section is introductory whereas it is more thorough in § 5.1 where derivations and discussions are included there. This is also reflected in the exercises which are essentially numerical in this section while they are generally of analytical nature in § 5.1.

Solved Problems

1. List a number of features that characterize Lorentz spacetime coordinate transformations.

   Answer: The Lorentz transformations of spacetime coordinates are characterized by the following features:
   • They are transformations between inertial frames of reference.
   • Spatial and temporal coordinates are mingled in the equations of the transformations of both spatial and temporal coordinates.
   • Maxwell's equations, as well as the laws of classical mechanics, are invariant under these transformations.
   • Spacetime intervals are invariant under these transformations, but the space intervals and the time intervals are not.
   • The Lorentz transformations converge practically to the Galilean transformations at low speeds, i.e. when $v \ll c$ and hence $\beta \simeq 0$ and $\gamma \simeq 1$.
   • These transformations are limited to speeds v below the characteristic speed of light c due to the presence of the Lorentz γ factor which becomes singular when $v = c$ and imaginary when $v > c$.
   • These transformations are also characterized by preserving the constancy of the speed of light between all inertial observers (see § 4.2.2 and § 9.3.2).[109]

Exercises

1. The given formulation of Lorentz transformations is based on a state of standard setting between the involved reference frames. What this means?
2. The Lorentz spacetime coordinate transformations were proposed to replace which transformations? What is the original motive for this proposal?

---

[109] This is linked to preserving the invariance of the spacetime interval.

3. Make a brief comparison between the Lorentzian and Galilean transformations of spatial and temporal coordinates highlighting their common features as well as their differences.
4. What is the significance of the convergence of Lorentz transformations to the Galilean transformations at low speeds?

4.2.1 Lorentz Spacetime Coordinate Transformations

Based on the above-described standard setting and assumptions, the Lorentz transformations of the coordinates of spacetime from an inertial frame O to another inertial frame O' are given by:

$$x' = \gamma(x - vt) = \gamma(x - \beta ct) \tag{90}$$
$$y' = y \tag{91}$$
$$z' = z \tag{92}$$
$$t' = \gamma\left(t - \frac{vx}{c^2}\right) = \gamma\left(t - \frac{\beta x}{c}\right) \tag{93}$$

where the primed and unprimed symbols belong to frames O' and O respectively, and β and γ are the speed ratio and Lorentz factor as defined earlier in Eq. 5. The last equation may also be written as:

$$ct' = \gamma\left(ct - \frac{vx}{c}\right) = \gamma(ct - \beta x) \tag{94}$$

so that all the coordinates of spacetime have the physical dimension of length. This will also facilitate the use of a unified symbolic notation (including tensor notation) for the spatial and temporal coordinates, i.e. $\mathbf{x} = (x^0, x^1, x^2, x^3)$ or $\mathbf{x} = (x^1, x^2, x^3, x^4)$ where $x^0 = x^4 = ct$.[110]

The above equations can be inverted to give the Lorentz transformations in the opposite direction (i.e. from frame O' to frame O), that is:[111]

$$x = \gamma(x' + vt') = \gamma(x' + \beta ct') \tag{95}$$
$$y = y' \tag{96}$$
$$z = z' \tag{97}$$
$$t = \gamma\left(t' + \frac{vx'}{c^2}\right) = \gamma\left(t' + \frac{\beta x'}{c}\right) \tag{98}$$

The last set of equations (i.e. primed to unprimed) looks identical to the previous set of equations (i.e. unprimed to primed) except for the exchange of the primed and unprimed symbols and the reversal of the sign of v (or β). This is because the two inertial frames are equivalent and the relative velocity between the two frames is equal in magnitude and

---

[110] In fact, the use of x^0 or x^4 to represent t rather than ct is also common in the literature. However, the use of a unified symbolic notation will be more appropriate if x^0 or x^4 are used to represent ct since all the coordinate symbols x^μ will then represent the same physical dimension.

[111] These may be called the inverse Lorentz transformations.

4.2.1 Lorentz Spacetime Coordinate Transformations

opposite in direction; hence, all is needed for this inversion is to interchange the primed and unprimed symbols of the spacetime coordinates and replace v with $-v$. For the same reason, the shift of transformations in the opposite direction (i.e. from primed to unprimed or the other way around) by exchanging the primed and unprimed symbols and reversing the sign of v (or β) is general and hence it applies to all types of transformation (e.g. velocity transformations) and not only to the spacetime coordinate transformations.

Solved Problems

1. Why the symbols v and β do not have a primed version?
   Answer: The symbols v and β do not have a primed version because the relative speed is the same for both frames and consequently the primed and unprimed status of these variables is represented by the sign of v (or β) which stands for the 1D velocity.

2. Explain the sign convention of the relative velocity v according to the standard setting between two inertial frames O and O'.
   Answer: In brief, v is the velocity of frame O' with respect to frame O along the common x-x' axis. Hence, v should be positive when O' moves (as seen from O) in the positive direction of the common x-x' axis and negative when O' moves in the negative direction of the common x-x' axis. Alternatively, v should be negative when O moves (as seen from O') in the positive direction of the common x-x' axis and positive when O moves in the negative direction of the common x-x' axis.

3. A given event is seen in an inertial frame O to have spacetime coordinates $\mathbf{x} = (5, 9, 13, -2)$ where the temporal coordinate ct is the first. What are the spacetime coordinates of this event as seen from another inertial frame O' which is in a state of standard setting with O where the relative velocity is $v = 0.55c$?
   Answer: Using the Lorentz spacetime coordinate transformations from frame O to frame O', we have:

$$ct' = \gamma(ct - \beta x) = \frac{5 - 0.55 \times 9}{\sqrt{1 - 0.55^2}} \simeq 0.0599$$

$$x' = \gamma(x - \beta ct) = \frac{9 - 0.55 \times 5}{\sqrt{1 - 0.55^2}} \simeq 7.4836$$

$$y' = y = 13$$

$$z' = z = -2$$

   Hence, $\mathbf{x}' \simeq (0.0599, 7.4836, 13, -2)$.

4. O and O' are two inertial observers in a state of standard setting. Two events, V_1 and V_2, are seen by O to be co-positional, while they are seen by O' to be separated by a spatial distance $\Delta x' = x'_2 - x'_1 = 10^6$ m and by a time interval $\Delta t' = t'_2 - t'_1 = 0.1$ s. What is the time interval between these events in frame O?
   Answer: Because the two events are co-positional in frame O then we should have: $x_1 = x_2$, that is:

$$x_1 = x_2$$
$$\gamma(x'_1 + vt'_1) = \gamma(x'_2 + vt'_2)$$

4.2.1 Lorentz Spacetime Coordinate Transformations

$$x_1' + vt_1' = x_2' + vt_2'$$
$$x_1' - x_2' = v(t_2' - t_1')$$
$$v = -\frac{x_2' - x_1'}{t_2' - t_1'}$$
$$v = -\frac{10^6}{0.1} = -10^7$$

Hence, the time interval between these events in frame O is:

$$t_2 - t_1 = \gamma\left(t_2' + \frac{vx_2'}{c^2}\right) - \gamma\left(t_1' + \frac{vx_1'}{c^2}\right)$$
$$= \gamma\left[(t_2' - t_1') + \frac{v}{c^2}(x_2' - x_1')\right]$$
$$= \frac{0.1 - (10^7/c^2) \times 10^6}{\sqrt{1 - (-10^7/c)^2}}$$
$$\simeq 9.9944 \times 10^{-2}\,\text{s}$$

We note that this question is an example of the relativity of co-positionality (see § 7.5) since the two events are seen co-positional from O frame but not from O' frame.

Exercises

1. Write the Lorentz spacetime coordinate transformations between two inertial frames, O and O', which are in a state of standard setting with a relative velocity $v = -0.6c$.
2. O and O' are two inertial frames in a state of standard setting with $v = -0.4c$. An event V_1 is observed from O to occur at $x_1 = 33.9$ and $t_1 = 2 \times 10^{-9}$. What are x_1' and t_1'? If another event V_2 occurred at $x_2' = 60$ and $t_2' = 8 \times 10^{-8}$, what are x_2 and t_2?
3. An event is observed from two inertial frames O and O', which are in a state of standard setting, to be at $\mathbf{x} = (12, 6, -21, 3)$ and $\mathbf{x}' = (10.4745, 1.3093, -21, 3)$ respectively where the temporal coordinates ct and ct' are the first. What is the relative speed between the two frames?
4. V_1 is an event seen in an inertial frame O to take place at $x_1 = \xi$, $y_1 = z_1 = 0$ and $ct_1 = \xi$, and V_2 is another event seen in O to take place at $x_2 = 3\xi$, $y_2 = z_2 = 0$ and $ct_2 = 2\xi$ where ξ is a positive number. However, V_1 and V_2 are seen in another inertial frame O' to take place at the same time. Assuming that O and O' are in a state of standard setting, what is the speed of O' relative to O? Comment on the significance of this exercise.
5. O and O' are two inertial observers in a state of standard setting with $v = -0.45c$. Two events, V_1 and V_2, are seen by O to be co-positional, while they are seen by O' to be separated by a time interval of 0.002. What is the spatial separation between V_1 and V_2 according to O'? Comment on the significance of this exercise.
6. O and O' are two inertial frames in a state of standard setting with $v = 0.3c$. An object moves in the xz plane of frame O with a constant velocity $u = 0.4c$ where its straight path makes an angle $\theta = \pi/6$ with the positive x axis. What are the equations of motion of the object in frame O' (i.e. x' and z' as functions of t')?

4.2.2 Velocity Transformations 138

7. An inertial observer O measures the spatial separation between two events, V_1 and V_2, to be $x_2 - x_1 = 10^8$ and the time interval to be $t_2 - t_1 = 1$. What is the proper time interval between V_1 and V_2?

4.2.2 Velocity Transformations

For the above two inertial observers O and O', the measured velocities, $\mathbf{u} = (u_x, u_y, u_z)$ and $\mathbf{u'} = (u'_x, u'_y, u'_z)$, of a given object in the two frames are Lorentz transformed from the frame of O to the frame of O' as follows:

$$u'_x = \frac{u_x - v}{1 - (vu_x/c^2)} \tag{99}$$

$$u'_y = \frac{u_y}{\gamma\left[1 - (vu_x/c^2)\right]} \tag{100}$$

$$u'_z = \frac{u_z}{\gamma\left[1 - (vu_x/c^2)\right]} \tag{101}$$

where the symbols are as explained before and γ is a function of v. As discussed earlier, the velocity v is positive/negative when O' moves, according to O, in the positive/negative direction of the common x-x' axis. These transformations, which can be obtained by differentiating the above Lorentz transformations of spacetime coordinates as will be seen in § 5.1.2, represent how the measurements of the velocity components of an object are related in the frames of O and O'.

As before and for the same reason, the inverse Lorentz transformations from O' to O are obtained by interchanging the primed and unprimed symbols of the velocity components of the object and replacing v with $-v$, that is:

$$u_x = \frac{u'_x + v}{1 + (vu'_x/c^2)} \tag{102}$$

$$u_y = \frac{u'_y}{\gamma\left[1 + (vu'_x/c^2)\right]} \tag{103}$$

$$u_z = \frac{u'_z}{\gamma\left[1 + (vu'_x/c^2)\right]} \tag{104}$$

Solved Problems

1. Discuss the general approach in tackling the velocity transformation questions in classical mechanics and in Lorentz mechanics.
   Answer: The approach in both mechanics is the same although the formulae are different. In brief, in the velocity transformation problems we have two inertial observers, O and O', and an observed object whose velocity is to be transformed from the frame of one observer to the frame of the other observer (or in other words we know its velocity in one frame and we want to find its velocity in the other frame). The two observers are in a state of relative uniform motion with a relative velocity $\mathbf{v}$. Moreover, the object has two velocities: one relative to O which we label as $\mathbf{u}$ and one relative to O'

4.2.2 Velocity Transformations

which we label as **u′**. Now, since the velocity is a vector in a 3D space it has three spatial components, and because we have three velocities (i.e. **v**, **u** and **u′**) we have nine velocity components. However, since we usually use standard setting then **v** is reduced to a single component which is its x component. i.e. v. So, in total we have seven quantities. As soon as we label the given data properly (i.e. what is v, u_x, etc.) then the only thing left to find the solution is to identify the appropriate formula or formulae that we should use to find the unknown or unknowns. What is left then is just mathematical manipulation and substitution of the numerical or symbolic data to find the final solution.

2. O and O' are two inertial observers in a state of standard setting with a relative velocity v where $0 \le |v| < c$. One of the observers observed a light signal propagating along the common x-x' axis with speed c (i.e. it can be in either direction and hence the velocity is $\pm c$). Assuming that the Lorentz velocity transformation formulae apply to light signal like any other object,[112] what is the speed of this signal from the frame of the other observer?
Answer: If O is the original observer then $u_x = \pm c$ and we have:

$$u'_x = \frac{u_x - v}{1 - \frac{vu_x}{c^2}} = \frac{\pm c - v}{1 - \frac{v(\pm c)}{c^2}} = \frac{\pm c - v}{1 - \frac{\pm v}{c}} = \frac{\pm c - v}{c \mp v} c = \frac{\pm (c \mp v)}{c \mp v} c = \pm c$$

while u'_y and u'_z are zero because u_y and u_z are zero.
Similarly, if O' is the original observer then $u'_x = \pm c$ and we have:

$$u_x = \frac{u'_x + v}{1 + \frac{vu'_x}{c^2}} = \frac{\pm c + v}{1 + \frac{v(\pm c)}{c^2}} = \frac{\pm c + v}{1 + \frac{\pm v}{c}} = \frac{\pm c + v}{c \pm v} c = \frac{\pm (c \pm v)}{c \pm v} c = \pm c$$

while u_y and u_z are zero because u'_y and u'_z are zero.
So, in all cases the speed of the signal is c in both frames.

3. O and O' are two inertial observers in a state of standard setting with a relative velocity v where $0 \le |v| < c$. One of the observers observed a light signal propagating along the y orientation with speed c (i.e. it can be in either direction and hence the velocity is $\pm c$). What is the speed of this signal from the frame of the other observer? Comment on the results.
Answer: If O is the original observer then $u_x = u_z = 0$ and $u_y = \pm c$ and we have:

$$u'_x = \frac{u_x - v}{1 - \frac{vu_x}{c^2}} = \frac{0 - v}{1 - 0} = -v$$

$$u'_y = \frac{u_y}{\gamma\left(1 - \frac{vu_x}{c^2}\right)} = \frac{\pm c\sqrt{1 - (v^2/c^2)}}{(1 - 0)} = \pm c\sqrt{1 - (v^2/c^2)}$$

$$u'_z = 0$$

Hence, the speed u' in O' frame is:

$$u' = \sqrt{(u'_x)^2 + (u'_y)^2 + (u'_z)^2} = \sqrt{v^2 + (c^2 - v^2) + 0} = \sqrt{c^2} = c$$

---
[112] This assumption should be valid at least as a limiting case.

4.2.2 Velocity Transformations

Similarly, if O' is the original observer then $u'_x = u'_z = 0$ and $u'_y = \pm c$ and we have:

$$u_x = \frac{u'_x + v}{1 + \frac{vu'_x}{c^2}} = \frac{0 + v}{1 + 0} = v$$

$$u_y = \frac{u'_y}{\gamma\left(1 + \frac{vu'_x}{c^2}\right)} = \frac{\pm c\sqrt{1 - (v^2/c^2)}}{(1 + 0)} = \pm c\sqrt{1 - (v^2/c^2)}$$

$$u_z = 0$$

Hence, the speed u in O frame is:

$$u = \sqrt{(u_x)^2 + (u_y)^2 + (u_z)^2} = \sqrt{v^2 + (c^2 - v^2) + 0} = \sqrt{c^2} = c$$

So, in all cases the speed of the signal is c in both frames.
Comment: the results indicate that the observed speed of light is invariant across inertial frames since it takes its characteristic value c in all these frames. However, the results also indicate that the velocity of light is dependent on the velocity of its source since it has an x velocity component $u'_x = -v$ in the first case and $u_x = v$ in the second case. This suggests that although the speed of light is independent of the speed of its source, its velocity is not. This rules out the possibility of a classical wave propagation model for light and hence the existence of a propagation medium (see § 1.13.2 and § 1.14) but it is consistent with a projectile propagation model (see § 1.13.1 and § 1.14).[113] These issues will be discussed in detail later where we will see that a Lorentzian projectile model is consistent with the invariance of the speed of light when we take account of the contraction of spacetime under the influence of motion. We also note that the projectile model is consistent with the existence of an absolute frame although it does not require such a frame (see § 3.8).

4. O and O' are two inertial observers in a state of standard setting with a relative velocity v where $0 \leq |v| < c$. One of the observers observed a light signal propagating with speed c in his xy plane along a straight line that makes an angle θ with the positive x axis. (a) What is the speed of this signal from the frame of the other observer? (b) What is the angle that this signal makes with the positive x axis in the frame of the other observer? (c) What you conclude from the results of this question?
Answer: Let assume that the original observer is O. The situation will be similar if the original observer is O'.
(a) If O is the original observer then $u_x = c\cos\theta$ and $u_y = c\sin\theta$. Hence:

$$u'_x = \frac{u_x - v}{1 - \frac{vu_x}{c^2}} = \frac{c\cos\theta - v}{1 - \frac{v\cos\theta}{c}} = \frac{c\cos\theta - v}{c - v\cos\theta}c$$

---

[113] This is about the propagation model and not about the nature of light and hence it does not contradict the fact that light has wave properties. In fact, this is completely consistent with the dual nature of light (i.e. wave-particle) according to modern physics. In brief, light is a wave phenomenon that propagates like particles.

4.2.2 Velocity Transformations

$$u'_y = \frac{u_y}{\gamma\left(1 - \frac{vu_x}{c^2}\right)} = \frac{c\sin\theta\sqrt{1 - v^2/c^2}}{1 - \frac{v\cos\theta}{c}} = \frac{c\sin\theta\sqrt{1 - v^2/c^2}}{c - v\cos\theta}c$$

$$u'_z = 0$$

Hence, the speed u' in O' frame is:

$$\begin{aligned}
u' &= \sqrt{(u'_x)^2 + (u'_y)^2 + (u'_z)^2} \\
&= \sqrt{\left(\frac{c\cos\theta - v}{c - v\cos\theta}c\right)^2 + \left(\frac{c\sin\theta\sqrt{1 - v^2/c^2}}{c - v\cos\theta}c\right)^2 + 0^2} \\
&= c\sqrt{\frac{(c\cos\theta - v)^2 + \left(c\sin\theta\sqrt{1 - v^2/c^2}\right)^2}{(c - v\cos\theta)^2}} \\
&= c\sqrt{\frac{(c^2\cos^2\theta - 2cv\cos\theta + v^2) + (c^2\sin^2\theta - v^2\sin^2\theta)}{(c - v\cos\theta)^2}} \\
&= c\sqrt{\frac{c^2(\cos^2\theta + \sin^2\theta) - 2cv\cos\theta + v^2(1 - \sin^2\theta)}{(c - v\cos\theta)^2}} \\
&= c\sqrt{\frac{c^2 - 2cv\cos\theta + v^2\cos^2\theta}{(c - v\cos\theta)^2}} \\
&= c\sqrt{\frac{(c - v\cos\theta)^2}{(c - v\cos\theta)^2}} \\
&= c
\end{aligned}$$

So, the speed is also c in O' frame.

(b) Regarding the angle θ' in O' frame, we have:

$$\theta' = \arctan\frac{u'_y}{u'_x} = \arctan\left(\frac{c\sin\theta\sqrt{1 - v^2/c^2}}{c\cos\theta - v}\right)$$

(c) The obvious conclusion from the results of this question (as well as the results of other previous and upcoming questions) is that according to the Lorentz velocity transformations the observed speed of light is invariant in all inertial frames and hence it is equal to its characteristic speed c in any frame. However, the observed velocity of light is frame dependent since the direction of the light path depends on the relative motion between the frames. This is because the velocity of light depends on the motion of its source since the velocity of light has a component from the motion of the source and this component depends on the relative motion of the source which is frame dependent. We should remark that in our answer we are assuming that the Lorentz velocity transformations apply to light as well as to massive objects at least in the limiting

4.2.2 Velocity Transformations 142

case where the speed approaches c from below. We should also remark that the above conclusion about the invariance of the observed speed of light across all inertial frames under the Lorentz velocity transformations applies to light signals in a more general setting where the light velocity has components in all directions (i.e. x, y and z) and not only in the x or y directions or in the xy plane as in the previous and present questions. This is because the z velocity transformation is mathematically identical to the y velocity transformation according to the standard setting and hence the above formulation and results should equally apply in the more general setting with minor amendments but with more messy mathematics. Alternatively, the speed should not depend on the choice of coordinate system and hence any coordinate system can be reduced to one of the above settings (i.e. in the x or y directions or in the xy plane) to find that the speed is c and hence it is invariant.[114]

5. A radioactive nucleus, which is moving along the x axis with a constant velocity of $0.35c$ relative to the laboratory frame, decays by emitting a beta particle with a speed of $0.45c$ relative to the nucleus. What is the velocity of the beta particle relative to the laboratory (a) if the velocity of the beta particle is in the same direction of the nucleus motion and (b) if the velocity of the beta particle is in a perpendicular direction (i.e. as seen in the frame of nucleus) to the nucleus motion? Assume that the given velocities are the values following the emission (or assume that the mass of the nucleus is much bigger than the mass of the beta particle).

Answer: We label the laboratory frame with O and the nucleus frame with O'. Accordingly, we have two inertial frames which are in a state of standard setting with $v = 0.35c$.

(a) In this case we have $u'_x = 0.45c$ and hence the velocity of the beta particle in the laboratory frame is:

$$u_x = \frac{u'_x + v}{1 + \frac{vu'_x}{c^2}} = \frac{0.45c + 0.35c}{1 + \frac{0.35c \times 0.45c}{c^2}} = \frac{0.45 + 0.35}{1 + (0.35 \times 0.45)}c \simeq 0.6911c$$

(b) In this case we have $u'_x = u'_z = 0$ and $u'_y = 0.45c$ and hence we have:

$$u_x = \frac{u'_x + v}{1 + \frac{vu'_x}{c^2}} = \frac{0 + 0.35c}{1 + 0} = 0.35c$$

$$u_y = \frac{u'_y}{\gamma\left(1 + \frac{vu'_x}{c^2}\right)} = \frac{0.45c\sqrt{1 - 0.35^2}}{1 + 0} \simeq 0.4215c$$

$$u_z = 0$$

Hence, the beta particle has a speed u in the laboratory frame given by:

$$u = \sqrt{(u_x)^2 + (u_y)^2 + (u_z)^2} \simeq c\sqrt{0.35^2 + 0.4215^2 + 0^2} \simeq 0.5479c$$

---

[114] In fact, all we need for this reduction is a rotation in the 3D space which keeps speed unchanged.

4.2.2 Velocity Transformations

and it moves in the xy plane along a straight line making an angle θ with the positive x axis of the laboratory frame where:

$$\theta = \arctan\frac{u_y}{u_x} \simeq 0.8779\,\text{rad} \simeq 50.30°$$

Exercises

1. An Object is seen by an inertial observer O to have a velocity $\mathbf{u} = (u_x, u_y, u_z) = (0.3c, 0.1c, 0.4c)$. What is the speed of the object as seen by another inertial observer O' who is in a state of standard setting with O where $v = 0.2c$?
2. An elementary (mother) particle decays into two (daughter) particles of equal speed. In the rest frame of the mother particle, the speed of each one of the daughter particles is $0.75c$. What are the velocities of the daughter particles in a frame in which the mother particle is moving at a speed of $0.5c$? Assume in your answer that the daughter particles move along the same orientation as the motion of the mother particle.
3. Repeat the previous exercise but this time assume that the daughter particles, as seen from the rest frame of the mother particle, move in a perpendicular direction to the direction of the relative motion between the two frames.
4. O and O' are two inertial observers in a state of standard setting with $v = 0.7c$. An object moves in the xy plane of O frame with a constant speed $u = 0.15c$ where its straight path makes an angle $\theta = \pi/3$ with the positive x axis, i.e. the x and y components of its constant velocity are positive. What is the velocity of the object in O' frame? Find this velocity once as components along the coordinate axes and once as speed and direction.
5. The observed speed of light in transparent material media, like air and water, is given by c/n where c is the characteristic speed of light in free space and n is the refractive index of the particular medium which is assumed to be in a standstill state. According to the Fizeau formula, which he derived[115] from his experiments in which he measured the speed of light in running water, the observed speed of light c_m in a medium that is moving uniformly with speed u_m relative to the observer is given by:

$$c_m = \frac{c}{n} + k u_m$$

where k is the dragging coefficient which is a medium-dependent parameter. In the mid 19th century, Fizeau found that the dragging coefficient of water is $k_w \simeq 0.44$. Try to justify this finding (both formula and $k_w \simeq 0.44$) by using the Lorentzian velocity transformations. Also, comment on the implication of this finding.
6. O and O' are two inertial observers in a state of standard setting with $v = 0.5c$. O' sent a light ray in the positive y' direction. What is the speed and direction of this signal as seen from O frame?
7. O and O' are two inertial observers in a state of standard setting with $v = 0.3c$. A light signal is emitted by O' in a direction that makes an angle $\theta' = 45°$ with the positive x' axis where this signal is contained in the $x'y'$ plane. Find the velocity components of this signal and its speed and direction as seen from O frame. Comment on the result.

---
[115] We should also refer to Fresnel in this regard (see § 2.6).

8. Referring to the previous question, justify the relation $\theta < \theta'$ with some graphic illustration.
9. Discuss the implications of the previous exercises about the speed and velocity of light in inertial frames.

4.2.3 Velocity Composition

According to the Lorentz velocity transformations, the composition of velocities (i.e. the velocity of an object O_3, which is moving in a given inertial frame O_2, as observed from another inertial frame O_1) is given by:

$$u_{x31} = \frac{u_{x32} + u_{x21}}{1 + \frac{u_{x32}u_{x21}}{c^2}} \tag{105}$$

where u_{x31} is the velocity of O_3 relative to O_1 in the x direction and similarly for u_{x32} and u_{x21}. It is obvious that Eq. 105 will reduce to the Galilean composition rule of velocity addition (Eq. 50) at low speeds, i.e. when $u_{x32}u_{x21} \ll c^2$.[116] We also note that Eq. 105 is no more than another form of Eq. 102 if we consider that: $u_{x31} \equiv u_x$, $u_{x32} \equiv u'_x$ and $u_{x21} \equiv v$. Following a similar argument, we obtain the other velocity components, that is:

$$u_{y31} = \frac{u_{y32}}{\gamma\left(1 + \frac{u_{x32}u_{x21}}{c^2}\right)} \tag{106}$$

$$u_{z31} = \frac{u_{z32}}{\gamma\left(1 + \frac{u_{x32}u_{x21}}{c^2}\right)} \tag{107}$$

where the symbols have similar meanings as before (e.g. u_{y31} is the velocity of O_3 relative to O_1 in the y direction) and γ is a function of u_{x21}. Again, Eq. 106 is another form of Eq. 103 if we consider that: $u_{y31} \equiv u_y$, $u_{y32} \equiv u'_y$, $u_{x32} \equiv u'_x$ and $u_{x21} \equiv v$. This equally applies to Eq. 107 in its correspondence to Eq. 104 with the change of y to z in the symbols.

We remark that since all the velocity composition formulae (i.e. Eqs. 105-107) are based on the velocity transformation formulae (i.e. Eqs. 102-104) that are founded on a state of standard setting with v in the latter being replaced by u_{x21} in the former, the relative motion between O_2 and O_1 is restricted to the x direction and that is why the subscript 21 in all the composition formulae is restricted to the x dimension. This is important to note when we employ these formulae to solve symbolic or numeric problems.[117]

Another remark is that although the composition formulae are not desperately needed since they can be replaced by the transformation formulae, there are cases in which the

---

[116] This condition may be realized by a combination of the following two conditions: $u_{x32} \ll c$ and $u_{x21} \ll c$ although this is not necessary.

[117] We should note that the velocity transformations are conceptually more appropriate for transforming one velocity from one inertial frame to another inertial frame while the composition of velocities is conceptually more appropriate for measuring the velocity of two objects from a given inertial frame. This should have an impact on the inertiality of the objects where one (i.e. the object that shares with the frame of observation the subscript 21) should be "inertial" since it corresponds to the other inertial frame in the velocity transformations.

4.2.3 Velocity Composition

setting of the problem and how it is stated makes the composition formulae more intuitive and easier to use due to their spontaneous labeling that facilitates the automation of the process of assigning the data to the symbols. However, as explained in the previous remark, when the relative motions belong to more than one dimension then special care is required in assigning the subscript 21 since the relative motion between 1 and 2 is restricted to the x dimension.

In Figure 12, we compare the velocity composition formulae of classical mechanics (i.e. Galilean) and Lorentz mechanics in the x direction where these formulae are given respectively by Eqs. 50 and 105. As we see, while the Galilean composition formula produces straight contours, due to the simple addition of velocities, with no limit on the maximum velocity and hence the resultant velocity can exceed c, the Lorentzian composition formula produces curved contours with a limit on the maximum velocity and hence the plane $u_{x31} = c$ acts as an asymptote to the u_{x31} surface of the Lorentzian formula.

Similarly, in Figure 13, we compare the velocity composition formulae of classical mechanics (i.e. Galilean) and Lorentz mechanics in the y direction where the Galilean is given by:

$$u_{y31} = u_{y32} \tag{108}$$

while the Lorentzian is given by Eq. 106 where in this figure we assume $u_{x32} = 0$. A similar figure should apply to the velocity composition formulae in the z direction.

We remark that the symbols in the velocity composition formulae, whether Galilean or Lorentzian, like the symbols in their parent formulae (i.e. the transformation formulae) represent velocities and hence they have sign as well as magnitude where these signs are based on the employed coordinate system. Hence, the signs of these velocities should be considered in solving numerical and symbolic problems. The situation in some cases may not be complicated to require employing a coordinate system and hence some rules of thumb are used to determine the signs, e.g. the two velocities in the numerator on the right hand side of Eq. 105 have opposite signs when they are in opposite directions and the same sign when they are in the same direction relative to the observer for whom the relative velocity of the observed object is computed. For example, if the velocity of a projectile fired from a rocket launched from the Earth is in the same direction as the rocket (i.e. forward) then the velocity of the rocket with respect to the Earth (i.e. u_{x21}) and the velocity of the projectile with respect to the rocket (i.e. u_{x32}) should have the same sign, but if the velocity of the projectile is in the opposite direction to the rocket motion then these two velocities should have opposite signs. This is logical because these two velocities are in the same or opposite direction according to the Earth observer to whom the velocity of the projectile (i.e. u_{x31}) is to be calculated. It is also based on the previously-stated fact that the symbols in the above formulations represent vectors[118] and hence they have magnitude and direction (represented by sign) although they look like scalars. This logic is also supported by common sense where the speeds are added in the first case while they are subtracted in the second case. This common sense can be used when confusion occurs in determining if the signs should be the same or opposite although it is generally safer

---

[118] Or components of vectors which in this sense can be seen as 1D vectors.

4.2.3 Velocity Composition

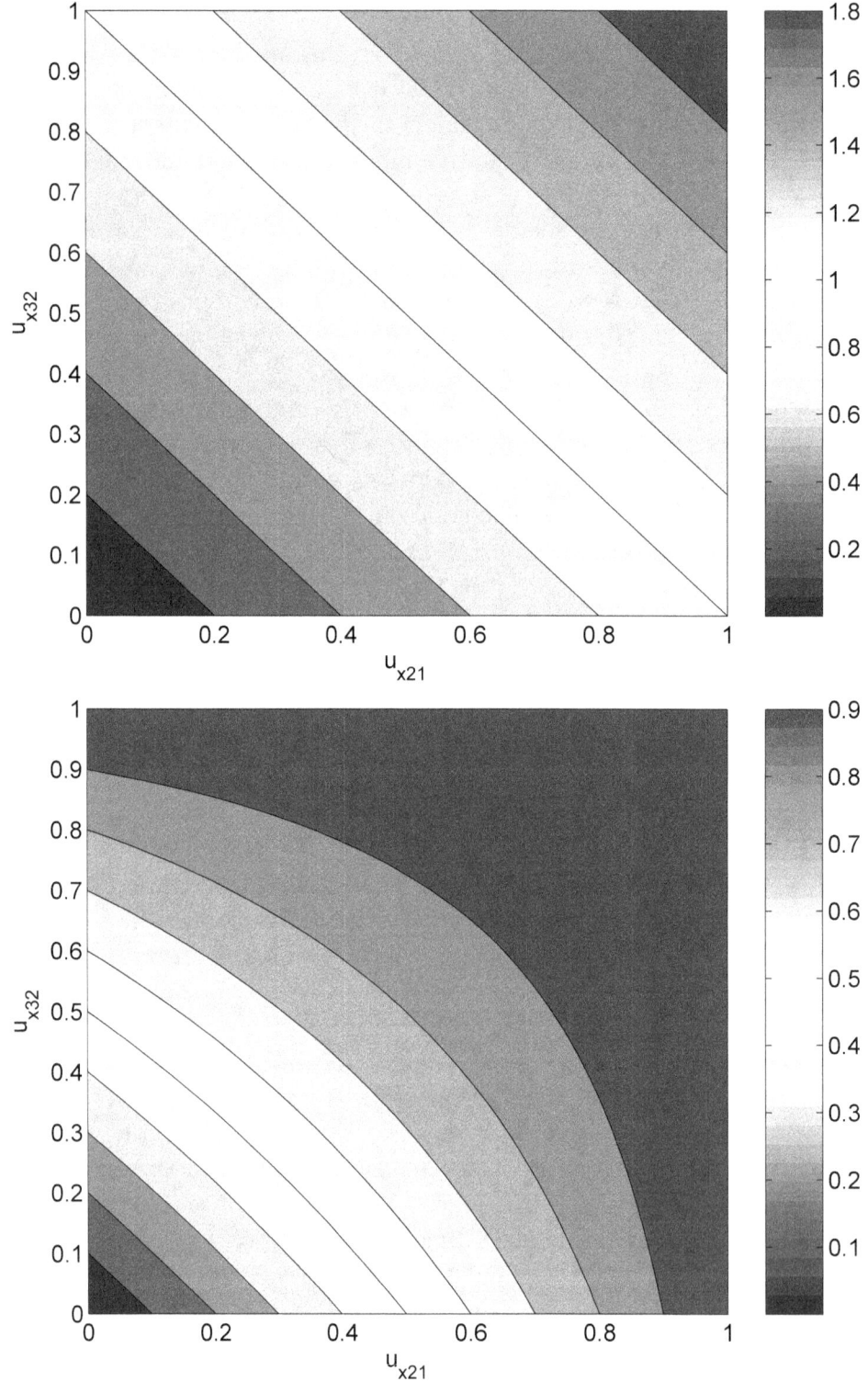

Figure 12: Contour plot of u_{x31} as a function of u_{x32} and u_{x21} for the Galilean velocity composition formula as given by Eq. 50 (upper subfigure) and for the Lorentzian velocity composition formula as given by Eq. 105 (lower subfigure). The axes and color bar in both plots are in units of c.

4.2.3 Velocity Composition

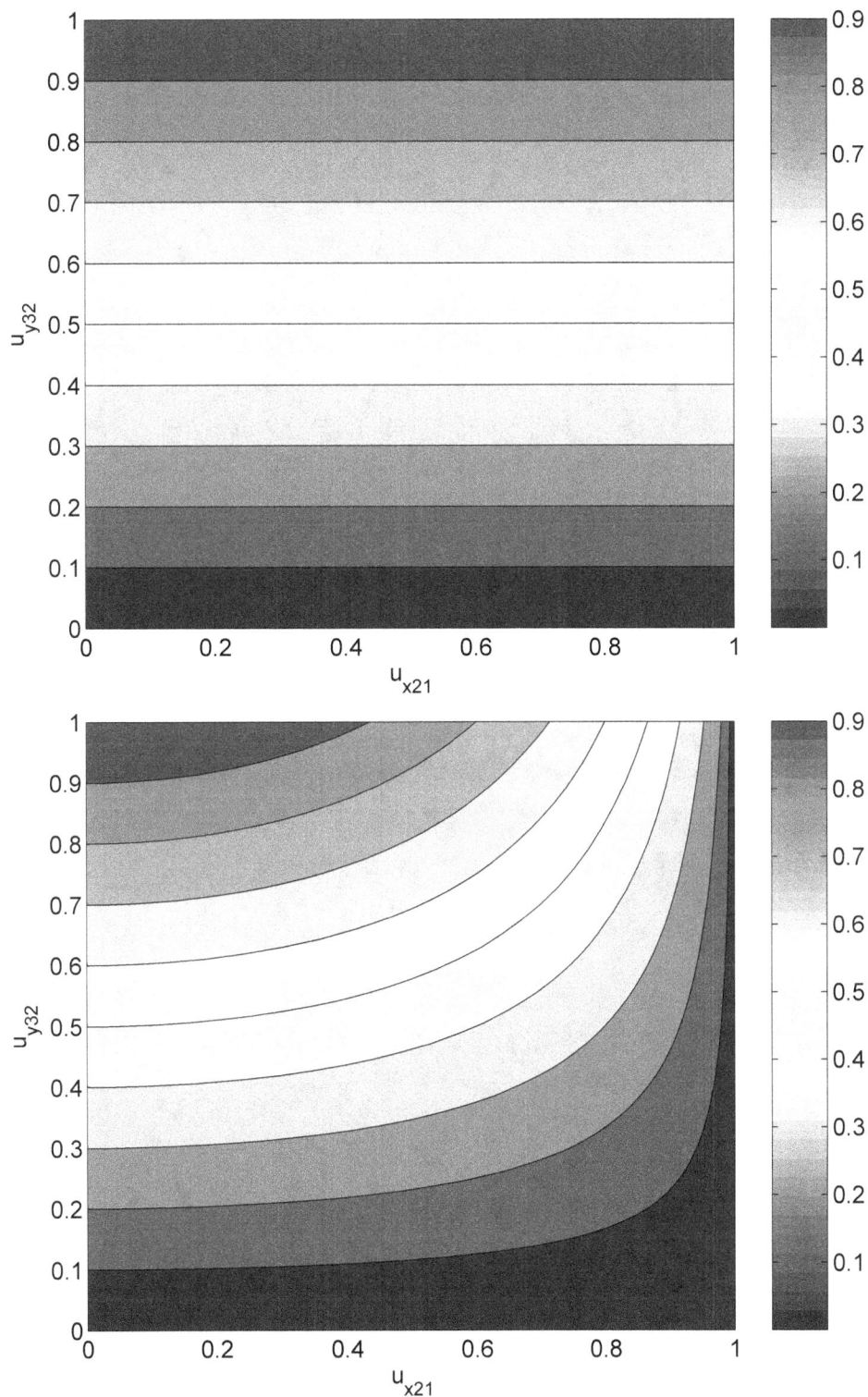

Figure 13: Contour plot of u_{y31} as a function of u_{y32} and u_{x21} for the Galilean velocity composition formula as given by Eq. 108 (upper subfigure) and for the Lorentzian velocity composition formula as given by Eq. 106 assuming $u_{x32} = 0$ (lower subfigure). The axes and color bar in both plots are in units of c.

4.2.3 Velocity Composition

to employ a coordinate system and hence automate the determination of signs and the solution process in general.

Another remark is that in solving problems related to the velocity composition formulae, whether Galilean or Lorentzian, we may need to compute one of the other two velocities (i.e. u_{21} or u_{32}) instead of u_{31} where in these symbols we drop the subscripts x, y and z to be general. In fact, this is just a matter of labeling and hence with a proper labeling we can solve any problem using the same formula. However, sometimes one labeling is more intuitive than another or the problem is set with certain labeling which is not supposed to be changed. In such cases, the formulae for the other velocities can be easily obtained from the given formulae for u_{31} by solving these formulae for the other velocities. The important thing in solving this sort of problems is to define u_{21}, u_{32} and u_{31} according to the given data. As soon as this is done, we can solve any problem either by the use of the above formulae directly if the unknown is u_{31} or by solving the formulae to one of the other velocities. As an example for solving the formulae to one of the other velocities, Eq. 105 can be solved for u_{x21} as follows:

$$u_{x31} = \frac{u_{x32} + u_{x21}}{1 + \frac{u_{x32}u_{x21}}{c^2}} \tag{109}$$

$$u_{x31}\left(1 + \frac{u_{x32}u_{x21}}{c^2}\right) = u_{x32} + u_{x21} \tag{110}$$

$$u_{x31} + u_{x31}\frac{u_{x32}u_{x21}}{c^2} = u_{x32} + u_{x21} \tag{111}$$

$$u_{x31} - u_{x32} = u_{x21} - u_{x31}\frac{u_{x32}u_{x21}}{c^2} \tag{112}$$

$$u_{x31} - u_{x32} = u_{x21}\left(1 - \frac{u_{x31}u_{x32}}{c^2}\right) \tag{113}$$

$$u_{x21} = \frac{u_{x31} - u_{x32}}{1 - \frac{u_{x31}u_{x32}}{c^2}} \tag{114}$$

Due to the symmetry of Eq. 105 with respect to u_{x21} and u_{x32}, the formula for u_{x32} can be easily obtained from the last formula (i.e. Eq. 114) by just exchanging these symbols in this formula.

Solved Problems

1. What are the implications of the fact that the symbols, like u_{x31} and u_{y32}, in the velocity composition formulae represent 1D vectors?[119]

    Answer: One of the implications is that they have direction (indicated by sign) as well as magnitude. Another implication is that the order of the indices is important in determining the sign, and hence the sign is reversed by reversing the order of the indices, e.g. $u_{x31} = -u_{x13}$.

2. O_2 is an inertial observer moving relative to another inertial observer O_1 with a speed ξc where $0 \leq \xi < 1$. O_2 emits a light signal which in his rest frame is measured to be

---

[119] We label them (here and elsewhere) as 1D vectors rather than components of vectors to be more clear about their vectorial nature where sign is not only of algebraic significance (like some scalars) but also of geometric significance due to the association with a set of basis vectors.

4.2.3 Velocity Composition

propagating with speed c. Assuming that all the motions have identical orientation, show that the speed of this light signal is also c in the frame of O_1. Also, discuss the case where $\xi = 1$.

Answer: Since all the motions have identical orientation we use the u_{x31} formula where we index the light signal with 3 while O_1 and O_2 keep their numerical labeling and hence we have: u_{x31} is the velocity of the light signal in O_1 frame, u_{x32} is the velocity of the light signal in O_2 frame, and u_{x21} is the velocity of O_2 relative to O_1. Hence, we have:

$$u_{x31} = \frac{u_{x32} + u_{x21}}{1 + \frac{u_{x32} u_{x21}}{c^2}} = \frac{(\pm c) + (\pm \xi c)}{1 + \frac{(\pm c)(\pm \xi c)}{c^2}} = \pm c$$

where all the four possible cases are considered in the $\pm$ signs of u_{x32} and u_{x21}, i.e. $++$, $--$, $+-$ and $-+$. Therefore, $|u_{x31}| = c$, i.e. the speed of the light signal is also c in the frame of O_1. So, in all the four cases about the direction of the velocities u_{x32} and u_{x21} we obtain the same speed c.

Regarding the case $\xi = 1$, although this case gives a consistent answer, it should be excluded to be consistent with the formalism of Lorentz mechanics which is restricted to speeds lower than c for massive objects. Yes, if light is considered as representing this case, it might be correct but then we should have a sensible definition of the frame of reference of light itself and if it is meaningful or not.

3. A scientist is observing a particle that decays by emitting an electron. The speed of the particle is $0.65c$ in the laboratory frame while the speed of the electron is $0.4c$ in the particle rest frame. What is the velocity of the electron in the laboratory frame (a) if the two motions are in the same direction, (b) if the two motions are in opposite directions and (c) if the two motions are in perpendicular directions?[120] Comment on this question.

Answer: We label the laboratory (which is the rest frame of the scientist), the particle and the electron with 1, 2 and 3. In (a) and (b) the two motions have identical orientation and hence the x velocity composition formula should be used, while in (c) the two motions have perpendicular directions and hence the y (or z) velocity composition formula should be used.

(a) We have $u_{x21} = 0.65c$ and $u_{x32} = 0.4c$. Hence:

$$u_{x31} = \frac{u_{x32} + u_{x21}}{1 + \frac{u_{x32} u_{x21}}{c^2}} = \frac{0.4c + 0.65c}{1 + \frac{0.4c \times 0.65c}{c^2}} \simeq 0.8333c$$

We can similarly assume $u_{x21} = -0.65c$ and $u_{x32} = -0.4c$ and hence we obtain $u_{x31} \simeq -0.8333c$.

(b) We have $u_{x21} = 0.65c$ and $u_{x32} = -0.4c$. Hence:

$$u_{x31} = \frac{u_{x32} + u_{x21}}{1 + \frac{u_{x32} u_{x21}}{c^2}} = \frac{-0.4c + 0.65c}{1 + \frac{-0.4c \times 0.65c}{c^2}} \simeq 0.3378c$$

---

[120] Expressions like "the two motions are in perpendicular directions", which are commonly used in the questions for ease and simplicity, mean as seen in one of the two frames which is the frame (labeled 2) in which the velocity of the observed object (labeled 3) has a single component (y or z component).

4.2.3 Velocity Composition

We can similarly assume $u_{x21} = -0.65c$ and $u_{x32} = 0.4c$ and hence we obtain $u_{x31} \simeq -0.3378c$.

(c) We have $u_{x21} = 0.65c$, $u_{x32} = u_{z32} = 0$ and $u_{y32} = 0.4c$. Hence:

$$u_{x31} = \frac{u_{x32} + u_{x21}}{1 + \frac{u_{x32}u_{x21}}{c^2}} = \frac{0 + 0.65c}{1 + 0} = 0.65c$$

$$u_{y31} = \frac{u_{y32}}{\gamma\left(1 + \frac{u_{x21}u_{x32}}{c^2}\right)} = \frac{0.4c\sqrt{1 - 0.65^2}}{1 + 0} \simeq 0.3040c$$

$$u_{z31} = 0$$

Hence, the speed of the electron in the laboratory frame is:

$$u_{31} = \sqrt{u_{x31}^2 + u_{y31}^2 + u_{z31}^2} \simeq c\sqrt{0.65^2 + 0.304^2 + 0^2} \simeq 0.7176c$$

and its movement is restricted to the xy plane where it makes an angle θ with the positive x axis which is given by:

$$\theta = \arctan\frac{u_{y31}}{u_{x31}} \simeq 0.4374\,\text{rad} \simeq 25.06°$$

We can also assume $u_{y32} = -0.4c$ (or indeed $u_{x21} = \pm 0.65c$ with $u_{y32} = \pm 0.4c$ considering all the four possibilities) and solve the problem similarly.

Comment: we solved this type of problems previously (see § 4.2.2) by using the velocity transformation formulae. This shows that the velocity transformation formulae and the velocity composition formulae are essentially the same. So, the velocity composition formulae are no more than a notationally different form of the velocity transformation formulae. The use of one form or the other is a matter of convenience in labeling, configuring and tackling the problem.

4. Re-solve part (a) of the previous question using different labeling to show that the labeling of the velocities in the velocity composition formulae does not affect the solution.
Answer: This time, we label the electron, the particle and the laboratory with 1, 2 and 3. So, $u_{x21} = -u_{x12} = -0.4c$ and $u_{x32} = -u_{x23} = -0.65c$ and hence the required velocity is $u_{x13} = -u_{x31}$. Accordingly, we have:

$$u_{x13} = -u_{x31} = -\frac{u_{x32} + u_{x21}}{1 + \frac{u_{x32}u_{x21}}{c^2}} = -\frac{-0.65c - 0.4c}{1 + \frac{(-0.65c)(-0.4c)}{c^2}} = \frac{0.65 + 0.4}{1 + 0.65 \times 0.4}c \simeq 0.8333c$$

as before. Other labeling schemes should also produce the same answer since the physical reality and truth should not depend on the conventions and labeling.

5. Verify that the formula of u_{x32} can be easily obtained from the formula of u_{x21} by just exchanging these symbols in the equation of u_{x21} due to the symmetry of the formula of u_{x31} in the symbols u_{x21} and u_{x32}.
Answer: On solving the formula of u_{x31} for u_{x32} we obtain:

$$u_{x31} = \frac{u_{x32} + u_{x21}}{1 + \frac{u_{x32}u_{x21}}{c^2}}$$

4.2.3 Velocity Composition

$$u_{x31}\left(1 + \frac{u_{x32}u_{x21}}{c^2}\right) = u_{x32} + u_{x21}$$

$$u_{x31} + u_{x31}\frac{u_{x32}u_{x21}}{c^2} = u_{x32} + u_{x21}$$

$$u_{x31} - u_{x21} = u_{x32} - u_{x31}\frac{u_{x32}u_{x21}}{c^2}$$

$$u_{x32}\left(1 - \frac{u_{x31}u_{x21}}{c^2}\right) = u_{x31} - u_{x21}$$

$$u_{x32} = \frac{u_{x31} - u_{x21}}{1 - \frac{u_{x31}u_{x21}}{c^2}}$$

On comparing the last equation with the equation of u_{x21} (Eq. 114) we see that they can be obtained from each other by just exchanging the symbols u_{x21} and u_{x32}.

Exercises

1. Make a comparison between the velocity transformation formulae and the velocity composition formulae.
2. We have two persons: person A who is standing on the street and watching a bus moving along a straight line with a constant velocity $v_B = 10$, and person C who is on the bus where C, according to A, is moving in the same direction as the bus. According to the bus rest frame, C is moving with a constant velocity $v_C = 2$. What is the velocity of C in the frame of A according to (a) classical mechanics and (b) Lorentz mechanics?
3. Repeat the previous question but now $v_B = 0.5c$ and $v_C = 0.2c$. Compare your results with the results of the previous question and comment.
4. Repeat the last question but now v_B and v_C are in opposite directions.
5. Show that when u_{x21} or/and u_{x32} in the Lorentz velocity composition formula for the x dimension approach c, u_{x31} also approaches c. What are the regions on the Lorentzian counter plot of u_{x31} that represent these cases?
6. Show that when u_{x21} or/and u_{x32} in the Lorentz velocity composition formula for the y dimension approach c, then u_{y31} should approach zero. Comment on the results.
7. Show that when u_{y32} in the Lorentz velocity composition formula for the y dimension approaches c, then u_{y31} can take any value between zero and c (i.e. $0 \leq u_{y31} \leq c$).[121]
8. Build a table showing u_{x31} as a function of the velocities u_{x21} and u_{x32} at and between the two limits of these velocities: zero and c.
9. Build a table showing u_{y31} as a function of the velocities u_{x21} and u_{x32} at and between the two limits of these velocities: zero and c.
10. Build a table showing u_{y31} as a function of the velocities u_{x21} and u_{y32} at and between the two limits of these velocities: zero and c.
11. Build a table showing u_{y31} as a function of the velocities u_{x32} and u_{y32} at and between the two limits of these velocities: zero and c.

---

[121] We note that in contexts like this, symbols like u_{y32} and u_{y31} may represent the magnitude of velocity. This is inline with our previously stated convention (see § 1.5) that symbols like these may represent 1D velocity and may represent speed.

12. Having three objects or observers (O, O_1 and O_2), the formula:

$$u_{x31} = \frac{u_{x32} + u_{x21}}{1 + \frac{u_{x32}u_{x21}}{c^2}} \qquad \text{may be simplified as:} \qquad u_3 = \frac{u_1 + u_2}{1 + \frac{u_1 u_2}{c^2}}$$

where u_1 is the velocity of O_1 relative to O, u_2 is the velocity of O_2 relative to O_1, and u_3 is the velocity of O_2 relative to O. What are the advantages and disadvantages of each form?
13. Compose two equal velocities to obtain a composite velocity that is $0.8c$ where all these velocities have the same orientation.[122]
14. Using the velocity composition formula, show that the composition of two velocities which are less than c is less than c. Assume that both velocities are in the positive x direction.

4.2.4 Acceleration Transformations

For the above-described two inertial observers O and O', the measured accelerations in the two frames are Lorentz transformed from O frame to O' frame as follows:

$$a'_x = \frac{a_x}{\gamma^3 \left(1 - \frac{vu_x}{c^2}\right)^3} \tag{115}$$

$$a'_y = \frac{c^2 a_y - vu_x a_y + vu_y a_x}{c^2 \gamma^2 \left(1 - \frac{vu_x}{c^2}\right)^3} \tag{116}$$

$$a'_z = \frac{c^2 a_z - vu_x a_z + vu_z a_x}{c^2 \gamma^2 \left(1 - \frac{vu_x}{c^2}\right)^3} \tag{117}$$

where the symbols are as explained before and γ is a function of v. These transformations can be obtained simply by differentiating the velocity transformation formulae (i.e. Eqs. 99-101) with respect to time as will be derived in § 5.1.3.

We note that, unlike classical mechanics, the transformation of acceleration in Lorentz mechanics requires knowledge of both velocity and acceleration in one frame, as well as the relative velocity between the two frames, to obtain the acceleration in the other frame. We also note that while in classical mechanics the acceleration is value-invariant across inertial frames (see Eqs. 51-53), in Lorentz mechanics the acceleration is frame dependent. However, if the acceleration vanishes in an inertial frame it should vanish in all inertial frames, as can be seen from the above transformations. Hence, the inertiality status in Lorentz mechanics is the same as the inertiality status in classical mechanics.

Solved Problems
1. Obtain the Lorentz acceleration transformations from O' frame to O frame.
   Answer: As usual, we obtain the opposite transformations by exchanging the primed

---

[122] In questions like this (which may be seen as ambiguous) it should be obvious that the orientation is determined by a primary inertial observer. Also, "equal velocities" should be sufficient to remove any ambiguity.

4.2.5 Length Transformation

and unprimed symbols and reversing the sign of the relative velocity v, that is:

$$a_x = \frac{a'_x}{\gamma^3 \left(1 + \frac{vu'_x}{c^2}\right)^3}$$

$$a_y = \frac{c^2 a'_y + vu'_x a'_y - vu'_y a'_x}{c^2 \gamma^2 \left(1 + \frac{vu'_x}{c^2}\right)^3}$$

$$a_z = \frac{c^2 a'_z + vu'_x a'_z - vu'_z a'_x}{c^2 \gamma^2 \left(1 + \frac{vu'_x}{c^2}\right)^3}$$

Exercises
1. O and O' are two inertial observers in a state of standard setting with relative velocity $v = 0.3c$. Observer O measures the velocity and acceleration of an object at a given instant of time to be $0.2c$ and 6 in the x direction and $0.1c$ and 4 in the z direction. What is the acceleration of this object at that time according to observer O' in these directions?
2. As stated in the text, while in classical mechanics the acceleration is value-invariant across inertial frames, in Lorentz mechanics the acceleration is frame dependent. Does this mean that the inertiality status (i.e. being inertial or non-inertial) of a frame could be different between these mechanics, i.e. a frame can be inertial in one of these mechanics but non-inertial in the other mechanics? If not, show that according to Lorentz mechanics if the acceleration vanishes in an inertial frame, it should vanish in all inertial frames.[123]
3. What other conclusion can be drawn from the argument in the previous question?

4.2.5 Length Transformation

In Lorentz mechanics, the length of a physical object transforms between proper and improper frames according to the length contraction formula. Accordingly, the proper length L_0 and the improper length L are related by the formula:

$$L_0 = \gamma L \tag{118}$$

where the contracted length is assumed to be oriented along the direction of the relative motion. As we will see in § 5.1.4 and § 7.2, length contraction effect is related to the Lorentz transformations of spatial coordinates. The reader is referred to § 4.1.1 for solved problems and exercises.

4.2.6 Time Interval Transformation

In Lorentz mechanics, the size of a time interval transforms between proper and improper frames according to the time dilation formula. Accordingly, the proper interval Δt_0 and

---
[123] The acceleration in this question belongs to a frame of reference rather than an object.

the improper interval Δt are related by the relation:

$$\Delta t_0 = \frac{\Delta t}{\gamma} \qquad (119)$$

As we will see in § 5.1.5 and § 7.3, time dilation effect is related to the Lorentz transformation of the temporal coordinate. The reader is referred to § 4.1.2 for solved problems and exercises.

4.2.7 Mass Transformation

As seen earlier (refer to § 4.1.3), there are two approaches in the literature of Lorentz mechanics about the mass: an old approach which considers mass as an extrinsic property and hence it varies between inertial observers, and a modern approach which, like classical mechanics, considers mass as an intrinsic property and hence it is independent of the observers. According to the old approach, the mass of an object transforms from its rest frame to a moving frame by the following relation:

$$m' = \gamma m \qquad (120)$$

where m and m' represent the mass in its rest frame and in the moving frame, while according to the modern approach it transforms invariantly as:

$$m' = m \qquad (121)$$

In this book we generally follow the new approach and hence we consider mass as an intrinsic property of massive objects that does not vary between inertial observers and frames although for convenience we may follow the old approach in some exceptional cases. Most solved problems and exercises about mass transformation can be found in § 4.1.3.

Solved Problems

1. Compare the mass in classical and Lorentz mechanics.
   Answer: In brief, in classical mechanics the mass in general is an intrinsic property while in Lorentz mechanics the rest mass is an intrinsic property but the non-rest mass may be an intrinsic or extrinsic property depending on the aforementioned conventions.

Exercises

1. Discuss mass transformation in Lorentz mechanics.

4.2.8 Frequency Transformation and Doppler Shift

The frequency of a light signal is transformed from the rest frame of the source of the signal to the frame of an observer according to the Doppler relation. The Doppler shift or Doppler effect is an observed frequency shift in the propagating waves due to the relative motion between the source of the waves and the observer. Doppler effect was

4.2.8 Frequency Transformation and Doppler Shift

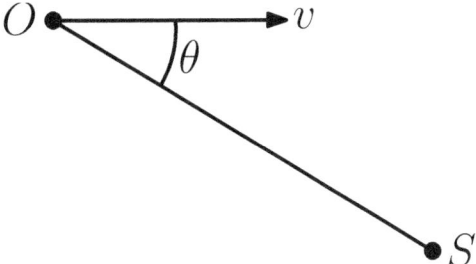

Figure 14: A schematic diagram demonstrating the setting of Doppler effect where S is the source and O is the observer who is moving with speed v relative to the frame of the source in the given direction.

originally considered for sound where the frequency of the emitted sound signal increases for an approaching source and decreases for a receding source. In Lorentz mechanics, the Doppler effect depends on the relative speed between the observer and the frame of the source and on the direction of this motion, i.e. it depends on the relative velocity between the observer and source.[124]

For a light source emitting a light signal whose frequency is measured in the rest frame of the source to be ν_0 and measured by an inertial observer in a frame moving with respect to the frame of the source with a speed v in a given direction to have a frequency ν, the Doppler frequency shift is given by:

$$\nu = \frac{\nu_0}{\gamma\left(1 - \frac{v}{c}\cos\theta\right)} \qquad (122)$$

where γ is a function of v and θ is the angle between the direction of the relative velocity between the two frames and the line of sight connecting the observer to the source as shown in Figure 14.

From the above equation we observe the following important special cases:

1. When the observer is heading towards the source (i.e. the direction of motion is along the line connecting the observer to the source), $\theta = 0$ and hence Eq. 122 becomes:

$$\nu = \frac{\nu_0}{\gamma\left(1 - \frac{v}{c}\right)} = \nu_0 \frac{\sqrt{1 - \frac{v^2}{c^2}}}{\left(1 - \frac{v}{c}\right)} = \nu_0 \sqrt{\frac{c+v}{c-v}} \qquad (123)$$

which means that the observed frequency is greater than the proper frequency (i.e. $\nu > \nu_0$ or blue shift) since $v > 0$.[125]

2. When the observer is moving transversely with respect to the source, $\theta = \pi/2$ and hence Eq. 122 becomes:

$$\nu = \frac{\nu_0}{\gamma} = \nu_0 \sqrt{1 - \frac{v^2}{c^2}} \qquad (124)$$

---

[124] We note that the observer in this context is localized like the source.

[125] The case $v = 0$ is a special case where the observer is at rest relative to the source and hence $\nu = \nu_0$. This can also be considered as a special case for the following two cases, i.e. $\theta = \pi/2$ and $\theta = \pi$, and indeed for the general case as given by Eq. 122.

4.2.8 Frequency Transformation and Doppler Shift

which means that the observed frequency is lower than the proper frequency (i.e. $\nu < \nu_0$ or red shift) since $\gamma > 1$.

3. When the observer is receding from the source (i.e. the direction of motion is along the line connecting the observer to the source), $\theta = \pi$ and hence Eq. 122 becomes:

$$\nu = \frac{\nu_0}{\gamma\left(1+\frac{v}{c}\right)} = \nu_0 \frac{\sqrt{1-\frac{v^2}{c^2}}}{\left(1+\frac{v}{c}\right)} = \nu_0 \sqrt{\frac{c-v}{c+v}} \qquad (125)$$

which means that the observed frequency is also lower than the proper frequency (i.e. $\nu < \nu_0$ or red shift) since $v > 0$.

Apart from the quantitative difference between the Doppler effect in classical mechanics and in Lorentz mechanics, the transverse Doppler effect (i.e. case 2 corresponding to $\theta = \pi/2$ in the above list) does not exist in classical mechanics and hence it is specific to Lorentz mechanics.[126] The relation between the observed wavelength λ and the proper wavelength λ_0 corresponding to the above Doppler frequency shift, can be easily obtained from Eq. 122 and the invariance of the speed of light across inertial frames (which can be inferred from the formalism of Lorentz mechanics as seen in § 4.2.2 and § 4.2.3). Accordingly, we have:

$$c = c \qquad (126)$$
$$\nu\lambda = \nu_0 \lambda_0 \qquad (127)$$
$$\lambda = \frac{\nu_0}{\nu}\lambda_0 \qquad (128)$$
$$\lambda = \gamma\left(1 - \frac{v}{c}\cos\theta\right)\lambda_0 \qquad (129)$$

Solved Problems

1. A light source of frequency $\nu_0 = 5 \times 10^{14}$ is located at the origin of coordinates of an inertial frame S. An inertial observer O is heading towards the light source with speed $v = 3.5 \times 10^7$. What is the frequency ν and the wavelength λ of the signal as observed by O?
   Answer: Since O is heading towards the source then we have:

$$\nu = \nu_0 \sqrt{\frac{c+v}{c-v}} \simeq 5 \times 10^{14} \times \sqrt{\frac{3 \times 10^8 + 3.5 \times 10^7}{3 \times 10^8 - 3.5 \times 10^7}} \simeq 5.6217 \times 10^{14}\,\text{Hz}$$

$$\lambda = \frac{c}{\nu} \simeq 533.64\,\text{nm}$$

i.e. the signal is blue shifted as it should be.

---

[126] In one of the exercises of this subsection, the classical formula is given where the velocities v_s and v_r are assumed to be along the line of sight and hence this case may be obtained from the given formula as a special case since the velocities along the line of sight are zero in this case. We should also remark that the transverse Doppler shift may be seen as experimental evidence in support of Lorenz mechanics and time dilation in particular.

4.2.8 Frequency Transformation and Doppler Shift

2. The speed of recession of a distant galaxy from the Earth is $v = 0.01c$. What is the observed frequency and wavelength of one of its emission lines whose proper wavelength is $\lambda_0 = 660$ nm?
   Answer: We have:
   $$\nu = \nu_0 \sqrt{\frac{c-v}{c+v}} = \frac{c}{\lambda_0}\sqrt{\frac{c-v}{c+v}} \simeq \frac{3 \times 10^8}{660 \times 10^{-9}} \times \sqrt{\frac{c - 0.01c}{c + 0.01c}} \simeq 4.5002 \times 10^{14} \text{ Hz}$$
   $$\lambda = \frac{c}{\nu} \simeq \frac{3 \times 10^8}{4.5002 \times 10^{14}} \simeq 666.63 \text{ nm}$$

   The line is red shifted as it should be.

3. An inertial observer O is following the trajectory $\mathbf{r} = (x, y, z) = (10^7 t, 10^8, 0)$ as seen from an inertial frame S where a light source is positioned at the origin of coordinates of S. The wavelength of the emitted light as measured in S is $\lambda_0 = 500$ nanometer. What are the frequency ν and the wavelength λ of the signal as observed by O at time: (a) $t = -2000$, (b) $t = 0$ and (c) $t = 10$?
   Answer: The speed of the observer O relative to frame S is:
   $$v = |\mathbf{v}| = \left|\frac{d\mathbf{r}}{dt}\right| = \left|(10^7, 0, 0)\right| = 10^7$$

   while the angle θ as a function of time is:
   $$\theta = \begin{cases} \arctan \frac{y}{-x} = \arctan \frac{10^8}{-10^7 t} = \arctan \frac{10}{-t} & (t \neq 0) \\ \pi/2 & (t = 0) \end{cases}$$

   where $0 \leq \theta \leq \pi$. The proper frequency is:
   $$\nu_0 = \frac{c}{\lambda_0} \simeq \frac{3 \times 10^8}{500 \times 10^{-9}} = 6 \times 10^{14}$$

   (a) At $t = -2000$ we have:
   $$\theta = \arctan \frac{10}{-(-2000)} = \arctan \frac{1}{200} \simeq 0.005 \text{ rad}$$

   and hence:
   $$\nu = \frac{\nu_0}{\gamma\left(1 - \frac{v}{c}\cos\theta\right)} \simeq \frac{6 \times 10^{14} \times \sqrt{1 - (10^7/c)^2}}{1 - (10^7/c) \times \cos(0.005)} \simeq 6.2034 \times 10^{14} \text{ Hz}$$
   $$\lambda = \frac{c}{\nu} \simeq 483.60 \text{ nm}$$

   These results are reasonable because the observer is essentially approaching the source (i.e. its velocity has an approaching component) and hence we should have a blue shift.
   (b) At $t = 0$ we have:
   $$\theta = \frac{\pi}{2} \text{ rad}$$

4.2.8 Frequency Transformation and Doppler Shift

and hence:

$$\nu = \frac{\nu_0}{\gamma\left(1 - \frac{v}{c}\cos\theta\right)} \simeq \frac{6 \times 10^{14} \times \sqrt{1 - (10^7/c)^2}}{1 - (10^7/c) \times 0} \simeq 5.9967 \times 10^{14}\,\text{Hz}$$

$$\lambda = \frac{c}{\nu} \simeq 500.28\,\text{nm}$$

These results are reasonable because the observer is moving transversely (corresponding to the second special case in the text) and hence we should have a red shift.

(c) At $t = 10$ we have:

$$\theta = \arctan\frac{10}{-10} = \arctan(-1) = \frac{3\pi}{4}\,\text{rad}$$

and hence:

$$\nu = \frac{\nu_0}{\gamma\left(1 - \frac{v}{c}\cos\theta\right)} \simeq \frac{6 \times 10^{14} \times \sqrt{1 - (10^7/c)^2}}{1 - (10^7/c) \times \cos\left(\frac{3\pi}{4}\right)} \simeq 5.8586 \times 10^{14}\,\text{Hz}$$

$$\lambda = \frac{c}{\nu} \simeq 512.07\,\text{nm}$$

These results are reasonable because the observer is essentially receding from the source (i.e. its velocity has a receding component) and hence we should have a red shift.

Exercises

1. A light source of frequency $\nu_0 = 6.6 \times 10^{14}$ is located at the origin of coordinates of an inertial frame S. An inertial observer O is receding from the light source with speed $v = 10^8$. What is the frequency ν and the wavelength λ of the signal as observed by O?

2. An electromagnetic transmitter of wavelength $\lambda_0 = 10^{-8}$ is located at the origin of coordinates of an inertial frame S. An inertial observer O observed that the transmitted signal has a constant wavelength $\lambda = 9 \times 10^{-9}$. What are the speed v and the angle θ of the relative motion between the observer and the transmitter?

3. An electromagnetic transmitter of wavelength $\lambda_0 = 10^{-7}$ is located at the origin of coordinates of an inertial frame S. An observer O observed that the transmitted signal has a constant wavelength $\lambda = 1.1 \times 10^{-7}$. What are the speed v and the angle θ of the relative motion between the observer and the transmitter? Assume that the speed of the observer is constant and $\theta \neq \pi$.

4. The sodium D_2 line of a star is observed from the Earth to be red shifted by 11 nm. What is the velocity of the star relative to the Earth?

5. A radio transmitter of frequency $\nu_0 = 10^8$ is located at the origin of coordinates of an inertial frame S. An inertial observer O whose speed in S is $v = 0.6c$ measured the frequency at $t = 0$ to be $\nu = 8 \times 10^7$. What is the velocity of O relative to the transmitter at $t = 0$?

4.2.9 Charge Density and Current Density Transformations

6. An observer at the origin of coordinates of an inertial frame O measured the frequency of a radio signal coming from an inertial spaceship S_1 to be $\nu_1 = 10^9$. If S_1 is receding from the observer along the positive x direction with a speed $v_1 = 0.5c$ relative to O, (a) What is the proper frequency ν_0 of the radio signal? (b) What is the frequency ν_2 as measured by a second inertial spaceship S_2 that recedes from the observer along the negative x direction with a speed $v_2 = 0.4c$ relative to O?

7. The equation of classical Doppler frequency shift is given by:

$$\nu = \frac{c + v_r}{c + v_s} \nu_0$$

where ν and ν_0 are the received (improper) and emitted (proper) frequency, while v_r and v_s are the velocity of the receiver and source relative to the medium of propagation along the line of sight. The sign of v_r is positive/negative when the receiver is approaching/receding from the source, while the sign of v_s is positive/negative when the source is receding/approaching the receiver. Show that the Lorentzian Doppler frequency converges to this classical limit when (a) The receiver is stationary with respect to the medium and the source is receding from the receiver. (b) The source is stationary with respect to the medium and the receiver is receding from the source.

4.2.9 Charge Density and Current Density Transformations

The electric charge density ρ and the electric current density $\mathbf{j}$ are defined as:

$$\rho \equiv \frac{Q}{V} \quad \text{and} \quad \mathbf{j} \equiv \rho \mathbf{u} \tag{130}$$

where Q is the electric charge, V is the volume that contains the charge[127] and $\mathbf{u}$ is the velocity of the charge relative to the observer. Now, let ρ and $\mathbf{j}$ be defined as above in an inertial frame O. From another inertial frame O' that moves with velocity $\mathbf{u}$ relative to O we will have:

$$\rho' = \rho_0 \quad \text{and} \quad \mathbf{j}' = 0 \tag{131}$$

where ρ_0 is the charge density as measured in the rest frame (i.e. proper charge density). This is because there is no relative motion between frame O' and the charge and hence this charge density represents the proper charge density and the current should vanish in the absence of relative motion between the charge and O'. On comparing the above equations, the electric charge density and the electric current density can be seen to transform between these frames (i.e. non-rest frame O and rest frame O' of the charge) by the following equations:

$$\rho = \gamma \rho' = \gamma \rho_0 \quad \text{and} \quad \mathbf{j} = \gamma \rho_0 \mathbf{u} \tag{132}$$

---

[127] We note that if the electric charge density is not uniform then Q and V should be interpreted in an infinitesimal sense. We also note that the charge can be positive or negative and hence Q and ρ have sign as well as magnitude.

where γ is a function of $u = |\mathbf{u}|$.

Solved Problems

1. A charged object has a charge density of 0.01 in its rest frame. What is the charge density and the current density as seen from an inertial frame O that is moving with a constant velocity $(-0.3c, 0, 0)$ relative to the object?
   Answer: We are given $\rho_0 = 0.01$; moreover if O is moving with velocity $(-0.3c, 0, 0)$ relative to the object, then the velocity of the object relative to O is $\mathbf{u} = (u_x, u_y, u_z) = (0.3c, 0, 0)$. Accordingly, the charge density ρ and the current density $\mathbf{j}$ as seen from frame O are:

$$\rho = \frac{\rho_0}{\sqrt{1 - (|\mathbf{u}|/c)^2}} = \frac{0.01}{\sqrt{1 - 0.3^2}} \simeq 0.0105 \, \text{C/m}^3$$

$$\mathbf{j} = \frac{\rho_0 \mathbf{u}}{\sqrt{1 - (|\mathbf{u}|/c)^2}} = \frac{0.01 \times (0.3c, 0, 0)}{\sqrt{1 - 0.3^2}} \simeq (9.4346 \times 10^5, 0, 0) \, \text{C/(m}^2\text{.s)}$$

Exercises

1. A charged object whose charge density in its rest frame is $\rho_0 = 6$ is seen from an inertial frame O_1 to have a charge density $\rho_1 = 10$. What is the charge density ρ_2 and the current density $\mathbf{j}_2$ as seen from another inertial frame O_2 which is moving uniformly relative to O_1 with a velocity of $0.2c$? Assume that the motion of O_1 and O_2 relative to the object is in the positive x direction.

4.3 Physical Relations

4.3.1 Newton's Second Law

As indicated previously, Newton's second law of motion, which states that force is equal to the time derivative of linear momentum, keeps its form in Lorentz mechanics if we use the Lorentzian expression for momentum as given by Eq. 81. Hence, we have:

$$f = \frac{dp}{dt} = \frac{d}{dt}(\gamma m u) \tag{133}$$

Most solved problems and exercises about Newton's second law can be found in § 4.1.7.

Solved Problems

1. Is there a contradiction between the fact that the domain of Lorentz mechanics is inertial frames where the velocities are constant, and dealing with concepts and physical quantities like acceleration and force where the physical objects have non-constant velocities?
   Answer: Although Lorentz mechanics is restricted to inertial frames and hence the observer should be inertial, the observed object can be subject to a force and hence it is accelerating and its velocity is not constant. Therefore, there is no contradiction.

Exercises

1. Show that all Newton's laws of motion are valid in Lorentz mechanics. What you conclude?
2. Compare Newton's second law in classical mechanics and in Lorentz mechanics.

4.3.2 Mass-Energy Relation

As indicated previously, in Lorentz mechanics the mass and energy are equivalent according to the Poincare mass-energy relation:

$$E_0 = mc^2 \qquad (134)$$

where E_0 is the rest energy and m is the mass of the object. This is unlike the traditional stand in classical mechanics where mass and energy are regarded as totally different entities. Consequently, the mass and energy have two independent conservation laws in classical mechanics, while in Lorentz mechanics they are combined in a single conservation law of mass-energy (see § 4.4). However, in § 5.3.2 we will see that some arguments for the equivalence between mass and energy are based on the principles of classical physics and hence the Poincare equivalence relation may not be proprietary to Lorentz mechanics.

We should remark that the term "binding energy" is used frequently in this context where this term may be defined as the potential energy that holds a composite physical system together. So, if we need a force to bring the components of the system together (e.g. bringing together two positive charges that repel each other), then energy is needed to bring this system together and this energy will be released when the system is fragmented. On the other hand, if we need a force to hold the components of the system apart (e.g. a positive charge and a negative charge that attract each other) then energy is released when the system is assembled and this energy will be needed to fragment the system. However, due to the equivalence between mass and energy, the binding energy of a physical system can be defined in this context in a more general way as the difference between the rest energy of the system in its bound state and the total rest energies of its individual components in their unbound state.[128] We note that some differentiate between the *reaction energy* where energy should be added to bind the system (and hence the mass of the composite system is greater than the mass of its components) and the *binding energy* where energy is released by the process of binding the system (and hence the mass of the composite system is less than the mass of its components).[129] The readers should therefore be vigilant about these conventions regarding the sign and terminology. However, for simplicity we define the binding energy generically as above to include all sign conventions and terminologies.

We should also remark that due to the mass-energy equivalence, the mass and energy may be expressed in units of each other with suitable conversion factors. For example, in

---

[128] The difference could be defined in one way or the other depending on the convention. However, this affects the sign (or the sense) of the binding energy but not the magnitude.

[129] This labeling may also be reversed by some. Moreover, other conventions and terminologies may also be found.

4.3.2 Mass-Energy Relation 162

atomic and particle physics the mass of particles may be expressed in units of eV/c^2 where eV stands for electron volt. Hence, we can state the mass of electron as $m_e = 0.511$ MeV/c^2 where MeV stands for mega electron volt, i.e. 10^6 eV. It is noteworthy that 1 eV is the energy acquired by an electron in passing through a potential difference of 1 volt, and hence $1\,\text{eV} \simeq 1.60218 \times 10^{-19}$ J which is numerically equal to the charge of electron.

Solved Problems

1. Why we label the formula $E_0 = mc^2$ as the Poincare relation?
   Answer: We label this formula as the Poincare mass-energy relation because we believe that Poincare is the originator of this formula and hence he should get the main credit for it although it seems that the formula has been developed over a considerable period of time and has been suggested in one form or another (mainly as a qualitative relation expressing the intimate link between mass and energy) even before Poincare. This issue will be discussed further in § 5.3.2.

2. What is the mass of electron, proton and neutron? Express each in units of kilogram and atomic mass unit (amu).
   Answer:
$$\begin{aligned} m_e &\simeq 9.1094 \times 10^{-31}\,\text{kg} \simeq 5.4858 \times 10^{-4}\,\text{amu} \\ m_p &\simeq 1.6726 \times 10^{-27}\,\text{kg} \simeq 1.0073\,\text{amu} \\ m_n &\simeq 1.6749 \times 10^{-27}\,\text{kg} \simeq 1.0086\,\text{amu} \end{aligned}$$

3. What is the rest energy of electron, proton and neutron? Express each in units of MeV and joule.
   Answer:
$$\begin{aligned} E_{0e} &= m_e c^2 \simeq 0.5110\,\text{MeV} \simeq 8.1985 \times 10^{-14}\,\text{J} \\ E_{0p} &= m_p c^2 \simeq 938.27\,\text{MeV} \simeq 1.5053 \times 10^{-10}\,\text{J} \\ E_{0n} &= m_n c^2 \simeq 939.57\,\text{MeV} \simeq 1.5074 \times 10^{-10}\,\text{J} \end{aligned}$$

4. The kinetic energy of a particle is half its rest energy. What is the speed of the particle?
   Answer: We have:
$$\begin{aligned} E_k &= 0.5 E_0 \\ mc^2(\gamma - 1) &= 0.5 mc^2 \\ \gamma &= 1.5 \\ \sqrt{1-\beta^2} &= 2/3 \\ 1-\beta^2 &= 4/9 \\ \beta^2 &= 5/9 \\ u &= c\sqrt{5}/3 \end{aligned}$$

Exercises

1. It is common in some branches of physics to use energy units as mass units (with proper conversion factors) and vice versa. Why?
2. Find the equivalent energy of the mass of neutron (i.e. its rest energy) giving your answer in SI units.
3. Find the energy required to combine an electron with a proton to produce a neutron.
4. A parent particle with mass m_P decays, while being at rest, into two daughter particles: A with mass m_A and B with mass m_B. If the velocity of A is $u_A = -0.4c$ and the velocity of B is $u_B = 0.6c$ (where both velocities are along a single orientation), what is the mass of the parent particle in terms of m_A and m_B?
5. The temperature of a block of aluminum of mass $m_b = 10$ kg was increased by $\Delta T = 1000$ K. What is the relative increase in mass?
6. The total energy of a particle of mass m is to be increased to 1.8 of its rest value. What is the required speed to realize this?
7. Object A of mass m_A and speed $u_A = 0.9c$ makes a perfectly inelastic collision with object B of mass $m_B = 2m_A$ which is at rest. What is the mass m_t of the single object that results from this collision in terms of m_A?
8. A radioactive nucleus of mass 109 amu (atomic mass unit) decays and releases $E_r = 7.4724 \times 10^{-11}$ J of energy. What is the total mass that is left after the decay?
9. A proton at rest is accelerated through a potential difference $\Delta V = 10^7$ V. What is its speed?

4.3.3 Momentum-Energy Relation

In Lorentz mechanics, the magnitude of momentum p and the total energy E of an object with mass m are linked through the following relation:[130]

$$E^2 = p^2 c^2 + m^2 c^4 \qquad (135)$$

As we will see in § 5.3.3, this relation can be easily derived from the expressions of the total energy and linear momentum, i.e. $E = \gamma mc^2$ and $p = \gamma mu$, as well as from other methods. It is noteworthy that from Eq. 135 we have the following two important special cases:

1. For a massive object at rest, $p = 0$ and hence $E = mc^2 = E_0$, that is its total energy is its rest energy, a relation that has already been established in § 4.3.2 (also refer to § 5.3.2) and hence this is a valid consistency check.
2. For a massless particle such as photon, $m = 0$ and hence $E = pc$. In fact, the relation: $E = pc$, which correlates the energy of radiation to its momentum, can be obtained from Maxwell's equations of electromagnetism regardless of Lorentz mechanics and the momentum-energy relation.[131]

---

[130] Here, as well as in some other similar locations in the book, we use E instead of E_t for total energy to follow the common form of this equation.
[131] However, as indicated in other parts of the book, Maxwell's equations may be seen as a Lorentzian physical theory and hence the relation $E = pc$ could be seen as Lorentzian even if it is obtained from Maxwell's equations although this may be debated since Maxwell's equations are historically a

4.3.3 Momentum-Energy Relation 164

Solved Problems

1. Based on the above momentum-energy relation, and following the style of expressing mass units as eV/c^2 which is based on the mass-energy equivalence relation, what units that involve energy may be used for momentum?
   Answer: The above momentum-energy relation may be expressed in a generic form as: $p = E/c$. Hence, units like eV/c or MeV/c may be used as units of momentum.

2. Express the unit of momentum eV/c in standard SI units.
   Answer: We have:
   $$\frac{\text{eV}}{c} \simeq \frac{1.60218 \times 10^{-19}}{3 \times 10^8} \simeq 5.3406 \times 10^{-28} \text{ kg.m/s}$$

3. What is the common conclusion from the energy and momentum relations, i.e. $E = \gamma mc^2$ and $p = \gamma mu$?
   Answer: The common conclusion is that no massive object can reach (or exceed) the speed of light to avoid having infinite energy and momentum. However, as indicated before, this should be considered within the validity domain of Lorentz mechanics and according to its current formalism.

4. What is the mass of an object whose momentum is $p = 10^4$ MeV/c and whose total energy is $E = 10^5$ MeV?
   Answer: From the momentum-energy relation we have:
   $$\begin{aligned} E^2 &= p^2 c^2 + m^2 c^4 \\ m^2 c^4 &= E^2 - p^2 c^2 \\ m &= \sqrt{\frac{E^2 - p^2 c^2}{c^4}} \\ m &= \sqrt{\frac{(10^5 \text{ MeV})^2 - (10^4 \text{ MeV}/c)^2 c^2}{c^4}} \\ m &= \sqrt{\frac{10^{10} \text{ MeV}^2 - 10^8 \text{ MeV}^2}{c^4}} \\ m &= \sqrt{10^{10} - 10^8} \text{ MeV}/c^2 \\ m &\simeq 9.9499 \times 10^4 \text{ MeV}/c^2 \\ m &\simeq 1.7713 \times 10^{-25} \text{ kg} \end{aligned}$$

5. What is the magnitude of the momentum of an electron whose kinetic energy is $E_k = 0.5$ MeV?

---

classical theory and their compatibility with the Lorentzian formulation does not qualify them to be Lorentzian. In fact, the emergence of Maxwell's equations before the emergence of Lorentz mechanics may indicate that these equations are not necessarily based on the framework of Lorentz mechanics (although they are compatible with it) and hence they may be based on another theory that could emerge in the future (possibly as extension or replacement to Lorentz mechanics). However, this may also put a question mark on their qualification as a classical theory.

4.3.4 Work-Energy Relation

Answer: From the momentum-energy relation we have:

$$E^2 = p^2c^2 + m_e^2c^4$$
$$(E_0 + E_k)^2 = p^2c^2 + E_0^2$$
$$p^2 = \frac{(E_0 + E_k)^2 - E_0^2}{c^2}$$
$$p^2 \simeq \frac{(0.511\,\text{MeV} + 0.5\,\text{MeV})^2 - 0.511^2\,\text{MeV}^2}{c^2}$$
$$p^2 = 0.761\,\text{MeV}^2/c^2$$
$$p \simeq 0.872\,\text{MeV}/c$$

Exercises

1. Some may argue that since we have $E = \gamma mc^2$ and $p = \gamma mu$, then from the relation $E = pc$ for a massless particle we have:

$$E = pc$$
$$\gamma mc^2 = \gamma muc$$
$$c = u$$

i.e. the speed of massless particles must be c. Assess this argument.

2. Assess the following argument which may be found in some textbooks: for a massless object, such as a photon, that is supposed to move with a speed lower than c, the expressions for total energy (i.e. $E = \gamma mc^2$) and momentum (i.e. $p = \gamma mu$) will vanish because $m = 0$, i.e. $E = p = 0$. Now, since this is untrue because E and p are not zero, then this should indicate that no massless object can move with a speed lower than c to avoid this contradiction.

3. Let assume that according to the formalism of Lorentz mechanics, no massive object can reach or exceed c and a massless object must not move with a speed less than c. What about massless objects moving with speeds greater than c? Is there a relation in Lorentz mechanics that implies the impossibility of exceeding c for massless objects?

4. The total energy of an electron is 0.53 MeV. What is the magnitude of its momentum?

5. What is the total energy of a neutron whose momentum is $p = 50\,\text{MeV}/c$?

6. Find the kinetic energy of a proton whose momentum is 150 MeV/c.

7. What is the advantage of expressing the energy in terms of momentum rather than velocity?

4.3.4 Work-Energy Relation

In Lorentz mechanics, as in classical mechanics, the change in kinetic energy ΔE_k of a massive object under the influence of an external force $\mathbf{f}(\mathbf{r})$ by displacing the object from position $\mathbf{r}_1$ to position $\mathbf{r}_2$ is equal to the work done W on the object by that force and hence it is given by the following integral:

$$\Delta E_k = W = \int_{\mathbf{r}_1}^{\mathbf{r}_2} \mathbf{f} \cdot d\mathbf{r} \qquad (136)$$

where the Lorentzian definition of force is to be used.

Solved Problems

1. Write the work-energy relation, which is given above in its vector form, in its scalar form[132] considering the x component of force and displacement. Explain all symbols.
   Answer: The scalar form of the work-energy relation considering the x component of force and displacement is:
   $$\Delta E_k = W = \int_{x_1}^{x_2} f_x \, dx$$
   where ΔE_k is the change in the kinetic energy, W is the work done by the force, f_x is the x component of the force, dx is the infinitesimal displacement in the x direction and x_1 and x_2 are the initial and final positions between which the x displacement took place.

2. Find the work done by a force given by $f_x = 3x^2$ in moving an object from position $x_1 = 3.5$ to position $x_2 = 7.8$.
   Answer: Using the given relation in the answer of the last question, we have:
   $$\begin{aligned} W &= \int_{x_1}^{x_2} f_x \, dx \\ &= \int_{3.5}^{7.8} 3x^2 \, dx \\ &= \left[x^3 \right]_{3.5}^{7.8} \\ &= 7.8^3 - 3.5^3 \\ &= 431.677 \, \text{J} \end{aligned}$$

Exercises

1. A force is given by: $\mathbf{f} = (x, y^2, \sqrt{z})$. What is the work done by this force in moving a particle from position $\mathbf{r}_1 = (1.2, -2.7, 5.3)$ to position $\mathbf{r}_2 = (-3.4, 2\pi, 3.9)$? What is the change in the kinetic energy of the particle in moving between these positions?

4.4 Conservation Laws

The commonly known conservation laws of mass, energy and momentum apply in Lorentz mechanics as in classical mechanics with the use of the Lorentzian forms of the conserved quantities as defined earlier. However, there are some modifications where the two independent laws of conservation of mass and conservation of energy in classical mechanics are merged in Lorentz mechanics into a single law of conservation of mass-energy due to the equivalence between mass and energy according to the Poincare relation.[133] We

---

[132] Or rather: scalar-like form or 1D vector form.

[133] Accordingly, the conservation of mass and the conservation of energy (i.e. in their classical meaning that exclude Lorentzian mass and rest energy) as such do not exist in Lorentz mechanics although they still apply when there is no conversion between mass and energy.

4.4 Conservation Laws

should also note that there are some other modifications in the form of some of these conservation laws where these modifications are based on the merge of space and time in Lorentz mechanics into spacetime. For example, energy and momentum are separate entities in classical mechanics and hence they have two different conservation laws, but in Lorentz mechanics they may be treated as components of the same vector representing its temporal and spatial dimensions and hence they may be merged in a single conservation law. Some aspects of this issue will be discussed further in § 6.5.5.

We note that if we follow the old convention about mass in Lorentz mechanics then mass may be considered as a conserved quantity that incorporates the kinetic energy, as was indicated for example in the exercises of § 4.3.2 (also see the exercises of § 5.2.3).[134] This reveals that the old and modern conventions about mass in Lorentz mechanics are just convenient ways of representing the same physical facts where according to the former the mass (i.e. Lorentzian mass) or energy (i.e. total energy) is the conserved quantity while according to the latter the mass-energy is the conserved quantity. We also note that the conservation laws are commonly used to solve many problems in Lorentz mechanics, as in classical mechanics, and hence they are very useful accessories in tackling symbolic and numeric problems. Examples of this use have been given previously and more examples will be given in the solved problems and exercises of this section as well as in other upcoming sections.

As in classical physics, charge is also conserved in Lorentz mechanics where the principle of conservation of charge may be stated mathematically by the continuity equation which is given by:

$$\frac{\partial \rho}{\partial t} + \nabla \cdot \mathbf{j} = 0 \tag{137}$$

where ρ is the electric charge density and $\mathbf{j} \equiv (j_x, j_y, j_z)$ is the electric current density 3-vector (see § 4.2.9 and § 6.5.7). It is noteworthy that the continuity equation can be expressed more compactly as an inner product between the nabla 4-operator $\Box \equiv \left(\frac{1}{c}\frac{\partial}{\partial t}, \frac{\partial}{\partial x}, \frac{\partial}{\partial y}, \frac{\partial}{\partial z}\right)$ and the electric current density 4-vector $\mathbf{J} \equiv (c\rho, j_x, j_y, j_z)$ (see § 6.2 and § 6.5.7), that is:[135]

$$\Box \cdot \mathbf{J} = 0 \tag{138}$$

Solved Problems

1. What the conservation of total energy in Lorentz mechanics means?

    Answer: It means that although the kinetic energy and the mass energy (including all non-kinetic forms of energy) of a closed physical system may not be conserved individually, the sum of all these forms of energy is conserved, i.e. it remains constant

---

[134] Accordingly, the difference between mass and total energy is the conversion factor c^2 which is a constant. In brief, what is conserved in Lorentz mechanics is the total energy $E_t = \gamma m_0 c^2$ which essentially is the same as the Lorentzian mass (or non-rest mass according to the old convention) $m = \gamma m_0$ where the rest mass m_0 incorporates all forms of non-kinetic energy according to the mass-energy equivalence while the Lorentz factor γ represents the kinetic aspect of the total energy.

[135] This scalar equation may be stated in tensor notation as: $\partial_\mu J^\mu = 0$ which reveals its invariant nature across frames (refer to § 6).

4.4 Conservation Laws

(e.g. as a function of time or position that does not affect its potential) and hence it does not increase or decrease despite possible changes from one form to another. At the heart of this unifying conservation law is the mass-energy equivalence which is governed by the Poincare equation.

2. Two particles of identical mass m and identical speed u that move in opposite directions collide and form a single particle. What is the rest energy of the single particle assuming that no mass or energy is lost to the environment (e.g. as sound or radiation)?
Answer: The initial momentum is zero and hence from the conservation of momentum the final momentum should also be zero, i.e. the speed of the single particle is zero. If we symbolize the rest energy of the single particle with E_{0s}, then from the conservation of total energy before and after collision we have:

$$\frac{mc^2}{\sqrt{1-(u/c)^2}} + \frac{mc^2}{\sqrt{1-(u/c)^2}} = E_{0s}$$

$$E_{0s} = \frac{2mc^2}{\sqrt{1-(u/c)^2}}$$

3. A particle of mass m and speed $u = 0.3c$ collides with an identical particle which is at rest and they coalesce. What are the mass, energy and momentum of the composite particle in terms of m and c?
Answer: If we label the mass and speed of the composite particle with M and v then from the conservation of momentum we have:

$$\frac{mu}{\sqrt{1-(u/c)^2}} = \frac{Mv}{\sqrt{1-(v/c)^2}}$$

and from the conservation of total energy we have:

$$\frac{mc^2}{\sqrt{1-(u/c)^2}} + mc^2 = \frac{Mc^2}{\sqrt{1-(v/c)^2}}$$

From the energy equation we obtain:

$$M = \sqrt{1-(v/c)^2}\left(\frac{m}{\sqrt{1-(u/c)^2}} + m\right)$$

On substituting this into the momentum equation we obtain:

$$\frac{mu}{\sqrt{1-(u/c)^2}} = \left(\frac{m}{\sqrt{1-(u/c)^2}} + m\right)v$$

4.4 Conservation Laws

$$v = \frac{u}{\sqrt{1-(u/c)^2}} \left(\frac{1}{\sqrt{1-(u/c)^2}} + 1 \right)^{-1}$$

$$v = \frac{0.3c}{\sqrt{1-0.3^2}} \left(\frac{1}{\sqrt{1-0.3^2}} + 1 \right)^{-1}$$

$$v \simeq 0.1535c$$

and hence:

$$M = \sqrt{1-(v/c)^2} \left(\frac{m}{\sqrt{1-(u/c)^2}} + m \right) \simeq 2.0240m$$

$$E = \frac{Mc^2}{\sqrt{1-(v/c)^2}} \simeq 2.0483mc^2$$

$$p = \frac{Mv}{\sqrt{1-(v/c)^2}} \simeq 0.3145mc$$

4. Analyze the conservation of mass and the conservation of energy in Lorentz mechanics considering the mass-energy equivalence relation.
   Answer: As indicated before, the conservation of mass and the conservation of energy (i.e. in their classical sense that exclude Lorentzian mass and rest energy[136]) as such do not exist in Lorentz mechanics. Therefore, the combination of these two conservation principles may be expressed in one of the following three variants:
   • The conservation of mass-energy where "mass" and energy have their classical meaning.
   • The conservation of Lorentzian mass, i.e. $m = \gamma m_0$ with m and m_0 being the Lorentzian and rest mass.
   • The conservation of total energy, i.e. $E_t = \gamma m_0 c^2$.
   As indicated earlier, the first variant may be more appropriate for the modern convention about mass while the second and third variants may be more appropriate for the old convention about mass.[137]

Exercises
1. Discuss the conservation laws of mass and energy in classical and Lorentz mechanics.
2. A bound electron captured a 2 MeV photon. Neglecting the recoil of the atom to which the electron is bound and assuming that the electron is initially at rest, find the speed of the electron after capture assuming that the external forces on the electron are negligible.

---
[136] Rest energy here means the energy content of the mass of the object in the classical sense of mass which represents the material part of the object.
[137] In fact, the third variant should be suitable to both conventions depending on the role of γ and if it belongs to the mass or to the whole expression of total energy as indicated earlier.

3. A particle A of mass m and kinetic energy $E_{kA} = 2mc^2$ collides with a stationary particle B of mass $1.5m$ and they coalesce. What are the mass, energy and momentum of the composite particle in terms of m and c assuming that there is no loss of mass or energy to the environment in this process?

4.5 Restoring Classical Formulation at Low Speed

As explained earlier, when the speed is very low compared to the characteristic speed of light (i.e. $v \ll c$), the speed ratio β approaches 0 while the Lorentz factor γ approaches 1. Accordingly, the formulation of Lorentz mechanics will converge to the formulation of classical mechanics at this speed limit and hence classical mechanics becomes a valid approximation to Lorentz mechanics at low speeds. In the following, we present a number of examples to demonstrate this fact where we present the Lorentzian form on one side and the approached approximate classical form on the other side.

1. Momentum:
$$p = \gamma m u \quad \rightarrow \quad p \simeq m u \qquad (139)$$

This is because $\gamma \simeq 1$ when $u \ll c$.

2. Kinetic energy:
$$E_k = mc^2 \left[\gamma - 1\right] \quad \rightarrow \quad E_k \simeq mc^2 \left[\left(1 + \frac{1}{2}\frac{u^2}{c^2}\right) - 1\right] = \frac{1}{2} m u^2 \qquad (140)$$

This is because when $u \ll c$, the Lorentz factor γ will reduce to the above form, which is a truncated binomial or Taylor power series of γ, since the quadratic and higher terms in β^2 will be vanishingly small and hence they are negligible (refer to Eqs. 13 and 85).

3. Velocity transformation in the x direction:
$$u'_x = \frac{u_x - v}{1 - \beta \frac{u_x}{c}} \quad \rightarrow \quad u'_x \simeq u_x - v \qquad (141)$$

This is because when $v \ll c$, $\beta \simeq 0$ and hence the denominator approaches 1. The reader is referred to the exercises for more details.

4. The formulae of Lorentz mechanics for length contraction and time dilation will reduce to their classical limit, i.e. $L' \simeq L$ and $\Delta t' \simeq \Delta t$ since the Lorentz factor γ, which scales length and time in the length contraction and time dilation formulae, approaches 1.

In fact, the convergence of Lorentz mechanics formulae to their corresponding classical mechanics forms when $v \ll c$ should be a verification test that must be conducted on any derived formula of Lorentz mechanics and hence no Lorentzian formula should be accepted if it does not reduce to its classical form at the low-speed limit. This is because classical mechanics is a very well tested and highly reliable theory at low speeds and hence no correct theory should depart from classical mechanics in its domain of validity.

Solved Problems

4.5 Restoring Classical Formulation at Low Speed

1. An inertial observer measures the speed of a massive object with mass $m = 1$ to be $u = 100$. What is the kinetic energy of this object according to Lorentz mechanics and according to classical mechanics? Comment on the result.
   Answer:
   According to Lorentz mechanics we have:

   $$E_{kl} = mc^2 \left(\gamma - 1\right) \simeq 1 \times \left(3 \times 10^8\right)^2 \times \left(\frac{1}{\sqrt{1 - \left(\frac{100}{3 \times 10^8}\right)^2}} - 1\right) \simeq 4996 \, \text{J}$$

   According to classical mechanics we have:

   $$E_{kc} = \frac{1}{2} m u^2 = 0.5 \times 1 \times 100^2 = 5000 \, \text{J}$$

   Comment: as we see, even at this rather high speed in comparison to the normal classical speeds that are commonly found in everyday life, the difference between the classical and Lorentzian kinetic energy is very small. The relative difference between the two is:

   $$\Delta E_{kr} = \frac{E_{kc} - E_{kl}}{E_{kc}} \simeq 0.0008$$

   i.e. about 0.08% which is negligible. Hence, the classical formula for kinetic energy is very good approximation at the low-speed limit, i.e. $u \ll c$.

2. Repeat the previous question considering this time the linear momentum of the object.
   Answer:
   According to Lorentz mechanics we have:

   $$p_l = \gamma m u = \frac{1 \times 100}{\sqrt{1 - \left(\frac{100}{3 \times 10^8}\right)^2}} \simeq 100.000000000006 \, \text{kg.m/s}$$

   According to classical mechanics we have:

   $$p_c = m u = 1 \times 100 = 100 \, \text{kg.m/s}$$

   Comment: the two results are virtually identical. The magnitude of the relative difference between the two is:

   $$\Delta p_r = \left| \frac{p_c - p_l}{p_c} \right| \simeq 5.55 \times 10^{-14}$$

   which is extremely small. This confirms the fact that the Lorentzian formulation converges to the classical formulation when $u \ll c$.

Exercises

1. Give examples (other than those given in the text) for the convergence of the Lorentzian formulation to the classical formulation at the low-speed limit, i.e. $v \ll c$.

2. Discuss the conditions for the validity of the classical formula of velocity transformation in the x direction. Also discuss the validity of the approximation in the y and z direction.
3. Find the speed condition for which the relative difference between the Lorentzian and classical velocity transformation relations for the x dimension does not exceed 0.05. Comment on the question.

4.6 Restrictions at High Speed

As discussed before, the formulation of Lorentz mechanics is restricted to speeds lower than the characteristic speed of light c. This is because when $v = c$ the Lorentz γ factor becomes singular, while when $v > c$ the Lorentz factor becomes imaginary. However, these limitations apply only within the domain of validity of Lorentz mechanics which is inertial frames. Moreover, these restrictions do not rule out the possibility of a more general formulation in the future that lifts these restrictions such as having a more general mechanical theory to which Lorenz mechanics is a special case, like classical mechanics in its relation to Lorentz mechanics. So in brief, the possibility of having physical speeds that reach or exceed c cannot be ruled out automatically because of the restrictions on the current formulation of Lorentz mechanics. Furthermore, the formulation-based speed restrictions should apply to massive objects and hence they cannot be extended automatically to massless objects without further evidence (refer to the solved problems and exercises for further discussion).

Solved Problems

1. Discuss the possibility of exceeding the characteristic speed of light c by massless objects.
   Answer: The restriction on the speed of physical objects by the requirement of the Lorenz γ factor to be finite (i.e. non-singular) does not apply to light which propagates at its characteristic speed c. Hence, the requirement of the Lorenz γ factor to be real (i.e. non-imaginary) should not apply to massless objects without further evidence. Therefore, denying the possibility of exceeding c by massless objects cannot be established by the requirement of the formalism of Lorenz mechanics.

Exercises

1. Give some examples for the speed restrictions on massive objects that are based on the current formalism of Lorentz mechanics.
2. Discuss the issue of the speed of physical objects within the framework of Lorentz mechanics and beyond.
3. An electron is seen from an inertial frame O_1 to have a momentum $p_1 = 2\,\text{MeV}/c$ and from another inertial frame O_2 to have a momentum $p_2 = 4\,\text{MeV}/c$. What is the relative speed between O_1 and O_2 assuming that O_1 and O_2 are in a state of standard setting and the momenta are referred to their common x axis? Comment on the result.
4. Can we infer from the velocity composition and velocity transformation formulae that the speed of massless objects is restricted to c (i.e. the speed cannot be less than or greater than c)? What about the restriction on the speed of massive objects to be less

than c?

Chapter 5
Derivation of Formalism

In this chapter, we derive the formalism of Lorentz mechanics (or rather the most common part of this formalism) with no reference to any particular interpretation such as special relativity. Accordingly, we only assume the validity of the Lorentz transformations for the four coordinates of spacetime as given by Eqs. 90-93.[138] These transformations can be taken as scientifically supported postulates through the experimental evidence in support of the direct consequences and derived formulations of these transformations. Hence, we do not postulate the relativity principle or the constancy of the observed speed of light as in special relativity. As we will see, all the consequences and implications of Lorentz mechanics can be obtained from the basic Lorentz transformations, and hence the whole formalism of special relativity is contained in Lorentz mechanics, so if we extract this formalism from special relativity, what remains is only the philosophical interpretation of special relativity and its epistemological framework, as well as some redundant attachments, which have no experimental verification or proof.

We think that our approach of postulating the Lorentz transformations of spacetime coordinates directly is more scientific and impartial. In fact, we can compare Lorentz mechanics with quantum mechanics and ask why these two branches of physics are treated differently. As we know, in quantum mechanics the subject starts from the formalism and ends with the interpretation, while in Lorentz mechanics (as dominated by special relativity) the subject starts from the interpretation and ends with the formalism which is seen as a result of the interpretation. Our belief is that for the sake of consistency and impartiality, if not for anything else, we should also start in Lorentz mechanics from the formalism of this subject (as represented by the Lorentz spacetime coordinate transformations and their formal consequences) and end with the interpretation where special relativity, as well as any other theory, can be presented as a possible interpretation for the formalism of Lorentz mechanics.

Accordingly, the two postulates of special relativity should be seen as part of the philosophical and epistemological interpretation of Lorentz mechanics and its formalism instead of being the basis of this formalism. Consequently, the formalism will not be imprisoned to any particular interpretation and its postulates. Moreover, no part of Lorentz mechanics other than its bare formalism will get the legitimacy and endorsement of experimental evidence. So in brief, special relativity is a scientifically redundant theory, like the theory

---

[138] In fact, we also assume the invariance of the rest mass and the conservation of the Lorentzian mass when we come to the dynamical part of Lorentz mechanics. Other similar assumptions, like the (empirically-established) invariance of electric charge across inertial frames, are also assumed in some derivations. These assumptions are generally established by experimental evidence although some are also embedded within a theoretical structure where they are shown to be derivable from other established principles.

of parallel worlds in quantum mechanics, although it may be philosophically significant and hence all we need in physics is the formalism of Lorentz mechanics on the basis of its endorsement by experimental evidence.

In the following sections, we follow the above plan and hence we investigate the formal structure of Lorentz mechanics by taking the Lorentz spacetime coordinate transformations as postulates while all the other transformations and formulations of Lorentz mechanics should be derived or obtained directly or indirectly from these basic coordinate transformations. We should remark that the presentation and derivation of the formalism of Lorentz mechanics are not thorough as we limit our attention to the main results. The primary objective is to outline the methods and provide the commonly needed formulations not to provide a completely structured theory and results. We should also remark that most of the formalism of Lorentz mechanics is based on employing a standard setting between inertial frames, as explained in § 1.5.

Solved Problems

1. Discuss the possible advantages and disadvantages of a philosophically-based scientific theory versus a formalism-based scientific theory.

   Answer: It may be claimed that while a philosophically-based theory of Lorentz mechanics, like special relativity theory, provides a better insight in the formalism and makes more sense of it, a formalism-based theory of Lorentz mechanics gives less insight as the formalism in itself does not provide any support to any specific interpretation although it may indicate a sort of intuitive interpretation or insight. However, as discussed in the text, a formalism-based theory is more objective and less susceptible to errors and bias. Moreover, it allows more freedom in interpreting the theory and probing more possibilities of its significance and implications. This will result in more diverse and rich investigation routes. A quick inspection to the history of Lorentz mechanics clearly shows that the rigid framework of special relativity and its taboos, such as considering c as a restricted and ultimate speed even for massless objects, has imposed many limits on the scientific research so far and excluded many potential routes of investigation. A formalism-based scientific theory will also separate the philosophical contemplation from the scientific investigation and hence it keeps each part away from the superfluous (and potentially harmful) influence of the other. Moreover, a postponed interpretation of a formalism-based scientific theory will come as a hindsight in the theory and its philosophical and epistemological infrastructure and hence no additional insight will be lost.

Exercises

1. Can we consider the limitation that is imposed in Lorentz mechanics on the physical speeds (i.e. $v < c$) as evidence for the proposal that c is a restricted speed to light and it is an ultimate speed for all physical objects?

5.1 Physical Transformations

We have two main types of transformations: the main transformations which are the Lorentz spacetime coordinate transformations, and the subsidiary transformations which are derived, directly or indirectly, from the main transformations. As stated earlier, the main transformations, which are investigated in § 5.1.1, are taken as postulates for the whole formalism of Lorentz mechanics and hence they are not supposed to be derived from other postulates. The justification for postulating these coordinate transformations is the experimental evidence in support of the formalism of Lorentz mechanics and its scientific and logical consequences (see § 8). However, for the sake of completeness we presented in § 12.4 some methods for deriving the main transformations if they are not postulated.

Regarding the subsidiary transformations which are derived directly or indirectly from the main transformations, we will investigate and derive the most common of these subsidiary transformations in the subsequent subsections. It is important to note that in the derivation of these transformations c stands, as anywhere else, for the characteristic speed of light (not the observed speed of light) and hence it is constant, so the reader should not be confused to believe that by treating c as a constant we are adopting the second postulate of special relativity since the essence of this postulate is the constancy of the observed speed of light.[139] We should also note that β and γ are also constant in general when they are functions of the speed between inertial frames because v (like c) is constant in any given problem although it is variable from one problem to another. However, they may also be variable when they are functions of the speed of the observed object like γ in the formula of the Lorentzian energy and momentum, i.e. $E = \gamma m c^2$ and $p = \gamma m u$. In brief, β and γ are functions of speed (i.e. v or u) and hence they are constant when the speed is constant and variable when the speed is variable.

Exercises

1. Why we prefer to postulate the Lorentz spacetime coordinate transformations and derive the other transformations and formulations of Lorentz mechanics from these postulates instead of deriving the Lorentz spacetime coordinate transformations from other postulates as it is the case in special relativity where they are derived from the postulate of relativity and the postulate of constancy of speed of light in all inertial frames?

5.1.1 Lorentz Spacetime Coordinate Transformations

Based on the previously stated assumptions, the Lorentz spacetime coordinate transformations between two inertial frames, O' and O, which are in a state of standard setting with a uniform relative velocity v are given by:

$$x' = \gamma (x - vt) = \gamma (x - \beta c t) \qquad (142)$$
$$y' = y \qquad (143)$$

---

[139] In more technical terms, the essence of the second postulate of special relativity is the invariance of the observed speed of light across all inertial frames where it takes the constant value c which symbolizes the characteristic speed of light (i.e. $c = 299792458$ m/s).

5.1.1 Lorentz Spacetime Coordinate Transformations

$$z' = z \tag{144}$$

$$t' = \gamma \left(t - \frac{vx}{c^2} \right) = \gamma \left(t - \frac{\beta x}{c} \right) \tag{145}$$

where the primed and unprimed symbols correspond to frames O' and O respectively and γ is the Lorentz factor which is a function of v as defined earlier in Eq. 5. The last equation may also be written as:

$$ct' = \gamma \left(ct - \frac{vx}{c} \right) = \gamma \left(ct - \beta x \right) \tag{146}$$

so that all the coordinates of the 4D spacetime manifold have the same physical dimension of length; moreover a more unified and homogeneous symbolic system (i.e. x^0, x^1, x^2, x^3 or x^1, x^2, x^3, x^4) can be used where the temporal and spatial coordinates are equally represented.

The above equations can be inverted to give the Lorentz transformations in the opposite direction (i.e. from frame O' to frame O), that is:

$$x = \gamma \left(x' + vt' \right) = \gamma \left(x' + \beta ct' \right) \tag{147}$$

$$y = y' \tag{148}$$

$$z = z' \tag{149}$$

$$t = \gamma \left(t' + \frac{vx'}{c^2} \right) = \gamma \left(t' + \frac{\beta x'}{c} \right) \tag{150}$$

As we noted earlier, since the two inertial frames are equivalent and the relative velocity between the two frames is equal in magnitude and opposite in direction, all is needed for this inversion is to interchange the primed and unprimed symbols of the spacetime coordinates and replace v with $-v$.

We note that the above Lorentz transformations correspond to a single event[140] as observed by O and O'. Sometimes, it is required to transform the time interval or the space separation between two events as observed by O and O'. In such cases, the temporal and spatial variables in the above Lorentz transformations are replaced by the difference between these variables for the two events. For instance, if it is required to find the time interval between two events V_A and V_B as measured by O' knowing the spacetime coordinates of the two events as measured by O, we subtract the related temporal coordinates of the two events in the Lorentz time transformation from O to O', that is:

$$t'_B - t'_A = \gamma \left(t_B - \frac{vx_B}{c^2} \right) - \gamma \left(t_A - \frac{vx_A}{c^2} \right) = \gamma \left[(t_B - t_A) - \frac{v}{c^2}(x_B - x_A) \right] \tag{151}$$

where the subscripts A and B correspond to V_A and V_B respectively. In fact, Eq. 145 may be seen as a special case for the last equation that corresponds to $t'_A = t_A = x_A = 0$ which is inline with the state of standard setting where V_A occurred at the common origin of space and time. This fact may also be demonstrated by writing the last equation as:

$$\Delta t' = \gamma \left[\Delta t - \frac{v}{c^2} \Delta x \right] \tag{152}$$

---

[140] Or rather: correspond to two events where one of these events is at the origin of spacetime.

where the similarity between the last equation and Eq. 145 is more obvious since the two equations are identical apart from the Δ symbol which is just a notational artifact.

An important feature of Lorentz transformations is that when there is a relative motion between two inertial frames then both space (in the x dimension which is the dimension of motion according to the state of standard setting) and time coordinates are defined in each frame in terms of both space and time coordinates in the other frame, as seen in Eqs. 142 and 145 and in Eqs. 147 and 150. Hence, space and time are entangled in the formalism of Lorentz mechanics to form a single spacetime manifold. Another important feature of Lorentz transformations is the Lorentz γ factor which makes the spacetime coordinates dependent on the speed of the relative motion between the frames. Both these features are Lorentzian and hence they do not exist in the classical Galilean transformations (although there is a temporal factor in the spatial x transformation).

Solved Problems
1. Find a mathematical expression for the spatial separation between two events V_A and V_B in the x dimension as measured by O knowing the spacetime coordinates of the two events as measured by O' where O and O' are in a state of standard setting.
   Answer: Similar to what we did in the text to find the time interval, we subtract the related spatial coordinates of the two events in the Lorentz x transformation from O' to O, that is:
   $$x_B - x_A = \gamma \left(x'_B + vt'_B\right) - \gamma \left(x'_A + vt'_A\right) = \gamma \left[\left(x'_B - x'_A\right) + v \left(t'_B - t'_A\right)\right]$$

Exercises
1. What is the physical significance of the Lorentz spacetime coordinate transformations and to which physical phenomena they are related?
2. Discuss the procedural aspects in solving the problems related to the Lorentz spacetime coordinate transformations.
3. Write the expressions for the space and time intervals between two events V_A and V_B as measured by O' knowing the spacetime coordinates of the events as measured by O where O and O' are in a state of standard setting.
4. Discuss if the mingling of space and time into spacetime is a complete novelty of Lorentz mechanics.

5.1.2 Velocity Transformations

These transformations are derived from the Lorentz spacetime coordinate transformations, as given in Eqs. 142-145, by taking the first derivative of the three spatial coordinates with respect to the temporal coordinate. These derivatives can be obtained either by the chain rule of differentiation or by taking the differentials and treating them as algebraic quantities. The latter method is normally easier than the former but it may be seen as less rigorous. Therefore, we will use the first method in the following derivations although for the sake of diversity and demonstration we will use the second method in the solution of some exercises.

5.1.2 Velocity Transformations

As we have three velocity components in the primed system, i.e. u'_x, u'_y and u'_z, we should derive three formulae. The transformations of the three velocity components in the unprimed system, i.e. u_x, u_y and u_z, can be derived similarly from Eqs. 147-150, but this is redundant because we can obtain these formulae from the formulae of u'_x, u'_y and u'_z by exchanging the primes in the symbols with reversing the sign of the relative velocity between the two frames, as explained and justified earlier.

A. Derivation of u'_x: using the basic definition of velocity and the chain rule of differentiation, we have:

$$u'_x = \frac{dx'}{dt'} = \frac{dx'}{dt}\frac{dt}{dt'} \tag{153}$$

Now, from Eq. 142 we obtain:

$$\frac{dx'}{dt} = \frac{d}{dt}[\gamma(x - vt)] = \gamma\left(\frac{dx}{dt} - v\right) = \gamma(u_x - v) \tag{154}$$

Similarly, from Eq. 150 we have:

$$\frac{dt}{dt'} = \frac{d}{dt'}\left[\gamma\left(t' + \frac{vx'}{c^2}\right)\right] = \gamma\left(1 + \frac{v}{c^2}\frac{dx'}{dt'}\right) = \gamma\left(1 + \frac{vu'_x}{c^2}\right) \tag{155}$$

Hence:

$$u'_x = \frac{dx'}{dt}\frac{dt}{dt'} \tag{156}$$

$$u'_x = \gamma(u_x - v)\gamma\left(1 + \frac{vu'_x}{c^2}\right) \tag{157}$$

$$u'_x = \frac{(u_x - v)\left(1 + \frac{vu'_x}{c^2}\right)}{1 - \beta^2} \tag{158}$$

$$u'_x - \beta^2 u'_x = u_x + \frac{v}{c^2}u_x u'_x - v - \beta^2 u'_x \tag{159}$$

$$u'_x = u_x + \frac{v}{c^2}u_x u'_x - v \tag{160}$$

$$u'_x - \frac{v}{c^2}u_x u'_x = u_x - v \tag{161}$$

$$u'_x = \frac{u_x - v}{1 - \frac{vu_x}{c^2}} \tag{162}$$

which is the same as Eq. 99.

B. Derivation of u'_y: using the basic definition of velocity and the chain rule, we have:

$$u'_y = \frac{dy'}{dt'} = \frac{dy'}{dt}\frac{dt}{dt'} \tag{163}$$

Now, from Eq. 143 we obtain:

$$\frac{dy'}{dt} = \frac{dy}{dt} = u_y \tag{164}$$

while $\frac{dt}{dt'}$ is given by Eq. 155. Hence:

$$
\begin{aligned}
u'_y &= \frac{dy'}{dt}\frac{dt}{dt'} \qquad (165)\\
&= u_y\gamma\left(1 + \frac{vu'_x}{c^2}\right)\\
&= u_y\gamma\left(1 + \frac{v}{c^2}\left[\frac{u_x - v}{1 - \frac{vu_x}{c^2}}\right]\right)\\
&= u_y\gamma\left(1 + \left[\frac{\frac{vu_x}{c^2} - \frac{v^2}{c^2}}{1 - \frac{vu_x}{c^2}}\right]\right)\\
&= u_y\gamma\left(\frac{1 - \frac{vu_x}{c^2} + \frac{vu_x}{c^2} - \frac{v^2}{c^2}}{1 - \frac{vu_x}{c^2}}\right)\\
&= u_y\gamma\left(\frac{1 - \frac{v^2}{c^2}}{1 - \frac{vu_x}{c^2}}\right)\\
&= \frac{u_y\gamma}{\gamma^2\left(1 - \frac{vu_x}{c^2}\right)}\\
&= \frac{u_y}{\gamma\left(1 - \frac{vu_x}{c^2}\right)}
\end{aligned}
$$

that is:
$$u'_y = \frac{u_y}{\gamma\left(1 - \frac{vu_x}{c^2}\right)} \qquad (166)$$

which is the same as Eq. 100.

C. Derivation of u'_z: this can be easily obtained from the derivation of u'_y by replacing the subscript y with the subscript z because the transformations of these coordinates are identical in form as seen in Eqs. 143 and 144.

Exercises
1. Derive the Lorentz velocity transformations from the unprimed frame to the primed frame using this time the differential method instead of the chain rule of differentiation.

5.1.3 Acceleration Transformations

These transformations are derived from the Lorentz spacetime coordinate transformations, as given in Eqs. 142-145, by taking the second derivative of the three spatial coordinates with respect to the temporal coordinate. This is equivalent to taking the first time derivative of the velocity transformation formulae which are derived in the previous subsection.

A. Derivation of a'_x: using the definition of acceleration and the chain rule of differentiation, we have:

$$a'_x = \frac{du'_x}{dt'} = \frac{du'_x}{dt}\frac{dt}{dt'} \qquad (167)$$

5.1.3 Acceleration Transformations

Now, from Eq. 162 and the quotient rule of differentiation we have:

$$\begin{aligned}
\frac{du'_x}{dt} &= \frac{d}{dt}\left[\frac{u_x - v}{1 - \frac{vu_x}{c^2}}\right] \\
&= \frac{a_x\left(1 - \frac{vu_x}{c^2}\right) - \left(-\frac{v}{c^2}a_x\right)(u_x - v)}{\left(1 - \frac{vu_x}{c^2}\right)^2} \\
&= \frac{a_x - a_x\frac{vu_x}{c^2} + a_x\frac{vu_x}{c^2} - \frac{v^2}{c^2}a_x}{\left(1 - \frac{vu_x}{c^2}\right)^2} \\
&= \frac{a_x - \frac{v^2}{c^2}a_x}{\left(1 - \frac{vu_x}{c^2}\right)^2} \\
&= \frac{a_x\left(1 - \frac{v^2}{c^2}\right)}{\left(1 - \frac{vu_x}{c^2}\right)^2} \\
&= \frac{a_x}{\gamma^2\left(1 - \frac{vu_x}{c^2}\right)^2}
\end{aligned}$$ (168)

while $\frac{dt}{dt'}$ is obtained in Eq. 155. Hence:

$$\begin{aligned}
a'_x &= \frac{du'_x}{dt}\frac{dt}{dt'} \\
&= \frac{a_x}{\gamma^2\left(1 - \frac{vu_x}{c^2}\right)^2}\left[\gamma\left(1 + \frac{vu'_x}{c^2}\right)\right] \\
&= \frac{a_x\left(1 + \frac{vu'_x}{c^2}\right)}{\gamma\left(1 - \frac{vu_x}{c^2}\right)^2} \\
&= \frac{a_x\left(1 + \frac{v}{c^2}\left[\frac{u_x - v}{1 - \frac{vu_x}{c^2}}\right]\right)}{\gamma\left(1 - \frac{vu_x}{c^2}\right)^2} \\
&= \frac{a_x\left(1 - \frac{vu_x}{c^2} + \frac{vu_x}{c^2} - \frac{v^2}{c^2}\right)}{\gamma\left(1 - \frac{vu_x}{c^2}\right)^3} \\
&= \frac{a_x\left(1 - \frac{v^2}{c^2}\right)}{\gamma\left(1 - \frac{vu_x}{c^2}\right)^3} \\
&= \frac{a_x}{\gamma^3\left(1 - \frac{vu_x}{c^2}\right)^3}
\end{aligned}$$ (169)

that is:

$$a'_x = \frac{a_x}{\gamma^3\left(1 - \frac{vu_x}{c^2}\right)^3}$$ (170)

which is the same as Eq. 115.

5.1.3 Acceleration Transformations

B. Derivation of a'_y: using the definition of acceleration and the chain rule, we have:

$$a'_y = \frac{du'_y}{dt'} = \frac{du'_y}{dt}\frac{dt}{dt'} \tag{171}$$

Now, from Eq. 166 and the quotient rule of differentiation we have:

$$\begin{aligned}
\frac{du'_y}{dt} &= \frac{d}{dt}\left[\frac{u_y}{\gamma\left(1-\frac{vu_x}{c^2}\right)}\right] \\
&= \frac{a_y\gamma\left(1-\frac{vu_x}{c^2}\right)-\gamma\left(-\frac{va_x}{c^2}\right)u_y}{\gamma^2\left(1-\frac{vu_x}{c^2}\right)^2} \\
&= \frac{a_y - a_y\frac{vu_x}{c^2} + \frac{va_x}{c^2}u_y}{\gamma\left(1-\frac{vu_x}{c^2}\right)^2} \\
&= \frac{c^2 a_y - vu_x a_y + vu_y a_x}{c^2\gamma\left(1-\frac{vu_x}{c^2}\right)^2}
\end{aligned} \tag{172}$$

while $\frac{dt}{dt'}$ is obtained in Eq. 155. Hence:

$$\begin{aligned}
a'_y &= \frac{du'_y}{dt}\frac{dt}{dt'} \\
&= \left[\frac{c^2 a_y - vu_x a_y + vu_y a_x}{c^2\gamma\left(1-\frac{vu_x}{c^2}\right)^2}\right]\gamma\left(1+\frac{vu'_x}{c^2}\right) \\
&= \left[\frac{c^2 a_y - vu_x a_y + vu_y a_x}{c^2\left(1-\frac{vu_x}{c^2}\right)^2}\right]\left(1+\frac{vu'_x}{c^2}\right) \\
&= \left[\frac{c^2 a_y - vu_x a_y + vu_y a_x}{c^2\left(1-\frac{vu_x}{c^2}\right)^2}\right]\left(1+\frac{v}{c^2}\left[\frac{u_x-v}{1-\frac{vu_x}{c^2}}\right]\right) \\
&= \left[\frac{c^2 a_y - vu_x a_y + vu_y a_x}{c^2\left(1-\frac{vu_x}{c^2}\right)^2}\right]\left(\frac{1-\frac{vu_x}{c^2}+\frac{vu_x}{c^2}-\frac{v^2}{c^2}}{1-\frac{vu_x}{c^2}}\right) \\
&= \left[\frac{c^2 a_y - vu_x a_y + vu_y a_x}{c^2\left(1-\frac{vu_x}{c^2}\right)^3}\right]\left(1-\frac{v^2}{c^2}\right) \\
&= \frac{c^2 a_y - vu_x a_y + vu_y a_x}{c^2\gamma^2\left(1-\frac{vu_x}{c^2}\right)^3}
\end{aligned} \tag{173}$$

that is:

$$a'_y = \frac{c^2 a_y - vu_x a_y + vu_y a_x}{c^2\gamma^2\left(1-\frac{vu_x}{c^2}\right)^3} \tag{174}$$

which is the same as Eq. 116.

C. Derivation of a'_z: this can be obtained from the derivation of a'_y by replacing the subscript y with the subscript z because the transformations of the velocities u'_y and u'_z are identical in form as can be seen from Eqs. 100 and 101.

Solved Problems

1. Someone may ask: if Lorentz mechanics is restricted to inertial frames and observers then why we should care about acceleration and its transformation? What is your answer to this question?
   Answer: Lorenz mechanics is restricted to inertial frames and inertial observers and not restricted to "inertial objects". This means that although in the domain of Lorentz mechanics the frame of reference from which we observe the physical laws and phenomena should be inertial, the observed objects are not required to be "inertial" and hence they can be accelerating. Accordingly, it is legitimate to investigate the transformation of acceleration between inertial frames since this is within the domain of applicability of Lorentz mechanics.

Exercises

1. Derive the Lorentz acceleration transformations from the unprimed frame to the primed frame using this time the differential method instead of the chain rule of differentiation.

5.1.4 Length Transformation

Let have a stick whose two end coordinates in the direction of motion in the proper (unprimed) and improper (primed) frames are x_1, x_2 and x'_1, x'_2 respectively. Accordingly, the length of the stick in the two frames will be related by (refer to Eq. 147):

$$L_x = x_2 - x_1 = \gamma(x'_2 + \beta ct') - \gamma(x'_1 + \beta ct') = \gamma(x'_2 - x'_1) = \gamma L'_x \tag{175}$$

that is:

$$L_x = \gamma L'_x \tag{176}$$

which is the well known length contraction formula where L_x and L'_x here represent the proper and improper length. So in brief, length transformation is based on space coordinate transformations. As discussed earlier (see for example the exercises of § 4.1.2), the Lorentzian length transformation applies to simultaneous events and hence in the above derivation we assumed $t'_1 = t'_2 = t'$ because the measurements of the spatial coordinates are simultaneous.

Solved Problems

1. Show that length contraction effect applies only in the direction of motion and not in the perpendicular directions to the direction of motion.
   Answer: In the y and z directions which are perpendicular to the direction of motion according to the standard setting we have:

$$\begin{aligned} L_y &= y_2 - y_1 = y'_2 - y'_1 = L'_y \\ L_z &= z_2 - z_1 = z'_2 - z'_1 = L'_z \end{aligned}$$

i.e. the length in the y and z directions which are perpendicular to the direction of motion are the same in both frames and hence they are not affected by the motion. This applies to any direction in the yz plane since any length in that plane will be resolved in y and z components and since these components are not affected by length contraction then the resultant length will not be affected.

Exercises

1. Find a general formula that correlates the total length of an object in one inertial frame O to its total length in another inertial frame O' where O and O' are in a state of standard setting and the object is arbitrarily oriented in any direction.

5.1.5 Time Interval Transformation

Using the Lorentz time transformation as given by Eq. 145, we see that the time intervals are transformed between the proper (unprimed) and improper (primed) frames by the following formula:

$$\Delta t' = t'_2 - t'_1 = \gamma \left(t_2 - \frac{\beta x}{c} \right) - \gamma \left(t_1 - \frac{\beta x}{c} \right) = \gamma (t_2 - t_1) = \gamma \Delta t \qquad (177)$$

which is the well known time dilation formula where Δt and $\Delta t'$ here represent the proper and improper time interval. As discussed earlier (see for example the exercises of § 4.1.2), the Lorentzian time interval transformation applies to events which are co-positional in the transformed frame and hence in the above derivation we assumed $x_1 = x_2 = x$ because the measurements of the temporal coordinates take place at the same position in that frame.

We note that in the length transformation we started from $L = x_2 - x_1$ (instead of $L' = x'_2 - x'_1$ which will produce an opposite formula of "length dilation"), while in the time interval transformation we started from $\Delta t' = t'_2 - t'_1$ (instead of $\Delta t = t_2 - t_1$ which will produce an opposite formula of "time contraction"). The reason is that length contraction and time dilation require these different approaches since length contraction is the effect that is seen by the rest frame observer of the length of the moving stick in the moving frame, while time dilation is the effect that is felt by the moving frame observer of the time interval in his frame.[141] Having ignored these historical factors in labeling and conceptualizing these effects, we will have equivalent effects where the resting observer will observe the same effect (i.e. contraction of length and time interval due to the contraction of spacetime coordinates by the motion) when he observes events in the moving frame. An

---

[141] In fact, there is another interpretation of time dilation where it is seen as the elongation of time as observed by the improper observer when he compares his own clock with the moving clock. Accordingly, we may recast the above phrase to read "since length contraction is the effect that is seen by the rest frame observer of the length of the *moving stick* in the moving frame, while time dilation is the effect that is observed by the rest frame observer of the time interval as measured by *his clock* as compared to the clock of the moving frame". However, these differences do not introduce any fundamental change to our point about the difference in perspective between time dilation and length contraction.

5.1.5 Time Interval Transformation

alternative to the above approach (i.e. starting from $\Delta t' = t'_2 - t'_1$ instead of $\Delta t = t_2 - t_1$) is to change the labeling of the frames by interchanging the primed and unprimed symbols (i.e. reversing the labeling of the standard setting).

We also note that this derivation method of the time dilation formula is based directly on the Lorentz spacetime coordinate transformations and not on any subsidiary method like light clock and hence any potential criticisms to these subsidiary methods (see e.g. § 9.7 and § 9.7.1) will not affect our derivation. In fact, this also applies to the length contraction formula which may also be derived as a lemma from the time dilation formula with the aid of light clock, as discussed in the solutions of the exercises of § 9.7 (also see the exercises of the present section and the exercises of § 4.1.2). So in brief, length transformation and time interval transformation are based on the Lorentz spacetime coordinate transformations which are the fundamental principles of Lorentz mechanics with no need for help from any subsidiary method or thought experiment or abstract device.

Solved Problems

1. Provide more clarification about the difference in perspective between length contraction and time dilation that resulted in using $L = x_2 - x_1$ for the first and $\Delta t' = t'_2 - t'_1$ for the second.
   Answer: Let have an observer O who from his rest frame watches a moving observer O'. When O watches a stick and a clock in O' frame, he will see both the length and the time interval of O' frame shorter (i.e. he observes length contraction and time contraction). For example, a stick which measures 1 meter in O frame will be seen by O to measure 0.5 m when it is in O' frame, i.e. length contraction. Similarly, when the clock of O records a time interval of 1 minute between two events an identical clock in O' frame will be seen by O to record 0.5 minute during that interval since it runs slower, i.e. time contraction. However, because O is aware of this time contraction in O' frame, he concludes that when his clock (i.e. O clock) records 1 minute and O' clock also records 1 minute then that time interval of O' clock is dilated, i.e. it represents elongated time. So, what O sees as time dilation in O' frame represents the knowledge of O that O' experiences a longer time during an interval of a given size (i.e. 1 minute) since the 1 minute of O' observer is equivalent to 2 minutes of O observer. We remark that we usually label this difference in perspective as proper-improper perspective which can be explained by the fact that by using the primed to unprimed transformation in the length contraction and using the unprimed to primed transformation in the time dilation we are changing our perspective although in both cases the proper is marked as unprimed and the improper is marked as primed. Anyway, this is just a label that we use and the reader should be aware of regardless of being an expressive label or not. More clarifications about this issue will follow.
2. Provide a simple argument based on the invariance of spacetime interval under the Lorentz transformations (refer to § 3.9.4) to show that length contraction and time dilation are essentially identical effects (i.e. they both represent the contraction of spacetime coordinates under the influence of motion) although their labels (which are based on different perspectives) may suggest otherwise.

Answer: The invariance of spacetime interval in its finite form across frames may be expressed as:
$$\left(\Delta x^0\right)^2 - \left[\left(\Delta x^1\right)^2 + \left(\Delta x^2\right)^2 + \left(\Delta x^3\right)^2\right] = 0$$
where Δ represents changes in time and space across different inertial frames which are in relative motion.[142] The important thing to note in the above equation is that the temporal and spatial terms have opposite signs. Accordingly, if the expression on the left hand side of this equation should vanish identically, then $\left(\Delta x^0\right)^2$ and $\left[\left(\Delta x^1\right)^2 + \left(\Delta x^2\right)^2 + \left(\Delta x^3\right)^2\right]$ should change in the same sense.[143] To put this argument in a more formal way, we can write the above equation as:
$$c^2 = \frac{\left(\Delta x^1\right)^2 + \left(\Delta x^2\right)^2 + \left(\Delta x^3\right)^2}{\Delta t^2}$$
Now, since c is the characteristic (not necessarily the observed) speed of light which is a constant, then the spatial coordinates in the numerator and the temporal coordinate in the denominator should change in the same sense to have a constant c.

Exercises
1. Use a single argument to derive both the Lorentzian length transformation and the Lorentzian time interval transformation. Comment on the results.
2. Repeat the previous question to show "time contraction" and "length dilation". Also, comment on the results.
3. Considering the procedural aspect of time measurement, what distinguishes proper time interval from improper time interval?
4. Provide more clarifications about the conditions that should be satisfied to justify using time dilation and length contraction formulae instead of the Lorentz spacetime transformations.

5.1.6 Mass Transformation

Because we follow the modern convention about mass (where it is considered an intrinsic invariant property), we do not need to derive the transformation (which is based on the old convention) of mass between proper and improper frames.[144] In fact, as indicated

---

[142] We note that the above equation is not the equation of the spacetime interval of light signal. However, with slight modifications the argument can be based on using the spacetime interval of light signal in the above equation since the invariance of spacetime interval is equivalent to the invariance of the speed of light due to the use of this speed for calibrating spacetime coordinates in all frames.

[143] The essence of this argument is that: ignoring the difference in sign between the temporal and spatial coordinates, these coordinates should change by the same amount in the same sense (i.e. both increase or decrease) to ensure that the above expression remains zero. Also, refer to the exercises of § 3.9.4 for a similar argument but that argument is based on the difference in sign although it is the same in essence.

[144] Yes, we need to postulate the invariance of mass (which is the same as the rest mass according to the modern convention that we follow) across all frames (i.e. the transformation $m = m_0$ or more generally $m' = m$).

5.1.7 Frequency Transformation and Doppler Shift   187

earlier and will be investigated further in the future (see § 5.4), the invariance of the *rest mass*[145] across frames and the conservation of the *Lorentzian mass*[146] are postulated where these are justified by experimental evidence (also see the exercises).

Exercises
1. Clarify the issue of the invariance of mass according to the old and modern conventions of Lorentz mechanics.
2. Following the old convention about mass, provide an argument based on the conservation of momentum to show that the mass is transformed from its rest frame to a moving frame according to $m = \gamma m_0$.

5.1.7 Frequency Transformation and Doppler Shift

In this subsection, we derive the Lorentz frequency transformation for light where we use in our derivation the definition of frequency and the standard transformations of Lorentz mechanics. Referring to Figure 14, we see that the instantaneous relative velocity between the source S of the light signal and the observer O along their line of sight is $v\cos\theta$. Now, let assume that O observed a light signal emitted by S where the speed of this signal in O frame is c and its wavelength is λ. According to the definition of frequency of light, the frequency ν of the signal in O frame is given by:

$$\nu = \frac{c}{\lambda} \tag{178}$$

Now, according to our previous findings in § 4.2.2 and § 4.2.3, the speed of light transforms invariantly across all inertial frames where it takes its characteristic value c. Moreover, according to the Lorentz spacetime coordinate transformations (see Eq. 90), λ transforms between S frame and O frame as:

$$\lambda = \gamma(x_2 - t_2 v \cos\theta) - \gamma(x_1 - t_1 v \cos\theta) \tag{179}$$
$$= \gamma[(x_2 - x_1) - (t_2 - t_1)v\cos\theta] \tag{180}$$
$$= \gamma(x_2 - x_1)\left(1 - \frac{t_2 - t_1}{x_2 - x_1}v\cos\theta\right) \tag{181}$$

where γ is a function of v.[147] Now, $x_2 - x_1$ is the wavelength in S frame and hence $x_2 - x_1 = \lambda_0$; moreover $t_2 - t_1$ and $x_2 - x_1$ belong to the light signal and hence $\frac{t_2 - t_1}{x_2 - x_1} = \frac{1}{c}$.

---
[145] Since we follow the modern convention then "rest mass" is the same as "mass". We note that "rest mass" here means the combination of matter and any form of non-kinetic energy.
[146] This conservation is equivalent to the conservation of total energy.
[147] We should note that the above transformation does not correspond exactly to Eq. 90 because γ is a function of v while t (as represented by t_1 and t_2) is multiplied by $v\cos\theta$ rather than v. This should be justified by the fact that the γ factor belongs to the spacetime (since it affects both space and time) while the factor that multiplies t represents the effect of the relative motion along the line of sight which is restricted to the spatial x dimension and hence it should be restricted to the spatial component along this dimension, i.e. $v\cos\theta$. More technically, Eq. 90 is based on a standard setting where v is restricted to the x dimension while in the above transformation v is not restricted to the x dimension (as can be seen from Figure 14) and hence the above transformation is not based on a standard setting.

Accordingly, the last equation becomes:

$$\lambda = \gamma \lambda_0 \left(1 - \frac{v}{c} \cos\theta\right) \qquad (182)$$

On substituting λ from the last equation into Eq. 178 we obtain:

$$\nu = \frac{c}{\lambda} = \frac{c}{\gamma \lambda_0 \left(1 - \frac{v}{c}\cos\theta\right)} = \frac{\nu_0}{\gamma\left(1 - \frac{v}{c}\cos\theta\right)} \qquad (183)$$

where $\nu_0 = c/\lambda_0$ is the frequency of the signal in S frame (which we regard as the proper frame of the signal). The last equation is the same as Eq. 122. As noted earlier, the speed of light transforms invariantly across inertial frames and hence it is the same for both O and S frames.

5.1.8 Charge Density and Current Density Transformations

As seen in § 4.2.9, the electric charge density ρ and the electric current density $\mathbf{j}$ are defined as:

$$\rho \equiv \frac{Q}{V} \qquad \text{and} \qquad \mathbf{j} \equiv \rho \mathbf{u} \qquad (184)$$

If ρ and $\mathbf{j}$ are defined as above in an inertial frame O then from another inertial frame O' which moves with velocity $\mathbf{u}$ relative to O we should have:

$$\rho' = \rho_0 \qquad \text{and} \qquad \mathbf{j}' = 0 \qquad (185)$$

where ρ_0 is the proper charge density. The first equation is justified by the fact that O' is a rest frame for the charge and hence the charge density takes its proper value, while the second equation is justified by the absence of relative motion between the charge and O' since it is a rest frame for the charge and hence the current should vanish in O'.

Now, if we go back to frame O then the dimension in the direction of $\mathbf{u}$ will be contracted by $\gamma(u) = \left[1 - (u/c)^2\right]^{-1/2}$ where $u \equiv |\mathbf{u}|$ and hence the proper volume (which is the volume as seen in the rest frame O') will be reduced by the factor γ, i.e.

$$V = \frac{V_0}{\gamma} = \frac{V'}{\gamma} \qquad (186)$$

where V and V' are the volume in O and O' while V_0 is the proper volume. Accordingly, the electric charge density will transform from O' to O as:

$$\rho = \frac{Q}{V} = \frac{Q}{V'/\gamma} = \gamma \frac{Q}{V'} = \gamma \rho' = \gamma \rho_0 \qquad (187)$$

where we assumed $Q = Q'$ because of the invariance of electric charge across inertial frames.[148] Similarly, the electric current density will transform from O' to O as:

$$\mathbf{j} = \gamma \rho_0 \mathbf{u} \qquad (188)$$

---

[148] In fact, the electric charge is conserved in each inertial frame and it is invariant across all inertial frames.

where in this transformation $\gamma\rho_0$ represents the transformation of the electric charge density from O' (where $\rho' = \rho_0$) to O, while $\mathbf{u}$ represents the transformation of the velocity from O' (where $\mathbf{u}' = \mathbf{0}$) to O.

Exercises
1. Provide more clarification about the transformation of electric current density.

5.2 Physical Quantities

We note first that the Lorentzian form of common physical quantities (i.e. length, time interval, mass, velocity, acceleration and work) have been established previously in § 4.1 and hence they do not need further discussion. Therefore, we discuss in the following subsections only those quantities that need further investigation (momentum, force and energy) to establish and justify their Lorentzian form.

5.2.1 Momentum

There are four requirements for an acceptable Lorentzian definition of momentum:
1. It should be reduced to its classical form (i.e. $p = mu$) at low speeds (i.e. $u \ll c$).
2. It should be form invariant under the Lorentz transformations across all inertial frames.
3. It should be conserved as in classical mechanics.
4. It should be compatible with Newton's second law, i.e. force is the time derivative of momentum.

The necessity for satisfying these requirements should be obvious. The first requirement is because classical mechanics is a well established theory at the low-speed limit and hence any correct formulation of physical laws (whether originating from Lorentz mechanics or from something else like quantum mechanics) should agree with the formulation of classical mechanics in its domain of validity (see § 4.5). As for the second requirement, form invariance of physical laws is a universally accepted principle in physics and hence it should be satisfied by any physical law (see § 1.8).[149] Regarding the third requirement, the conservation of momentum (as based originally on its classical definition that applies within the domain of validity of classical mechanics) is a well established principle in physics and has strong experimental and theoretical support and hence its validity within the classical domain and in the extended Lorentzian domain is required. As for the fourth requirement, Newton's second law (with its many auxiliary results and applications) is a well established and commonly used law and hence we should avoid reformulating large

---

[149] We note that the above "definition" of momentum is a physical law since it has certain observable physical consequences on its own and as part of the formulation of other physical laws that are related to momentum. This is similar to Newton's second law where it is a law and "definition" at the same time. In brief, this sort of physical laws can be regarded as definitions but not in an ordinary way, i.e. they are not arbitrary and based on sheer convention but they have real roots in the physical world since they capture real physical phenomena although they are based on a certain framework of conceptualization and formulation.

5.2.1 Momentum

part of physics by adopting a momentum definition that is incompatible with Newton's second law.[150]

Based on the above requirements, it can be shown that the following form of the definition of momentum satisfies all the four requirements:

$$p = \gamma m u \tag{189}$$

The first requirement is obviously satisfied by this form since when $u \ll c$ we have $\gamma \simeq 1$ and hence $p \simeq mu$ which is the well known classical form while the last three requirements will be established later.

Regarding the issue of value invariance of momentum across frames, we should note that the components of momentum in the perpendicular directions to the direction of relative motion between the frames of observation are value invariant across these frames and hence only the component in the direction of relative motion is variant (i.e. not value invariant). Accordingly, any potential disagreement between different inertial observers about the value of momentum is with respect to the x component of the momentum according to the state of standard setting which is assumed here. Regarding the y and z components, since there is no relative motion between the frames in these directions according to the standard setting, then all inertial observers should agree on the value of momentum in these directions. This issue will be discussed further later on.[151]

Solved Problems

1. Justify the Lorentzian definition of momentum and assess its compliance with the aforementioned requirements.

    Answer: There are several formal justifications for the Lorentzian form of linear momentum, as given by Eq. 189, and how it can be obtained from first principles. In the following, we outline some of these justifications. One of the simplest methods to obtain a Lorentzian form of momentum that satisfies the above requirements is to use the rest frame[152] of the object whose momentum is under consideration. Now, the classical definition of momentum (i.e. $p = m\frac{dx}{dt}$) varies in form between different frames since $\frac{dx}{dt}$ is not the same in all inertial frames due to the difference in the spacetime coordinates according to the Lorentz transformations. However, all inertial observers agree on the rest frame of the object and hence they agree on the definition of momentum in the rest frame. So, if we define the momentum in Lorenz mechanics by the same expression that is used in classical mechanics (i.e. $p = m\frac{dx}{dt}$) but with the understanding that this definition belongs to the rest frame of the object (and hence we write it as $p = m\frac{dx}{dt_0}$ to indicate proper time), then we have a definition that is universal and invariant across all inertial frames. Now, if we want to express this definition in terms of the improper time (i.e. the time of any frame other than the rest frame of the object) then we can

---

[150] In fact, the third and fourth requirements can be regarded as an instance of the same principle which the first requirement is based upon, i.e. the necessity of the convergence of Lorentz formulation to its classical limit (see the exercises).

[151] This issue has been addressed indirectly in a previous exercise about mass transformation (see § 5.1.6).

[152] Or rather the instantaneous rest frame.

5.2.1 Momentum 191

use the chain rule of differentiation, that is:

$$p = m\frac{dx}{dt_0} = m\frac{dx}{dt}\frac{dt}{dt_0} = mu\gamma \qquad (190)$$

where t is the improper time and the equality $\frac{dt}{dt_0} = \gamma$ is based on the time dilation formula (i.e. $\Delta t = \gamma \Delta t_0$) since the derivative $\frac{dt}{dt_0} = \gamma$ is just the infinitesimal form of the time dilation formula. As we saw, the above definition of momentum (i.e. $p = mu\gamma$) satisfies the first two of the above requirements since it converges to the classical form $p = mu$ at low speeds and it is the same across all inertial frames (including the rest frame of the object itself) since each frame takes in this definition its measured value of u and γ. Regarding the third and fourth requirements they will be dealt with later (refer to § 5.4, and to § 5.3.1 and § 6.5.6). The reader is also referred to § 5.4 for a more formal and rigorous presentation of this argument.

2. Demonstrate the invariance of the components of momentum in the perpendicular directions to the direction of relative motion between the frames of observation by making a simple argument for the case that the y (or z) component of the momentum is the same for two inertial observers who are in a state of standard setting.
Answer: Let have two observers, O and O', who are in a state of standard setting. Observer O' fires a bullet in the y direction towards a block of wood which is in his rest frame and the bullet penetrates the block creating a hole of length L. Now, since O and O' agree on the lengths in the y direction which is perpendicular to the direction of their relative motion they should agree on L. It is sensible to assume that the length of the hole is solely dependent on the momentum of the bullet in the y direction. Accordingly, since O and O' agree on the length of the hole they should agree on the momentum of the bullet in the y direction. The reader is also referred to the next question.

3. Challenge the argument of the previous question and assess this challenge.
Answer: Someone may challenge the argument of the previous question by the following: since O and O' agree on the momentum of the bullet in the y direction then we should have:

$$p_y = p'_y$$

$$\frac{mu_y}{\sqrt{1-(u_y/c)^2}} = \frac{mu'_y}{\sqrt{1-(u'_y/c)^2}}$$

$$\frac{u_y}{\sqrt{1-(u_y/c)^2}} = \frac{u'_y}{\sqrt{1-(u'_y/c)^2}}$$

where u_y and u'_y are the y components of the bullet velocity and hence we should have $u_y = u'_y$ which contradicts the Lorentz velocity transformation in the y direction.[153] However, this challenge is based on ignoring a γ factor that belongs to the motion in the x direction. To simplify the explanation and symbolism, let follow the old Lorentzian

---
[153] The equality $u_y = u'_y$ represents the physically viable case.

5.2.1 Momentum

formalism where the mass is frame dependent and hence we have a rest mass m_0, a mass in O frame m, and a mass in O' frame m'. Now, since the bullet in frame O' has only a y velocity component then we should have:

$$m' = \frac{m_0}{\sqrt{1 - \left(u'_y/c\right)^2}}$$

However, the bullet in frame O has both x and y velocity components and hence we have:

$$m = \frac{m_0}{\sqrt{1-(u/c)^2}} = \frac{m_0}{\sqrt{1-\frac{u_x^2+u_y^2}{c^2}}} = \frac{m_0}{\sqrt{1-\frac{v^2+u_y^2}{c^2}}}$$

where v is the relative speed between the two frames. On applying the Lorentzian velocity transformation to u_y in the argument of the square root in the last equation noting that $u'_x = 0$ we obtain:

$$1 - \frac{v^2 + u_y^2}{c^2} = 1 - \frac{v^2}{c^2} - \frac{u_y^2}{c^2}$$
$$= 1 - \frac{v^2}{c^2} - \frac{1}{c^2}\left(u'_y\sqrt{1-\frac{v^2}{c^2}}\right)^2$$
$$= \left(1-\frac{v^2}{c^2}\right) - \frac{1}{c^2}\left(u'_y\right)^2\left(1-\frac{v^2}{c^2}\right)$$
$$= \left(1-\frac{v^2}{c^2}\right)\left(1-\frac{\left(u'_y\right)^2}{c^2}\right)$$

On substituting from the last equation into the equation of m we obtain:

$$m = \frac{m_0}{\sqrt{1-\frac{v^2+u_y^2}{c^2}}} = \frac{m_0}{\sqrt{1-(v/c)^2}\sqrt{1-\left(u'_y/c\right)^2}} = \frac{m'}{\sqrt{1-(v/c)^2}}$$

Accordingly, we have:

$$p_y = mu_y = \frac{m'}{\sqrt{1-(v/c)^2}}u_y = \frac{m'}{\sqrt{1-(v/c)^2}}u'_y\sqrt{1-(v/c)^2} = m'u'_y = p'_y$$

The last equation establishes the claim that the y (and z) component of the momentum is the same for two inertial observers who are in a state of standard setting in a more formal way than the argument in the previous question and fills any gap in that argument. We note that the γ factor in the equations of the Lorentzian momentum (i.e. $p_y = mu_y$ and $p'_y = m'u'_y$) is absorbed in the mass symbols (i.e. m and m') since we are following here the old formalism about mass (or what we call the Lorentzian mass) and hence the rest mass is symbolized by m_0.[154]

Exercises
1. Make a simple argument to obtain the Lorentzian form of momentum with no need for formal derivation.
2. Make a simple argument (based on the validity of conservation of momentum in classical mechanics) to justify the third requirement of an acceptable Lorentzian definition of momentum.
3. Make a simple argument (based on the validity of Newton's second law in classical mechanics) to justify the fourth requirement of an acceptable Lorentzian definition of momentum.
4. What is the implication of the following statement: "The components of momentum in the perpendicular directions to the direction of relative motion between the frames of observation are value invariant across these frames and hence only the component in the direction of relative motion is variant".

5.2.2 Force

The definition of force in Lorentz mechanics is essentially the same as in classical mechanics, that is force is the time derivative of linear momentum where the momentum is as defined in Lorentz mechanics by Eq. 189.[155] Hence, the force in Lorentz mechanics is given by:

$$f = \frac{dp}{dt} = \frac{d}{dt}(\gamma m u) \tag{191}$$

where γ is a function of u and where f and u are assumed to be along the same direction (which is usually taken as the x direction) in this 1D form. In fact, the above equation is no more than the Lorentzian form of Newton's second law in its 1D version. Accordingly, we have:

$$f = \frac{d}{dt}(\gamma m u) \tag{192}$$

---

[154] We note that the formal argument in this question is based on the transformation of mass according to the old convention and hence this transformation should be established first by an independent argument to avoid circularity. Accordingly, the above formal argument should not be used simultaneously with the argument for the transformation of mass which was given in a previous exercise (see § 5.1.6) since that argument is based on the invariance of momentum in the y direction. We presented both arguments (in different contexts) for pedagogical purposes. Yes, it may be claimed that the previous argument is ultimately based on the invariance of length in the y direction and hence there is no circularity. However, even if we accept this claim we can argue that the formal argument is at least redundant unless we assume that its sole purpose is refuting the above challenge (through a formal demonstration) rather than establishing the invariance of momentum in the y direction which is supposedly established by the invariance of length in the previous argument.
[155] In fact, this is a result of choosing a suitable definition for momentum that satisfies the aforementioned requirements, as discussed in the previous subsection.

5.2.3 Energy

$$
\begin{aligned}
&= \frac{d}{dt}\left(\frac{mu}{\sqrt{1-\beta^2}}\right)\\
&= \frac{ma\sqrt{1-\beta^2} - mu\left(\frac{1}{2}\left[1-\beta^2\right]^{-1/2}\left[-2\beta\right]\left[a/c\right]\right)}{1-\beta^2}\\
&= \frac{ma\sqrt{1-\beta^2} + ma\left(1-\beta^2\right)^{-1/2}\beta^2}{1-\beta^2}\\
&= \frac{ma\left[\sqrt{1-\beta^2} + \left(1-\beta^2\right)^{-1/2}\beta^2\right]}{1-\beta^2}\\
&= \frac{ma\left(1-\beta^2+\beta^2\right)}{\left(1-\beta^2\right)^{3/2}}\\
&= \frac{ma}{\left(1-\beta^2\right)^{3/2}}\\
&= ma\gamma^3
\end{aligned}
$$

where $\beta = u/c$ and $a\,(=du/dt)$ is the acceleration in the same direction as f and u. The third line is based on the quotient rule of differentiation while the sixth line is based on multiplying the numerator and denominator by $\sqrt{1-\beta^2}$. We note that u, which is the velocity of the object, may be supposed to be constant since Lorentz mechanics is about inertial frames, but this is not the case because the observer is required to be inertial but not the object. So, u which is the velocity of the observed object is unlike v which is the relative velocity between the inertial frames. We also note that the above derivation is based on assuming that the mass of the object is constant, i.e. it does not increase or decrease due to an exchange with the environment of the massive object (refer to § 1.10 and § 4.1.7).[156]

Exercises

1. Write down the mathematical form of Newton's second law in Lorentz mechanics.

5.2.3 Energy

In the following we derive the Lorentzian form of kinetic energy using largely the classical formulations but with the use of the Lorentzian definitions of the quantities involved. From the classical relation between the work done by a force and the change in the kinetic energy as a result of this work we have:

$$W = \Delta E_k = E_{k2} - E_{k1} \tag{193}$$

where W is the work done, ΔE_k is the change in the kinetic energy and E_{k1} and E_{k2} are the initial and final kinetic energy. If we use the classical definition of work, with standard

---

[156] In fact, the validity of this derivation and its alike (where the rest mass is treated as constant) should depend on another assumption that is there is no conversion between kinetic and non-kinetic forms of energy. Hence, the above derivation and its alike may be seen as special cases or approximations.

5.2.3 Energy

definitions and common mathematical techniques, then we should have:

$$
\begin{aligned}
\Delta E_k &= W & (194) \\
&= \int_{x_1}^{x_2} f\, dx \\
&= \int_{x_1}^{x_2} \frac{d}{dt}(\gamma m u)\, dx \\
&= \int_{t_1}^{t_2} \frac{d}{dt}(\gamma m u) \frac{dx}{dt} dt \\
&= m \int_{t_1}^{t_2} \frac{d}{dt}(\gamma u)\, u\, dt \\
&= m \int_{u_1}^{u_2} u\, d(\gamma u)
\end{aligned}
$$

where f and u are the force and velocity in the x direction and where the limits of integration in the last step are justified by the fact that γ is a function of u. Hence, according to the last equation plus Eq. 16 we have:

$$\Delta E_k = m \int_{u_1}^{u_2} u\, d(\gamma u) = \left[mc^2 \gamma \right]_{u_1}^{u_2} = \frac{mc^2}{\sqrt{1-(u_2/c)^2}} - \frac{mc^2}{\sqrt{1-(u_1/c)^2}} \quad (195)$$

Now, if we have $u_1 = 0$, then $E_{k1} = 0$ and $\Delta E_k = E_{k2}$ which for notational simplicity can be written as $\Delta E_k = E_k$. Accordingly, from the last equation we obtain:

$$E_k = \Delta E_k = \frac{mc^2}{\sqrt{1-(u/c)^2}} - mc^2 = mc^2(\gamma - 1) \quad (196)$$

where for notational simplicity we replaced u_2 with u. The last equation, i.e. $E_k = mc^2(\gamma - 1)$, is the well known expression for the Lorentzian kinetic energy. Now, since $E_0 = mc^2$ (refer to § 4.3.2 and § 5.3.2), then from the last equation we immediately obtain the Lorentzian expression for the total energy, which is the sum of the rest and kinetic energy, that is:

$$E_t = E_0 + E_k = mc^2 + mc^2(\gamma - 1) = \gamma mc^2 \quad (197)$$

Solved Problems

1. Show that the Lorentzian formula for kinetic energy converges to the classical formula for kinetic energy at the low-speed limit, i.e. $u \ll c$.
   Answer: We have:

$$
\begin{aligned}
E_k &= mc^2(\gamma - 1) \\
&\simeq mc^2 \left(1 + \frac{1}{2}\beta^2 - 1 \right)
\end{aligned}
$$

$$= \frac{1}{2}mc^2\beta^2$$
$$= \frac{1}{2}mu^2$$

where the second line is justified by Eq. 13 because when $u \ll c$ the quartic and higher order terms in β will become vanishingly small and hence they can be ignored.

Exercises

1. Evaluate the following integral analytically using a method other than the integration by parts which is used in the text (see § 1.4):

$$\int_{x_1}^{x_2} \frac{d}{dt}(\gamma m u)\, dx$$

2. Discuss the generalization (or extension) of the concept of mass in Lorentz mechanics.

5.3 Physical Relations

5.3.1 Newton's Second Law

As seen earlier in § 4.3.1 and § 5.2.2, Newton's second law of motion will keep its basic form (i.e. $f = dp/dt$) in Lorentz mechanics with the use of the Lorentzian expression for momentum as given by Eq. 189. The justification is that the Lorentzian definition of momentum is chosen to satisfy this requirement. We note that in taking the time derivative of the momentum, γ and u are treated as variable while m is treated as constant if there is no exchange of mass between the object and its environment.

Solved Problems

1. Derive the Lorentzian form of the speed of an object of mass m and charge Q that conducts a uniform circular motion in a plane which is perpendicular to a uniform magnetic field of magnitude B.
   Answer: From the Lorentzian version of Newton's second law in its vector form we have:

$$\begin{aligned}\mathbf{f} &= \frac{d}{dt}(\gamma m \mathbf{u}) \\ &= m\frac{d}{dt}\left(\mathbf{u}\left[1 - \frac{\mathbf{u}\cdot\mathbf{u}}{c^2}\right]^{-1/2}\right) \\ &= m\left(\mathbf{a}\left[1 - \frac{\mathbf{u}\cdot\mathbf{u}}{c^2}\right]^{-1/2} + \mathbf{u}\left[1 - \frac{\mathbf{u}\cdot\mathbf{u}}{c^2}\right]^{-3/2}\frac{\mathbf{u}\cdot\mathbf{a}}{c^2}\right)\end{aligned}$$

where $\mathbf{f}$ is force, $\mathbf{u}$ is velocity and $\mathbf{a}$ is acceleration. Now, in circular motion the acceleration $\mathbf{a}$ is perpendicular to the velocity $\mathbf{u}$ and hence $\mathbf{u}\cdot\mathbf{a} = 0$. Accordingly, we have:

$$\mathbf{f} = m\mathbf{a}\left[1 - \frac{\mathbf{u}\cdot\mathbf{u}}{c^2}\right]^{-1/2} = \frac{m\mathbf{a}}{\sqrt{1 - u^2/c^2}}$$

where $u = |\mathbf{u}|$. On taking the modulus of both sides we obtain:

$$f = \frac{ma}{\sqrt{1 - u^2/c^2}}$$

Now, for magnetic force we have: $f = QuB$, and for uniform circular motion we have: $a = u^2/r$ where r is the radius of the circular path. On substituting these in the last equation we obtain:

$$QuB = \frac{m}{\sqrt{1 - u^2/c^2}} \frac{u^2}{r}$$

$$QrB\sqrt{1 - u^2/c^2} = mu$$

$$(QrB)^2 \left(1 - u^2/c^2\right) = m^2 u^2$$

$$(QrB)^2 - \frac{u^2 (QrB)^2}{c^2} = m^2 u^2$$

$$(QrB)^2 = u^2 \left(m^2 + \frac{(QrB)^2}{c^2}\right)$$

$$u^2 = (QrB)^2 \left(m^2 + \frac{(QrB)^2}{c^2}\right)^{-1}$$

$$u = QrB \left(m^2 + \frac{(QrB)^2}{c^2}\right)^{-1/2}$$

Exercises

1. Starting from the generic expression of Newton's second law in its vector form, develop and derive the Lorentzian version of Newton's second law in its 1D form.

5.3.2 Mass-Energy Relation

As discussed previously, in Lorentz mechanics the mass and energy are equivalent according to the Poincare relation:[157]

$$E_0 = mc^2 \tag{198}$$

where E_0 is the rest energy and m is the mass of the object. There are several methods, based on various arguments, for deriving the above formula. In the following bullet points we present some of these methods which can be found in the literature of Lorentz mechanics. Due to the importance of this relation, we present a rather significant number

---

[157] We note that the "mass-energy equivalence relation" may be used to refer to the relation $E = mc^2$ where E is the total energy and m is the mass according to the old convention (i.e. $m = \gamma m_0$). However, the essence of the mass-energy equivalence is contained in the relation $E_0 = mc^2$ (where m is the mass according to the modern convention) and this essence is passed to the relation $E = mc^2$ and hence we think this label is more appropriate to refer to the relation $E_0 = mc^2$. Nevertheless, this label can be used appropriately to refer to any one of these formulae.

5.3.2 Mass-Energy Relation

of arguments. We note that most of these arguments and derivations lack rigor and may be challenged. They are like demonstrations to the idea behind this formula rather than proofs. We also note that some of these derivations may be circular as they might depend on formulae or principles whose derivation depends on the derived formula or some of its assumptions and subsidiary formulae. Anyway, the validity of the Poincare mass-energy relation should be ultimately based on the experimental evidence and not necessarily on these arguments whose main objective is to rationalize this relation and embed it within a consistent theoretical structure.[158] As we will see, some of these methods are also based on purely classical arguments with no need for Lorentz mechanics and its formalism. A consequence of this is that the experimental evidence in support of the mass-energy equivalence relation is not necessarily a supporting evidence to Lorentz mechanics (see the end of this subsection and § 8.2).

• One method for deriving the Poincare mass-energy relation is based on the conservation of momentum. According to this method, we have a particle that moves uniformly at a constant speed u with respect to an inertial observer O. According to O, the momentum of the particle is initially p_1. The particle then emits two pulses of light simultaneously: one in the direction of its motion and one in the opposite direction. Each one of these pulses has energy $0.5E_p$ as seen in the frame of the particle, and hence the speed of the particle u does not change. Now, although the speed of the particle does not change, its momentum should change due to a change in its mass by Δm. Hence, following this emission the momentum of the particle according to O is p_2. So, according to the conservation of momentum in O frame we should have:

$$p_1 = p_2 + 0.5\frac{E_p}{c}\left(1 + \frac{u}{c}\right) - 0.5\frac{E_p}{c}\left(1 - \frac{u}{c}\right) \tag{199}$$

where the second and third terms on the right hand side represent the momenta of these light pulses[159] (see § 4.3.3 and § 5.3.3) with $\left(1 + \frac{u}{c}\right)$ and $\left(1 - \frac{u}{c}\right)$ being the blue and red shift factors of the light pulses in the approaching and receding directions relative to the observer (see the exercises). Hence, we have:

$$\Delta p = p_1 - p_2 = \frac{uE_p}{c^2} \tag{200}$$

Now, by using the classical formula for momentum noting that u did not change we obtain:

$$\Delta p = \Delta(mu) = u\Delta m = \frac{uE_p}{c^2} \tag{201}$$

that is:

$$E_p = c^2 \Delta m \tag{202}$$

---

[158] In other words, the main objective is to convert an empirical relation to an experimentally-endorsed theoretical relation.

[159] For massless objects, the momentum-energy relation becomes $E = pc$. This can also be obtained from Maxwell's equations.

5.3.2 Mass-Energy Relation

Now, if we consider the extreme case where the whole mass of the particle is converted to light then we have:
$$E_0 = mc^2 \qquad (203)$$
where the subscript 0 is used to indicate the rest (or mass) energy.

As we see, this derivation method lacks rigor. In the following points we put some question marks on the validity of this method and the argument on which it is based:

1. This derivation already assumes the equivalence between mass and energy (although its quantitative form may still be unknown) since it assumes there is a change in the momentum through the presumed change in mass following the emission of pulses and conversion of mass to light. So, the argument may be logical if its objective is to obtain the mathematical form of the equivalence relation following an assumption that such an equivalence does exist, but it is illogical if it is supposed to establish even the principle of equivalence.

2. It is assumed that the speed u does not change following the emission of light pulses as if the presumed change of momentum must totally occur through mass loss not through speed loss or through both.[160] In reality, the change of speed u in O frame may be justified by the difference of energy between the two pulses because the energy of these pulses is equal in the frame of the particle but not in the frame of O and that is why we have blue and red shift factors. The argument seems to suggest that the energy of the pulses is equal in the particle frame and hence the speed in O frame should not change, and this is not obvious.

3. We also note that this method uses the classical definition of momentum[161] and the classical factors for the blue and red shifts or an approximation of the Lorentzian factors (see the exercises of the present subsection) and hence if it is accepted it should be a questionable classical proof or at least a hybrid classical-Lorentzian proof (with possible compromising approximations and degrading assumptions) where in the latter case the validity of the whole argument may be challenged.

4. The argument requires the validity of the conservation of momentum principle which may be questioned if there is a presumed conversion of mass to energy. The classical form of this principle cannot be used because there is no mass-energy equivalence or conversion in classical mechanics (at least prior to the derivation of this relation) while the Lorentzian form of this principle may also lead to circularity in the argument.

5. The possibility of the extreme case scenario should be established first for a rigorous proof because even if we assume that part of the mass of a massive object can be converted to energy it is not obvious that the whole mass of the object can be converted since the physical mechanism of conversion may depend on the existence of a massive object for the process to occur. Hence, the final generalization is questionable and

---

[160] In fact, this should also require the invariance of the conservation of momentum because initially the momentum is supposed to be conserved in the frame of the particle and hence its conservation in O frame should depend on this invariance (see § 5.4).

[161] This may be justified by the condition $u \ll c$ but this will damage the generality and rigor of the argument anyway since the equivalence relation is not supposed to hold only in the classical limit.

5.3.2 Mass-Energy Relation

hence even the mathematical form of the equivalence relation will be questionable in this generalization.

6. This argument is totally based on a thought experiment and hence its validity is questionable. If there is any experimental support to the mass-energy equivalence, then this support is the evidence with no need for this argument although this argument may still be useful for demonstrative and pedagogical purposes. Yes, if the purpose of this argument is to rationalize the mass-energy equivalence relation and incorporate it within a proper theoretical framework (whether classical or Lorentzian) following its establishment by experimental evidence then the argument apparently failed to achieve this objective properly.

Although, a number of justifications may be proposed to fix some or all of these (as well as other) loopholes, the derivation still cannot be regarded as a real derivation that can rigorously prove this formula. So, any validation of this formula should rely either on experimental evidence or on a more rigorous method of derivation.

• Another method, which is similar to the previous one, is based on the conservation of energy. Let have a particle that moves at a constant speed u with respect to an inertial observer O. In its rest frame the particle has energy E_1 and in O frame it has energy $E_1 + E_{k1}$ where E_{k1} is its kinetic energy as seen by O. The particle then emits two pulses of light simultaneously: one in the direction of its motion and one in the opposite direction. Each one of these pulses has energy $0.5E_p$ in the particle frame; however in O frame the approaching/receding pulses will have energies $0.5E_p\gamma\left(1 \pm \frac{u}{c}\right)$ where the Lorentzian frequency shift factors are used (see § 5.1.7 and the exercises of this subsection). Although the speed of the particle did not change by its emission, its energy will change due to the energy loss in the emission and hence its energy will be E_2 in the particle frame and $E_2 + E_{k2}$ in O frame. Because of the conservation of energy, we should have:

$$E_1 = E_2 + E_p \tag{204}$$
$$E_1 + E_{k1} = E_2 + E_{k2} + \gamma E_p \tag{205}$$

where the first and second equations represent the conservation of energy in the particle frame and in O frame respectively. We note that the last term in the second equation is based on adding the energies of the two pulses, i.e. $0.5E_p\gamma\left(1 + \frac{u}{c}\right) + 0.5E_p\gamma\left(1 - \frac{u}{c}\right) = \gamma E_p$. On substituting from Eq. 204 into Eq. 205 we obtain:

$$E_2 + E_p + E_{k1} = E_2 + E_{k2} + \gamma E_p \tag{206}$$
$$E_p + E_{k1} = E_{k2} + \gamma E_p \tag{207}$$
$$E_{k1} - E_{k2} = \gamma E_p - E_p \tag{208}$$
$$\Delta E_k = (\gamma - 1) E_p \tag{209}$$
$$\frac{1}{2}\Delta m u^2 \simeq \frac{1}{2}\frac{u^2}{c^2} E_p \tag{210}$$
$$\Delta m \simeq \frac{1}{c^2} E_p \tag{211}$$
$$E_p \simeq c^2 \Delta m \tag{212}$$

5.3.2 Mass-Energy Relation

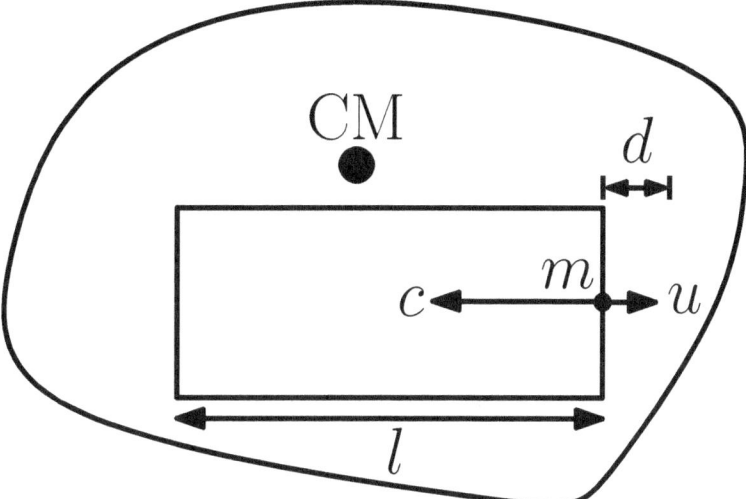

Figure 15: Schematic illustration of a cross section of a massive object that is used in the argument of Maxwell's equations to establish the mass-energy equivalence formula: $E_0 = mc^2$. We note that CM is the center of mass of the object.

where Δm is the fraction of the mass of the particle that is lost to the emission and where in Eq. 210 we used the approximation $(\gamma - 1) \simeq \frac{1}{2}\frac{u^2}{c^2}$ which is based on the binomial or Taylor expansion (see Eq. 85).[162] Now, if we consider the extreme case where the whole mass of the particle is converted to light then we have:

$$E_0 = mc^2 \tag{213}$$

where m is the mass of the particle. Again, we see that this method, like its predecessor, lacks rigor for similar reasons and hence it cannot establish this formula as a proof although it may be useful as demonstration to the idea behind the mass-energy equivalence.

• Another method is based on Maxwell's equations where light with energy E is shown to have a momentum $p = E/c$. Let assume that we have a massive object with an enclosure which can be assumed (with no loss of generality) to have a rectangular parallelepiped shape (refer to Figure 15). A light pulse is emitted from one face of the enclosure (say the face on the right hand side) and it is absorbed on the opposite face (the face on the left hand side) following a time interval Δt which is given by:

$$\Delta t = \frac{l}{c} \tag{214}$$

where l is the distance between the two faces.

By the conservation of momentum, the object should move to the right when the light pulse is emitted and hence we should have:

$$Mu = \frac{E}{c} \tag{215}$$

---

[162] We should also note that we used the classical definition of kinetic energy on the left hand side, and hence this method should be a hybrid classical-Lorentzian method.

5.3.2 Mass-Energy Relation

where M is the mass of the object excluding the light pulse, u is its speed due to the movement caused by the emission of the light pulse and E is the energy of the light pulse. Because of this movement, the object is displaced to the right a distance d and we have:

$$d = u\Delta t = \frac{El}{Mc^2} \tag{216}$$

where Eqs. 214 and 215 are used to substitute for Δt and u. Now, the object as a whole (i.e. including the light pulse) is an isolated system with no external forces acting on it and hence its center of mass does not move as a result of the emission. Accordingly, a mass m should have been transformed from the right to the left by the light pulse, that is:

$$Md - ml = 0 \tag{217}$$
$$Md = ml \tag{218}$$
$$M\frac{El}{Mc^2} = ml \tag{219}$$
$$E = mc^2 \tag{220}$$

which is the required result with $E \equiv E_0$. As we see, the method is based on a classical approach with no involvement of the formalism and principles of Lorentz mechanics.[163] We note that the equality $Md - ml = 0$ may be seen as an approximation, but this can be fixed by changing the definition of l in this derivation to mean the actual distance traversed by the light pulse instead of being the distance between the two faces.

• Another method is based on the time dilation triangle of Figure 16 (refer to § 9.7) where by Pythagoras theorem we have:

$$(c\Delta t')^2 = (c\Delta t)^2 + (u\Delta t')^2 \tag{221}$$

Now, if we multiply the two sides of this equation by $(mc)^2$ and use the time dilation formula with some algebraic manipulations, we obtain:

$$(mc^2\Delta t')^2 = (mc^2\Delta t)^2 + (mcu\Delta t')^2 \tag{222}$$
$$\left(mc^2\frac{\Delta t'}{\Delta t}\right)^2 = (mc^2)^2 + \left(mcu\frac{\Delta t'}{\Delta t}\right)^2 \tag{223}$$
$$(mc^2\gamma)^2 = (mc^2)^2 + (mcu\gamma)^2 \tag{224}$$
$$E^2 = m^2c^4 + p^2c^2 \tag{225}$$

where in Eq. 224 we used the time dilation formula (Eq. 177), while in Eq. 225 we used the Lorentzian formulae for total energy (Eq. 197) and momentum (Eq. 189). Now, when $u = 0$ the momentum will vanish (i.e. $p = 0$) and hence we obtain:

$$E_0 = mc^2 \tag{226}$$

---

[163] We are considering Maxwell's equations as classical. This may be challenged because although Maxwell's equations predate the emergence of Lorentz mechanics, they are Lorentzian in nature due to their form invariance under the Lorentz transformations and hence they may be seen as the first Lorentzian formulation of a physical theory.

5.3.2 Mass-Energy Relation

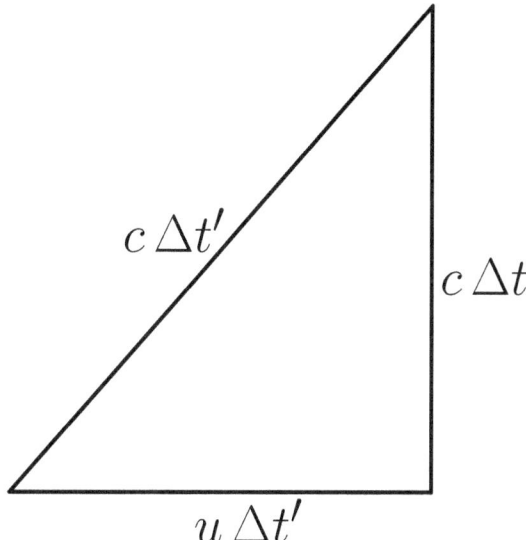

Figure 16: Time dilation triangle that is used in one of the methods for deriving the momentum-energy relation and hence the Poincare mass-energy relation.

which is the required result. The absurdity of this method is obvious since the definition of total energy is based on the equivalence between mass and energy although it may be sensible if the method is meant to find the exact mathematical expression of this equivalence following the assumption of this equivalence and assuming there is no dependence of the definition of total energy on the exact equivalence relation (i.e. no circularity). We note that Eq. 225 is the same as Eq. 135, which is the momentum-energy relation, and hence this derivation method may be considered as another way for deriving Eq. 135 if the employed formulae do not lead to circularity (see § 5.3.3). We also note that this method is based on the argument of the working principle of light clock which may be challenged (see § 9.7 and § 9.7.1) although the result can be obtained legitimately from the formalism. Moreover, it may lead to circularity in some cases as indicated earlier. Anyway, if this method of derivation is valid in principle (i.e. using the momentum-energy relation to derive the mass-energy relation), then all the methods that are used to derive the momentum-energy relation (see § 5.3.3) can be used to establish the mass-energy relation assuming that the latter relation is not used in the derivation of the momentum-energy relation (i.e. no circularity).

• Another method is based on the classical premise that the change in the kinetic energy ΔE_k is equal to the work done W, that is:

$$\Delta E_k = W = \int f \, dx \tag{227}$$

where f and x are the force and displacement in a one dimensional motion. Now, if we use the Lorentzian definition of force in the above integral (see Eq. 82), we obtain:

$$\Delta E_k = \int \frac{d}{dt}(\gamma m u) \, dx \tag{228}$$

5.3.2 Mass-Energy Relation

$$\begin{aligned}
&= m \int \frac{d}{dt}(\gamma u)\, dx \\
&= m \int \frac{d}{dt}(\gamma u)\, \frac{dx}{dt}\, dt \\
&= m \int u \frac{d}{dt}(\gamma u)\, dt \\
&= m \int u\, d(\gamma u) \\
&= mc^2 \gamma + C \\
&= \frac{mc^2}{\sqrt{1-(u/c)^2}} + C
\end{aligned}$$

where C is the constant of integration and where Eq. 16 is used in the analytical evaluation of the integral in the fifth line. If we now set $\Delta E_k = 0$ which corresponds to $u = 0$, then we have $C = -mc^2$, that is:

$$\Delta E_k = \frac{mc^2}{\sqrt{1-(u/c)^2}} - mc^2 = \left[\frac{1}{\sqrt{1-(u/c)^2}} - 1\right] mc^2 = \alpha mc^2 \qquad (229)$$

where $\alpha \geq 0$ is a dimensionless factor.

We note that this method of derivation may be sufficient to establish a proportionality relation between the kinetic energy of the object and its mass where c^2 plays a role in this proportionality but it is not sufficient to establish the equivalence relation between mass and energy (i.e. in its full sense that includes possible conversion) without further assumptions and efforts.[164] We also note that if this method of derivation is valid, then it should establish the equivalence between mass and kinetic energy and hence the generalization to any type of energy requires proof. Yes, the equivalence between different types of energy may be sufficient to establish this generalization with no extra effort. This sort of generalization may also be required for the other methods of derivation where mass is assumed to convert to a certain type of energy (e.g. radiation energy in the form of light pulse) and this generalization may also be provided by the equivalence between different types of energy. In fact, all the above methods (and indeed any other method) require some sort of generalization since all these methods are based on certain physical settings and processes and hence the equivalence between mass and energy which is supposedly

---

[164] It may be claimed by some that due to the conservation of total energy the kinetic energy in a given improper frame is equivalent to a certain amount of the rest mass in the proper frame, but this is based on a confusion between "conservation" and "invariance". In fact, the rest energy should be invariant but not the total (or kinetic) energy although the total energy is conserved in each frame. Moreover, it requires the equivalence between mass and kinetic energy in the definition of the total energy and this equivalence is supposed to be shown by this proof. We should also note that this method of derivation may be more appropriate to the old convention about mass where the kinetic energy is part of the total (or non-rest) mass.

5.3.2 Mass-Energy Relation

proved for that particular setting and process should be generalized to include all settings and processes.

We remark that the lack of rigor in most (if not all) of these arguments may allow a space for some modifications, e.g. adding a dimensionless numerical scale factor to the formula: $E_0 = mc^2$. Although the experimental evidence in support of the Poincare relation in the given form seems to be overwhelming, the difference between this formula and any modified version may be easy to explain if the deviation is rather small (i.e. the scale factor is close to unity like 0.99) since it may be possible to absorb the difference within other sources of experimental error and uncertainty.

We should also emphasize that if the mass-energy equivalence, as represented by the Poincare relation, is established (in principle at least) by a purely classical argument (as well as by a purely Lorentzian argument) then in this case it will not be proprietary to Lorentz mechanics and hence it will not be necessarily a supporting evidence to Lorentz mechanics (see § 8.2) because its validity within the Lorentzian framework may be based on its validity within the classical framework (since the latter is a special case of the former). Accordingly, the exact form of the Lorentzian framework may not be correct although it should still be a good approximation (as a minimum) to another framework. This should also be the case if the mass-energy equivalence relation can be established only by purely classical arguments (which seems unlikely although the Maxwell's equations argument is a good candidate if it is regarded as classical) where the minimum is based on the fact that classical mechanics is a special case of another theory (which could be, but not necessarily, Lorentz mechanics) since classical mechanics is not an exact theory according to the experimental evidence. In fact, if Lorentz mechanics failed to provide a valid theoretical justification for the mass-energy relation then this should indicate the necessity of amendment to Lorentz mechanics. So in brief, the mass-energy equivalence will necessarily be a supporting evidence to Lorentz mechanics only if it can be properly established by purely Lorentzian, but not classical, methods and arguments.

Finally, if the mass-energy equivalence relation cannot be established properly by any theoretical method (i.e. neither classical nor Lorentzian) then although this will not affect its empirical validity as a law of physics due to its overwhelming experimental evidence, it should indicate the necessity for a search for a different (existing or non-existing) theoretical structure or amendment of an existing theoretical structure (especially Lorentz mechanics) that can provide a theoretical justification for this law. Accordingly, the mass-energy equivalence will be evidence only to the theoretical structure that succeeds to provide a reliable theoretical justification.

Solved Problems
1. Discuss the credit for proposing and deriving the mass-energy formula: $E_0 = mc^2$.
   Answer: The roots of this formula go deep in the history of classical physics where some sort of equivalence between mass and energy has been contemplated in the old literature of classical physics but it did not go beyond a vague relation of qualitative nature to become a rigorous quantitative formula until the emergence of Maxwell's equations which played a driving role in the subsequent rise and development of Lorentz

mechanics. Based on the historical records that are available to the author, Poincare was the first to propose this formula in the form $m = E/c^2$ in 1900 in the context of electromagnetic radiation. There are also claims of an Italian scholar (known as Olinto De Pretto) who published a paper in 1903 in which he proposed this formula. Similar claims attribute the derivation of this formula to an Austrian scholar (known as Hasenohrl) in 1904 but with an added numerical factor. Einstein has also contributed to the development of this formula and generalizing its physical basis. There was also a later work around 1906 on this relation by Planck who apparently proposed a more general and sound argument than the one presented by Einstein. Other arguments and derivation methods have also been proposed by other scholars after Planck; these include Larmor, Pauli and Lenard. In fact, the research and debate about this formula are active even in these days as can be observed from the continuously emerging research papers on this subject. Accordingly, we believe that labeling this formula as Einstein formula (or Einstein mass-energy equation) is another example of the distortion of facts and falsification of history which is common in the literature of relativity. It is certain that Einstein was not the first to propose and derive this formula and he was not the last to work on the development, derivation and generalization of this formula. So, the formula should either be labeled by a more general term like "mass-energy formula" with acknowledgment to all those who contributed to this formula (or at least the main contributors), or it should be called Poincare relation because Poincare seems to be the originator of this formula in its modern form where all the subsequent work of generalization and providing physical and mathematical arguments for its derivation is based on the Poincare insight. For this reason, we label this formula in this book as the Poincare mass-energy equivalence relation.

Exercises

1. Using a Lorentzian argument, justify the factors $(1 + u/c)$ and $(1 - u/c)$, which appear in the first method for deriving the mass-energy relation, as the blue and red shift factors.
2. Using a classical argument, justify the factors $(1 \pm u/c)$, which appear in the first method for deriving the mass-energy relation, as the blue and red shift factors.
3. Show that $\gamma \left(1 \pm \frac{u}{c}\right)$, which appear in the second method for deriving the mass-energy relation, are the Lorentzian blue and red shift factors.

5.3.3 Momentum-Energy Relation

The relation between momentum and energy can be derived by a number of methods. In the following points we outline some of these methods.[165] We note that in the following we use E for the total energy, i.e. E_t. The reason for this replacement of symbols is to obtain the formula in its famous form, i.e.

$$E^2 = p^2 c^2 + m^2 c^4 \tag{230}$$

---

[165] Since momentum is vector while energy is scalar, the momentum-energy equation in the given form is a relation between the magnitude of momentum and energy.

5.3.3 Momentum-Energy Relation

- One method is based on the use of the total energy formula (Eq. 88) where we have:

$$\begin{align}
E^2 &= \gamma^2 m^2 c^4 \tag{231}\\
&= \gamma^2 m^2 c^4 - m^2 c^4 + m^2 c^4 \\
&= m^2 c^4 \left(\gamma^2 - 1\right) + m^2 c^4 \\
&= m^2 c^4 \left(\gamma^2 \frac{u^2}{c^2}\right) + m^2 c^4 \\
&= m^2 c^2 \gamma^2 u^2 + m^2 c^4 \\
&= p^2 c^2 + + m^2 c^4
\end{align}$$

where in the fourth line Eq. 6 is used while in the last line Eq. 189 is used.

- Another method is based on the use of the momentum formula (Eq. 189) where we have:

$$\begin{align}
p &= \gamma m u \tag{232}\\
p^2 &= \gamma^2 m^2 u^2 \tag{233}\\
p^2 c^2 &= \gamma^2 m^2 u^2 c^2 \tag{234}\\
p^2 c^2 &= \gamma^2 m^2 \beta^2 c^4 \tag{235}\\
p^2 c^2 &= \gamma^2 m^2 c^4 \left(1 - \frac{1}{\gamma^2}\right) \tag{236}\\
p^2 c^2 &= \gamma^2 m^2 c^4 - m^2 c^4 \tag{237}\\
p^2 c^2 &= E^2 - m^2 c^4 \tag{238}\\
E^2 &= p^2 c^2 + + m^2 c^4 \tag{239}
\end{align}$$

where in the fifth line Eq. 10 is used while in the seventh line Eq. 88 is used.

We note that the above momentum-energy relation may also be stated in different forms, e.g. the following widely used form:

$$E^2 = p^2 c^2 + E_0^2 \tag{240}$$

where Eq. 198 is used to express the rest energy. We also remark that a special case of the above momentum-energy equation is when the speed is zero and hence $p = 0$ where this equation reduces to the formula of the rest energy, i.e. Eq. 198. Another special case applies to massless objects where the momentum-energy equation reduces to the form: $E = pc$ because $m = 0$. Also, because the rest mass is an invariant quantity, the rest energy is also invariant. Therefore, the quantity $E^2 - p^2 c^2$ (which is equal to the rest energy squared) is also invariant.

Solved Problems

1. From inspecting the previous sections, try to find another method for deriving the momentum-energy relation.
   Answer: It is the method of time dilation triangle which we presented in § 5.3.2. The reader is also referred to the exercises of § 6.5.5 for another method of derivation which is based on the definition of the momentum 4-vector.

Exercises

1. Derive the momentum-energy relation of Lorentz mechanics by combining the momentum and total energy formulae.
2. What you observe from a close inspection of the above three methods (two given in the text and one in the previous exercise) for deriving the momentum-energy relation?

5.3.4 Work-Energy Relation

As seen earlier, in Lorentz mechanics (as in classical mechanics) the change in the kinetic energy of a massive object under the influence of an external force $\mathbf{f}(\mathbf{r})$ where the object is displaced by the force from position $\mathbf{r}_1$ to position $\mathbf{r}_2$ is equal to the work done on the object by that force and hence it is given by the following integral:

$$\Delta E_k = W = \int_{\mathbf{r}_1}^{\mathbf{r}_2} \mathbf{f} \cdot d\mathbf{r} \tag{241}$$

This may be derived in the following way although it may not be truly formal since in some of the previous derivations the work-energy relation in Lorentz mechanics is assumed where it is based on its classical form.

For simplicity, we assume a one dimensional motion along the x axis and hence we use f to symbolize the force in the x direction and x to symbolize the displacement along this direction. On combining the classical definition of work with the Lorentzian definition of force (see Eq. 82) we obtain:

$$\begin{aligned}
W &= \int_{x_1}^{x_2} f \, dx \tag{242} \\
&= \int_{x_1}^{x_2} m a \gamma^3 \, dx \\
&= m \int_{x_1}^{x_2} \gamma^3 \frac{du}{dt} \, dx \\
&= m \int_{t_1}^{t_2} \gamma^3 \frac{du}{dt} \frac{dx}{dt} dt \\
&= m \int_{u_1}^{u_2} \gamma^3 u \, du \\
&= mc^2 \int_{u_1}^{u_2} \gamma^3 \beta \, d\beta \\
&= mc^2 \int_{u_1}^{u_2} d\gamma \\
&= mc^2 \left[\gamma \right]_{u_1}^{u_2} \\
&= mc^2 \left[\frac{1}{\sqrt{1 - (u_2/c)^2}} - \frac{1}{\sqrt{1 - (u_1/c)^2}} \right]
\end{aligned}$$

where the differential form of Eq. 8 is used in the seventh line. Now, if we start from rest then $u_1 = 0$ and we have:

$$W = mc^2 \left[\frac{1}{\sqrt{1 - (u/c)^2}} - 1 \right] = mc^2 (\gamma - 1) \qquad (243)$$

where in the last equation we used u to symbolize u_2 for the purpose of notational simplicity. Now, according to Eq. 84 we have:

$$E_k = mc^2 (\gamma - 1) \qquad (244)$$

Moreover, we have $\Delta E_k = E_{k2} - E_{k1}$ where E_{k1} and E_{k2} are the initial and final kinetic energy with $E_{k1} = 0$ because $u_1 = 0$. So, we have:

$$\Delta E_k = E_{k2} = E_k \qquad (245)$$

where for notational simplicity we use E_k to symbolize E_{k2}. On comparing the last three equations we immediately obtain:

$$W = \Delta E_k \qquad (246)$$

which is the well known work-energy relation of classical mechanics that also applies in Lorentz mechanics. As indicated above, this may not be considered as a formal derivation. Instead, it may be seen as a consistency check (i.e. the various formulae of Lorentz mechanics are consistent with each other) and a demonstration of the correspondence between the classical formulation and the Lorentzian formulation.

5.4 Conservation Laws

Regarding the conservation laws of total energy and momentum, there are two main issues: the conservation of these quantities in a given frame of reference and the invariance of the conservation principles between different frames of reference under the Lorentz spacetime coordinate transformations. As we will see, the momentum and total energy are conserved in any given frame and their conservation principles are invariant, i.e. if the momentum or energy is conserved in one inertial frame, then it is conserved in all inertial frames. However, all these are dependent on the invariance of rest mass across all inertial frames and the conservation of what we call the "Lorentzian mass" (which is another term for the frame dependent mass as in the old formalism) in a particular inertial frame, as will be seen in the solved problems. Regarding the conservation of charge and its invariance across inertial frames, they are postulated (based on experimental evidence) as indicated earlier (see for example § 4.4, § 5 and § 5.1.8) and hence no formal derivation is required.

Solved Problems

1. Outline a plan for establishing the conservation principles of total energy and momentum.
   Answer: There are several methods for establishing the conservation principles of total

5.4 Conservation Laws

energy and momentum. In the following points we outline our preferred method for doing this:

- We start from two assumptions: the invariance of rest mass (which includes all forms of non-kinetic energy) across all inertial frames and the conservation of Lorentzian mass in a particular inertial frame. The first is justified by the fact that all forms of non-kinetic energy do not depend on the relative motion between the frames while the second can be justified empirically (e.g. by verifying this conservation in a laboratory frame).
- We show that the conservation of Lorentzian mass in a particular inertial frame is equivalent to the conservation of total energy in that frame.
- We show that the conservation of total energy in a particular inertial frame implies the conservation of momentum in that frame.
- We show that the conservation of total energy is invariant across all inertial frames, i.e. if the total energy is conserved in a particular frame it is conserved in all inertial frames.
- We show that the conservation of momentum is invariant across all inertial frames, i.e. if the momentum is conserved in a particular frame it is conserved in all inertial frames.

2. Define the "Lorentzian mass" of a massive object as γm where m is the frame independent mass (or rest or proper mass) of the object to show that if the Lorentzian mass of a given physical system is conserved in a given inertial frame then the total energy (i.e. sum of rest and kinetic energies) is also conserved in that frame. What you conclude?

Answer: Let have a system of n massive objects with proper masses $m_1, m_2, \cdots, m_n$. If the Lorentzian mass is conserved in a given inertial frame O then in this frame we should have:

$$\gamma_1 m_1 + \gamma_2 m_2 + \cdots + \gamma_n m_n = \sum_{i=1}^{n} \gamma_i m_i = \text{constant}$$

where the subscript of the γ factor indicates its dependency on the speed of the particular object, e.g. $\gamma_1 = (1 - u_1^2/c^2)^{-1/2}$ where u_1 is the speed of object 1 whose mass is m_1. On multiplying this equation by c^2 we obtain:

$$\gamma_1 m_1 c^2 + \gamma_2 m_2 c^2 + \cdots + \gamma_n m_n c^2 = \sum_{i=1}^{n} \gamma_i m_i c^2 = \text{constant}$$

that is:

$$E_1 + E_2 + \cdots + E_n = \sum_{i=1}^{n} E_i = \text{constant}$$

where E_i stands for the total energy of object i. So, we conclude that if the Lorentzian mass of the system is conserved then its total energy is also conserved.

Conclusion: the conservation of Lorentzian mass and the conservation of total energy (i.e. rest energy plus kinetic energy) are equivalent and they essentially represent the same fact with slightly different forms.

5.4 Conservation Laws

3. Show that the conservation of total energy in a particular inertial frame implies the conservation of momentum in that frame.
   Answer: The conservation of momentum applies in the absence of external forces (as may be inferred from Newton's second law). Now, in the absence of external forces no conversion between rest energy and kinetic energy can occur. Hence, the conservation of total energy, which is given by $E = mc^2\gamma$, means the rest mass (i.e. m) is constant and the speed (i.e. u of γ) is constant (and hence γ is also constant). Accordingly, the momentum, which is given by $p = \gamma m u$, should also be constant since it is the product of three constants, i.e. the momentum is conserved.[166]
   The claim that "in the absence of external forces no conversion between rest energy and kinetic energy can occur" may be challenged by the fact that such conversion can occur even in the absence of external forces, e.g. by decay of a radioactive particle. However, this challenge can be addressed by the fact that the kinetic energy in the claim is the external kinetic energy of the object (represented by the motion of its center of mass) as seen from the frame of observation and hence it does not include internal kinetic energy (in the frame of the object itself and its center of mass). In fact, this internal energy should be regarded as part of the rest energy (i.e. heat) since it is invariant across all inertial frames.

4. Show that if the total energy (or Lorentzian mass) is conserved in an inertial frame then it is conserved in all inertial frames (i.e. the conservation of total energy is Lorentz invariant). Refer to the exercises of § 1.9.2 for a similar Galilean case.
   Answer: The total energy consists of rest energy and kinetic energy. According to the second assumption (i.e. the conservation of Lorentzian mass in a particular inertial frame), the total energy is conserved in a given frame (say O). This implies that the change of kinetic energy and the change of rest mass in O are equal in magnitude and opposite in sense since the increase in one must equal the decrease in the other if the total energy should remain constant. Based on the objectivity of energy and the invariance of the rest energy (or rest mass), this should also apply to all other frames. Now, let assume that the kinetic energy in O is totally converted to rest energy, i.e. O becomes a rest frame. In any other frame O', the energy will then consist of two parts: rest energy which is equal to the total energy in O due to the invariance of rest energy, and kinetic energy that solely depends on the relative motion between O and O', and hence the kinetic energy in O' is constant since the relative motion between any two inertial frames is uniform. Accordingly, the total energy in O' is the sum of two parts: rest energy which is constant (according to the conservation of total energy in O), and kinetic energy which is also constant, and therefore the total energy in O' should also

---

[166] In fact, the above method demonstrates the conservation of the magnitude of the momentum and hence we may still need a proof for the "conservation of direction" since momentum is a vector. However, this is a marginal issue to our objective and hence we do not go through unnecessary discussions. Yes, if we assume that the conservation of momentum means the conservation of each single component of the momentum vector then the "conservation of direction" should follow from the conservation of momentum (when it is proved) with no need for an independent proof. However, this should depend on the nature of the method used to prove or demonstrate the conservation of momentum, and hence the above method may not be sufficient to establish the "conservation of direction".

5.4 Conservation Laws

be constant, i.e. conserved. Now, since O and O' are arbitrary, then we conclude that if the total energy is conserved in an inertial frame then it is conserved in all inertial frames, as required.

5. Using the Lorentz spacetime coordinate transformations and their subsidiaries and assuming the conservation of total energy (or Lorentzian mass) in any frame (as shown in the previous question), show that if the momentum is conserved in an inertial frame then it is conserved in all inertial frames (i.e. the conservation of momentum is Lorentz invariant) and comment on the result. Refer to the exercises of § 1.9.2 for the corresponding Galilean case.

Answer: Let have two inertial frames, O and O', in a state of standard setting with relative velocity v. We consider a general collision (elastic or inelastic) between two bodies of mass m_1 and m_2 which approach each other along the x axis with velocities u_1 and u_2 where they collide and form two bodies of mass m_3 and m_4 that depart with velocities u_3 and u_4.[167] The use of standard setting between the frames does not affect the generality of the argument, as explained previously. Similarly, considering only the x component of the momentum of the observed object under the state of standard setting between the two frames does not affect the generality of the argument in a sense that the y and z components of the momentum under this state are invariant as shown earlier (refer to § 5.2.1).[168] Accordingly, if the momentum is conserved in frame O then we should have:

$$\gamma_{u_1} m_1 u_1 + \gamma_{u_2} m_2 u_2 = \gamma_{u_3} m_3 u_3 + \gamma_{u_4} m_4 u_4 \qquad (247)$$

where the subscript on γ indicates its velocity dependence, i.e. it is a function of which velocity, e.g. $\gamma_{u_1} = \left[1 - (u_1/c)^2\right]^{-1/2}$.

Now, it can be shown that (see the exercises):

$$p = \gamma_u m u = \gamma_u m \frac{dx}{dt} = m \frac{dx}{d\tau}$$

where τ is the proper time parameter which is invariant under the Lorentz transformations. Accordingly, the above equation of the conservation of momentum in frame O

---

[167] The state of coalescence is a special case corresponding to $u_3 = u_4$ and hence it does not require a special attention.

[168] In other words, if we ignore the standard setting then the momentum components that are perpendicular to the direction of relative motion are invariant across frames and hence if they are conserved in one frame (as it is assumed) then they should be conserved in all frames (see § 1.8), so all we need to do is to show that the momentum component in the direction of relative motion is also conserved. We note that this issue has a link to the issue of the "conservation of direction" which was indicated earlier, so if the conservation of direction is established independently then the invariance (and hence conservation) of the perpendicular components plus the conservation of direction should lead to the conservation of the parallel component (and hence the conservation of momentum itself). On the other hand, the invariance (and hence conservation) of the perpendicular components plus the conservation of the parallel component (when it is proved) should lead to the conservation of direction. There are many details and possibilities about these issues that the reader may try to consider and work out.

5.4 Conservation Laws

(i.e. Eq. 247) can be written as:

$$m_1 \frac{dx_1}{d\tau_1} + m_2 \frac{dx_2}{d\tau_2} = m_3 \frac{dx_3}{d\tau_3} + m_4 \frac{dx_4}{d\tau_4} \qquad (248)$$

Now, m and τ are invariant under the Lorentz transformations while x transforms as:

$$x = \gamma_v (x' + vt') \qquad \rightarrow \qquad dx = \gamma_v (dx' + vdt')$$

Hence, Eq. 248 will become:

$$m_1 \gamma_v \left(\frac{dx'_1}{d\tau_1} + v \frac{dt'_1}{d\tau_1} \right) + m_2 \gamma_v \left(\frac{dx'_2}{d\tau_2} + v \frac{dt'_2}{d\tau_2} \right) = m_3 \gamma_v \left(\frac{dx'_3}{d\tau_3} + v \frac{dt'_3}{d\tau_3} \right) + m_4 \gamma_v \left(\frac{dx'_4}{d\tau_4} + v \frac{dt'_4}{d\tau_4} \right)$$

that is:

$$m_1 \left(\frac{dx'_1}{d\tau_1} + v \frac{dt'_1}{d\tau_1} \right) + m_2 \left(\frac{dx'_2}{d\tau_2} + v \frac{dt'_2}{d\tau_2} \right) = m_3 \left(\frac{dx'_3}{d\tau_3} + v \frac{dt'_3}{d\tau_3} \right) + m_4 \left(\frac{dx'_4}{d\tau_4} + v \frac{dt'_4}{d\tau_4} \right)$$

Now, it can be shown that (see the exercises):

$$\frac{dt'}{d\tau} = \frac{E'}{mc^2}$$

where E' is the total energy in frame O'. Hence, we have:

$$m_1 \left(\frac{dx'_1}{d\tau_1} + \frac{vE'_1}{m_1 c^2} \right) + m_2 \left(\frac{dx'_2}{d\tau_2} + \frac{vE'_2}{m_2 c^2} \right) = m_3 \left(\frac{dx'_3}{d\tau_3} + \frac{vE'_3}{m_3 c^2} \right) + m_4 \left(\frac{dx'_4}{d\tau_4} + \frac{vE'_4}{m_4 c^2} \right)$$

that is:

$$\left(m_1 \frac{dx'_1}{d\tau_1} + \frac{vE'_1}{c^2} \right) + \left(m_2 \frac{dx'_2}{d\tau_2} + \frac{vE'_2}{c^2} \right) = \left(m_3 \frac{dx'_3}{d\tau_3} + \frac{vE'_3}{c^2} \right) + \left(m_4 \frac{dx'_4}{d\tau_4} + \frac{vE'_4}{c^2} \right)$$

Now, since the total energy is conserved in all frames (as shown in the previous question based on the given assumptions) then we should have:

$$E'_1 + E'_2 = E'_3 + E'_4$$
$$\frac{vE'_1}{c^2} + \frac{vE'_2}{c^2} = \frac{vE'_3}{c^2} + \frac{vE'_4}{c^2}$$

and hence the terms involving v in the momentum equation of frame O' will cancel from both sides, that is:

$$m_1 \frac{dx'_1}{d\tau_1} + m_2 \frac{dx'_2}{d\tau_2} = m_3 \frac{dx'_3}{d\tau_3} + m_4 \frac{dx'_4}{d\tau_4}$$

As we see, the last equation is identical in form to Eq. 248, i.e. the momentum is conserved in frame O' (according to the last equation) if it is conserved in frame O

(according to Eq. 248).

Now, since O and O' are arbitrary inertial frames, then we conclude that if the Lorentzian momentum is conserved in an inertial frame then it is conserved in all inertial frames under the Lorentz transformations (i.e. the conservation of momentum is Lorentz invariant) where we used in this argument the result of the previous question, i.e. the conservation of the total energy (or Lorentzian mass) in all inertial frames. We note that the above proof can be easily extended to include any number of interacting objects where the above formulation can be generalized as $\sum_i \gamma_{u_i} m_i u_i = 0$ (with $i = 1, \cdots, n$ where n is the number of objects and u_i represents 1D velocity). It can also be extended to multiple directions where the components in each direction are considered separately by the above formulation (although this has no benefit since we have already shown that the components in the perpendicular directions to the relative motion are invariant and hence there is no advantage in orienting the coordinate system differently by violating the standard setting unless the setting of the problem requires this).

Comment: the results of this and previous questions indicate that the conservation of Lorentzian momentum is invariant between inertial frames under the Lorentz spacetime coordinate transformations if and only if[169] the Lorentzian total energy (or equivalently the Lorentzian mass) is conserved, and all these results are based on the given two assumptions, i.e. the invariance of rest mass across all inertial frames and the conservation of Lorentzian mass in a particular inertial frame.

6. Discuss the implication of the previous questions.

Answer: Starting from the assumptions that the rest mass is invariant across all inertial frames and the Lorentzian mass is conserved in a given inertial frame, we sequentially demonstrated that the total energy is conserved in the given frame, the momentum is conserved in the given frame, the conservation of total energy is Lorentz invariant across all inertial frames (and hence the total energy is conserved in all inertial frames), and the conservation of momentum is Lorentz invariant across all inertial frames (and hence the momentum is conserved in all inertial frames). This means that all these four important principles (i.e. the conservation of total energy, the conservation of momentum, the invariance of total energy conservation, and the invariance of momentum conservation) rely on the invariance of rest mass across all inertial frames and the conservation of the Lorentzian mass in a particular inertial frame.

Exercises

1. Prove the relation: $p = m \frac{dx}{d\tau}$ which is given in the solved problems.
2. Prove the relation: $\frac{dt'}{d\tau} = \frac{E'}{mc^2}$ which is given in the solved problems.
3. Show that the conservation of kinetic energy is Lorentz invariant across all inertial frames, i.e. if the kinetic energy is conserved in one inertial frame then it is conserved in all inertial frames (and if not it is not).[170]

---

[169] The "only if" part may be established by reversing the above argument (i.e. exchanging the role of total energy and momentum in the conservation assumption).

[170] This question is not about the conservation of kinetic energy, which is untrue in general, but about the invariance of this conservation, i.e. assuming it is conserved in a given inertial frame (as indicated

5.4 Conservation Laws

4. Discuss the issue that have been indicated in several places in the book, and in this chapter in particular, that is many of the arguments and proofs that are used to establish the theory of Lorentz mechanics are not sufficiently rigorous.[171]
5. Suggest a different plan for establishing the conservation principles of total energy and momentum and their invariance across all inertial frames.
6. Why we need to establish the conservation principles of total energy and momentum plus their invariance across all inertial frames (i.e. why it is not sufficient to establish these conservation principles without their invariance)?

---

by "if not it is not"). It should also be obvious that this question is not about the invariance of kinetic energy which is obviously untrue.

[171] In fact, this lack of rigor applies even to some arguments that we presented in this section. However, our purpose of these arguments is to outline the theoretical foundations of these principles rather than proving them rigorously, and hence they are more sensible and useful as pedagogical demonstrations than mathematical and theoretical substantiations. This fact applies to large parts of the theoretical structure of Lorentz mechanics and its alleged proofs, whether those presented in this book or in the wider literature of this subject, especially those related to the conservation laws which are distinguished by their problematic nature.

Chapter 6
Tensor Formulation of Lorentz Mechanics

This chapter is provided for the mathematically-oriented readers, especially those who are interested in tensor calculus. Apart from the difference in the mathematical formulation, there is nothing essentially new in this chapter. Hence, the readers who have no interest in tensor calculus can ignore this chapter with no loss of physical substance and indeed even essential mathematical substance as far as Lorentz mechanics is concerned. We should also note that because the present book is about Lorentz mechanics and not about tensor calculus, the presentation of tensors in this chapter is rather abrupt and may not be sufficiently rigorous in some parts to avoid lengthy explanations and justifications.

6.1 Preliminaries

Before we start our investigation of the tensor formulation of Lorentz mechanics we should draw the attention to the following points:

• As stated earlier, we generally follow the summation convention in the tensor formulation.

• The spacetime coordinates may be indexed by $0, 1, 2, 3$ or by $1, 2, 3, 4$ where 0 in the former and 4 in the latter represent the temporal coordinate (i.e. ct) while the other three indices represent the spatial coordinates. Although the latter indexing system may be more appropriate for tensor formulation since tensor indices usually start from 1, we follow the former indexing system in the present chapter due to its wide spread use and its more convenient notational utilization.

• The Latin indices represent spatial coordinates and variables and hence they range over $1, 2, 3$ while the Greek indices represent spacetime coordinates and variables and hence they range over $0, 1, 2, 3$ (or $1, 2, 3, 4$ if they are used).

• We define the proper time parameter τ as the absolute value (or length) of the spacetime interval σ divided by the characteristic speed of light c. Using the differential form (which

6.1 Preliminaries

is commonly employed in this chapter) we have:[172]

$$d\tau = \frac{\sqrt{|(d\sigma)^2|}}{c} \tag{249}$$

Since σ is Lorentz invariant (see § 3.9.4) and c is constant then τ is also Lorentz invariant. The last equation is dimensionally consistent since $\sqrt{|d\sigma^2|}$ has the physical dimension of length while c represents speed and hence τ has the physical dimension of time which is consistent with its "proper time" label. However, we may distinguish this rather more technical use of "proper time" from the more generic use of proper time as seen earlier.

• Following the commonly used terminology in the literature of tensor calculus formulation of Lorentz mechanics, we use 4-vector to refer to a vector (in its technical tensorial sense) in the 4D spacetime. Similarly, we use 3-vector to refer to a vector in the 3D ordinary "spatial" space. We may also use 3-tensor and 4-tensor to refer to tensors in 3D space and 4D spacetime. We also use terms like 3-operator, 4-operator, 3-*quantity* or 4-*quantity* where *quantity* is a given physical quantity (e.g. 4-force). In this context, the reader should be reminded that the dimension of the space of a tensor (which is represented by the range of its indices) is different from the rank of the tensor (which is represented by the number of its free indices). Technically, a 4-vector **A** transforms as:

$$A'^{\mu} = L^{\mu}_{\nu} A^{\nu} \tag{250}$$

where L^{μ}_{ν} is the mixed type Lorentz tensor (see § 6.4). This means that 4-vectors in a technical sense should satisfy certain invariance requirements under the Lorentz spacetime coordinate transformations.

• In general, the physical quantities that are based on the time derivative of spatial quantities, like velocity and acceleration, as defined previously are not tensors under the Lorentz transformations (i.e. they do not transform invariantly under these transformations). To make these quantities tensors, the 3D spatial quantities are replaced by 4D spacetime quantities and the t-derivatives are replaced by τ-derivatives. This leads to the replacement of the 3-vectors of the 3D "spatial" space with 4-vectors of the 4D spacetime where these 4-vectors transform invariantly between inertial frames by the Lorentz tensor (see § 6.4).

• In this chapter, we generally use lower case symbols to represent 3-vectors and upper case symbols to represent 4-vectors. For example, **u** is the spatial 3-velocity while **U** is the spacetime 4-velocity. However, there are some exceptions like **E** and **B** 3-vectors which kept their upper case symbols to keep inline with the common convention noting that they are not defined as 4-vectors. Another important exception is the coordinates of space

---

[172] The modulus sign is used to avoid excluding spacelike intervals (considering the reality of $d\tau$), although this may be redundant if we consider the physically possible causal relations and Lorentzian speed restrictions. Anyway, this is a minor issue and hence it can be ignored. We note that in the case of spacelike the interval may be interpreted as proper length (rather than proper time as in the case of timelike). Accordingly, proper time parameter is specifically associated with timelike intervals and hence the modulus sign becomes redundant.

and spacetime (or "position" vectors **r** and **x**) where both are symbolized with lower case (using different letters). We also changed the common symbol **A**, which stands for the electromagnetic potential 3-vector, to **a** to ensure the consistency of our labeling scheme and to avoid any confusion since **A** is used in this chapter to label the electromagnetic potential 4-vector. We should also remark that these symbols (i.e. **a** and **A** for electromagnetic vector potential) should not be confused with the symbols **a** and **A** which are used to label the acceleration 3- and 4-vectors.

• Most tensor formulations in this chapter are based on employing a rectangular Cartesian coordinate system for the underlying 3D space (i.e. the space that represents the spatial part of the 4D spacetime manifold) and hence ordinary derivatives are used. Accordingly, when a more general type of curvilinear coordinate system is employed, these derivatives should be replaced by tensor derivatives, i.e. ordinary total derivatives should be replaced by absolute (or intrinsic) derivatives while partial derivatives should be replaced by covariant derivatives.

• In general, the inner product of a 4-vector by itself (i.e. covariant by contravariant) produces an invariant scalar. Examples include the quadratic form of the spacetime interval (see § 3.9.4 and § 6.5.2), the modulus of velocity (see § 6.5.3) and the modulus of momentum (see § 6.5.5).

Solved Problems

1. Justify, briefly and descriptively, the labeling of τ as "proper time".
   Answer: The term "proper time" that is attached to this parameter originates from the fact that an observer in his rest frame will have constant spatial coordinates and hence $dx^1 = dx^2 = dx^3 = 0$. Therefore, he will have $d\tau = dt$, i.e. τ is equivalent to his time which is the proper time of his frame.

2. Make your answer to the previous question more formal by using the definition of the spacetime interval $d\sigma$ to find the mathematical definition of proper time $d\tau$.
   Answer: Using the definition of the spacetime interval in its infinitesimal form, we have:
   $$(d\sigma)^2 = (dx^0)^2 - (dx^1)^2 - (dx^2)^2 - (dx^3)^2$$
   Now, from the definition of the proper time interval $d\tau$ as the time between two events as measured in a frame in which the events occur at the same spatial location we should have: $dx^0 = cd\tau$ iff $dx^1 = dx^2 = dx^3 = 0$, that is:
   $$(d\sigma)^2 = (cd\tau)^2 - 0 - 0 - 0$$
   $$cd\tau = \sqrt{|(d\sigma)^2|}$$
   $$d\tau = \frac{\sqrt{|(d\sigma)^2|}}{c}$$
   which is the definition of proper time as given by Eq. 249.

Exercises

1. Highlight the role of tensor calculus in making the physical laws form invariant.

2. Briefly explain the summation convention that is commonly used in tensor calculus. What about the variance type (i.e. being covariant or contravariant) of the repeated index?
3. Show formally that τ really represents the proper time.

6.2 Useful Mathematics

The nabla 3-operator ∇, which is a 3D spatial vector operator, can be extended to the 4D spacetime manifold by adding the temporal coordinate where it is symbolized as $\Box$ and is given in its covariant form as:[173]

$$\partial_\mu = \frac{\partial}{\partial x^\mu} \equiv \left(\frac{\partial}{\partial x^0}, \nabla \right) \qquad (251)$$

and in its contravariant form as:[174]

$$\partial^\mu = \frac{\partial}{\partial x_\mu} \equiv \left(\frac{\partial}{\partial x_0}, -\nabla \right) \qquad (252)$$

where the nabla 3-operator ∇ represents the three spatial components (refer to § 6.5.1 about the definition of the spacetime 4-vector). The mostly used 3D coordinate system is the rectangular Cartesian for which the nabla 3-operator is:

$$\nabla = \left(\frac{\partial}{\partial x}, \frac{\partial}{\partial y}, \frac{\partial}{\partial z} \right) \qquad (253)$$

and hence the nabla 4-operator becomes $\left(\frac{1}{c}\frac{\partial}{\partial t}, -\frac{\partial}{\partial x}, -\frac{\partial}{\partial y}, -\frac{\partial}{\partial z} \right)$ in its contravariant form and $\left(\frac{1}{c}\frac{\partial}{\partial t}, \frac{\partial}{\partial x}, \frac{\partial}{\partial y}, \frac{\partial}{\partial z} \right)$ in its covariant form. So in brief, $\Box$ in the 4D spacetime corresponds to ∇ in the 3D "spatial" space.

Similarly, the Laplacian 3-operator ∇^2, which is a 3D spatial scalar operator, may be extended to the 4D spacetime manifold by adding the temporal coordinate where it is symbolized as $\Box^2$ and is given as an inner product of the nabla 4-operator by itself, i.e. covariant by contravariant. Accordingly, the Laplacian 4-operator (which is also known as d'Alembert operator or the d'Alembertian) is given by:[175]

$$\Box^2 = \frac{1}{c^2}\frac{\partial^2}{\partial t^2} - \nabla^2 \qquad (254)$$

where ∇^2 is the ordinary spatial Laplacian 3-operator as defined in the mathematical textbooks for various coordinate systems. The d'Alembertian my also be given by:

$$\Box^2 = -\frac{1}{c^2}\frac{\partial^2}{\partial t^2} + \nabla^2 \qquad (255)$$

---

[173] This labeling of covariant and contravariant may be reversed depending on the convention.
[174] We note that an upper index in the denominator of partial derivative is like a lower index in the numerator, and hence a lower index in the denominator (according to this notation) should be like an upper index in the numerator. So, the notation is consistent.
[175] The d'Alembertian is also commonly symbolized as $\Box$ rather than $\Box^2$.

where the sign is reversed which is equivalent to multiplication by -1 and hence it is essentially the same. The mostly used 3D coordinate system is the rectangular Cartesian for which the 3D spatial Laplacian is:

$$\nabla^2 = \frac{\partial^2}{\partial x^2} + \frac{\partial^2}{\partial y^2} + \frac{\partial^2}{\partial z^2} \tag{256}$$

and hence the d'Alembertian becomes:

$$\Box^2 = \frac{1}{c^2}\frac{\partial^2}{\partial t^2} - \frac{\partial^2}{\partial x^2} - \frac{\partial^2}{\partial y^2} - \frac{\partial^2}{\partial z^2} \tag{257}$$

In a more general tensor notation where the underlying spatial coordinate system could be curvilinear, the d'Alembertian is given by:

$$\Box^2 \equiv \partial^\mu \partial_\mu = g^{\mu\nu}\partial_\nu \partial_\mu = g_{\mu\nu}\partial^\nu \partial^\mu = \partial_\mu \partial^\mu \tag{258}$$

where $g^{\mu\nu}$ and $g_{\mu\nu}$ represent the contravariant and covariant metric tensor of the spacetime and $\mu, \nu = 0, 1, 2, 3$. In fact, this tensor form also applies to the above definition of the d'Alembertian (i.e. Eq. 257) where $g^{\mu\nu}$ and $g_{\mu\nu}$ are given by Eq. 261.

Finally, it can be shown (refer to the solved problems and exercises) that:

$$\frac{dt}{d\tau} = \gamma \qquad \text{and} \qquad \frac{d\tau}{dt} = \frac{1}{\gamma} \tag{259}$$

$$\frac{dx^i}{d\tau} = \gamma u^i \qquad (i = 1, 2, 3) \tag{260}$$

where the symbols are as defined previously. In fact, Eq. 259 is no more than a variant form of the time dilation formula.

Solved Problems

1. Show that the d'Alembertian operator is invariant under the Lorentz transformations but not invariant under the Galilean transformations.
   Answer: The reader is referred to § 12.3.3 where it is shown that the electromagnetic wave equation (which is no more than the d'Alembertian operator in one of its forms acting on a scalar field h) is Lorenz invariant but is not Galilean invariant. The invariance of the d'Alembertian operator under the Lorentz transformations may also be deduced from the fact that it is the inner product of the covariant and contravariant forms of the nabla 4-operator and hence it should be invariant.
2. Derive the following relations (which are just different forms of the time dilation formula) by using the definition of the proper time parameter:

$$\frac{dt}{d\tau} = \gamma \qquad \text{and} \qquad \frac{d\tau}{dt} = \frac{1}{\gamma}$$

   Answer: Let start from the following finite differential form of the definition of proper time parameter, that is:

$$\Delta\tau = \frac{\sqrt{|\Delta\sigma^2|}}{c}$$

6.3 Minkowski Metric Tensor

$$
\begin{aligned}
c\Delta\tau &= \sqrt{|\Delta\sigma^2|} \\
(c\Delta\tau)^2 &= \left|(\Delta x^0)^2 - (\Delta x^1)^2 - (\Delta x^2)^2 - (\Delta x^3)^2\right| \\
\left(\frac{c\Delta\tau}{\Delta x^0}\right)^2 &= \left|1 - \left(\frac{\Delta x^1}{\Delta x^0}\right)^2 - \left(\frac{\Delta x^2}{\Delta x^0}\right)^2 - \left(\frac{\Delta x^3}{\Delta x^0}\right)^2\right| \\
\frac{c\Delta\tau}{c\Delta t} &= \left[1 - \left(\frac{\Delta x^1}{c\Delta t}\right)^2 - \left(\frac{\Delta x^2}{c\Delta t}\right)^2 - \left(\frac{\Delta x^3}{c\Delta t}\right)^2\right]^{1/2} \\
\frac{\Delta t}{\Delta\tau} &= \left[1 - \left(\frac{\Delta x^1}{c\Delta t}\right)^2 - \left(\frac{\Delta x^2}{c\Delta t}\right)^2 - \left(\frac{\Delta x^3}{c\Delta t}\right)^2\right]^{-1/2}
\end{aligned}
$$

where the fifth line is justified by the restriction $u^2 < c^2$.[176] On taking the limit, we obtain the following infinitesimal differential form:

$$
\begin{aligned}
\frac{dt}{d\tau} &= \left[1 - \frac{1}{c^2}\left\{\left(\frac{dx^1}{dt}\right)^2 + \left(\frac{dx^2}{dt}\right)^2 + \left(\frac{dx^3}{dt}\right)^2\right\}\right]^{-1/2} \\
&= \left[1 - \frac{u^2}{c^2}\right]^{-1/2} \\
&= \gamma
\end{aligned}
$$

where u is the speed. The relation $\frac{d\tau}{dt} = \frac{1}{\gamma}$ can be similarly established from the fifth line of the above sequence of equations (or can be obtained from the relation $\frac{dt}{d\tau} = \gamma$ by following the rules of calculus).

Exercises
1. Briefly define the Laplacian 4-operator.
2. Using tensor notation and the generic definition of Laplacian in nD space (i.e. divergence of gradient), derive the mathematical expression for the d'Alembertian operator assuming an underlying rectangular Cartesian coordinate system for the spatial part.
3. Using tensor notation and assuming a rectangular Cartesian coordinate system for the spatial part, show that the d'Alembertian operator is invariant under the Lorentz transformations.
4. Show that:
$$\frac{dx^i}{d\tau} = \gamma u^i \qquad (i = 1, 2, 3)$$

6.3 Minkowski Metric Tensor

The metric tensor of Lorentz mechanics, which is commonly known as the Minkowski metric and may also be called the Lorentz metric, corresponding to a temporal coordinate

---

[176] This may justify the above-indicated restriction of proper time parameter to timelike intervals and hence the redundancy of the modulus sign.

6.3 Minkowski Metric Tensor

$x^0 = ct$ and spatial rectangular Cartesian coordinates x^1, x^2, x^3 is given by one of the following two forms:

$$[g_{\mu\nu}] = [g^{\mu\nu}] = \text{diag}\,[+1, -1, -1, -1] \tag{261}$$
$$[g_{\mu\nu}] = [g^{\mu\nu}] = \text{diag}\,[-1, +1, +1, +1] \tag{262}$$

where $\mu, \nu = 0, 1, 2, 3$. Hence the line element in the Minkowski spacetime, $d\sigma$, is given by one of the following two quadratic forms:

$$(d\sigma)^2 = +(dx^0)^2 - (dx^1)^2 - (dx^2)^2 - (dx^3)^2 \tag{263}$$
$$(d\sigma)^2 = -(dx^0)^2 + (dx^1)^2 + (dx^2)^2 + (dx^3)^2 \tag{264}$$

Because the temporal coordinate may also be indexed as x^4 while the spatial coordinates keep their former indices, the metric tensor can also take one of the following two forms:

$$[g_{\mu\nu}] = [g^{\mu\nu}] = \text{diag}\,[-1, -1, -1, +1] \tag{265}$$
$$[g_{\mu\nu}] = [g^{\mu\nu}] = \text{diag}\,[+1, +1, +1, -1] \tag{266}$$

where $\mu, \nu = 1, 2, 3, 4$ and hence the line element of the Minkowski spacetime will be given by one of the following two quadratic forms:

$$(d\sigma)^2 = -(dx^1)^2 - (dx^2)^2 - (dx^3)^2 + (dx^4)^2 \tag{267}$$
$$(d\sigma)^2 = +(dx^1)^2 + (dx^2)^2 + (dx^3)^2 - (dx^4)^2 \tag{268}$$

All these four forms are essentially equivalent since they differ only by a sign change or/and by a change in the order of coordinates with no essential change to the geometric properties of the spacetime that they represent.

Now, because the line element, in any one of the above four forms, is an invariant of the spacetime under the Lorentz transformations, we should have (see § 3.9.4):

$$(d\sigma)^2 = (d\sigma')^2 \tag{269}$$

where the unprimed and primed symbols correspond to two inertial observers, O and O', who are in a state of relative uniform motion. For example, using the form of Eq. 263 we have:

$$(dx^0)^2 - (dx^1)^2 - (dx^2)^2 - (dx^3)^2 = (dx'^0)^2 - (dx'^1)^2 - (dx'^2)^2 - (dx'^3)^2 \tag{270}$$

It should be obvious that the invariance of the line element in any one of the above forms implies the invariance of the other forms because all these forms differ from each other by a constant multiplicative factor of unity magnitude or by the indexing of the coordinates which is just reordering of the algebraic terms in the quadratic form. In § 3.9.4, we demonstrated the invariance of the line element (or what we called there the spacetime interval) under the Lorentz transformations, but not under the Galilean transformations, using a simplified set of symbols and hence there is no need to repeat here. As indicated

6.3 Minkowski Metric Tensor

earlier, the invariance of the line element of spacetime implies the invariance of the proper time, as represented by the parameter τ, because the definition of τ involves only the line element and the constant c and both are invariant.

We should also note that, depending on the purpose or convenience, the line element or spacetime interval may be expressed in a differential finite form as $\Delta\sigma$ or in a differential infinitesimal form as $d\sigma$ as well as in an ordinary finite form σ.[177] The meaning should be obvious in all cases and hence some of these labels may be used interchangeably for the purpose of simplicity and convenience with no fear of confusion. As noted earlier, the line element $d\sigma$ can be real or imaginary and this depends on the employed quadratic form of the line element as well as the relative size of the spatial and temporal coordinates. However, the modulus may be taken in some applications or contexts to obtain its real positive value (or "length") as we saw in the definition of τ. We should also note that the temporal variable, whether symbolized as x^0 or x^4, stands for ct and hence it has a physical dimension of length like the spatial variables x^1, x^2, x^3. Also, for the sake of convenience or mathematical uniformity the temporal coordinate of the spacetime may be defined by some as ict where i is the imaginary unit so that homogeneous coordinates can be used, and hence an "imaginary time" emerges which is no more than a notational handiness. Accordingly, all the diagonal elements of the metric tensor in any one of the above forms will have the same sign which is plus (+). Therefore, assuming an underlying spatial Cartesian coordinate system the line element $d\sigma$ will take one of the following two quadratic forms:[178]

$$(d\sigma)^2 = +(dx^0)^2 + (dx^1)^2 + (dx^2)^2 + (dx^3)^2 = dx^\mu dx^\mu \qquad (271)$$

$$(d\sigma)^2 = +(dx^1)^2 + (dx^2)^2 + (dx^3)^2 + (dx^4)^2 = dx^\nu dx^\nu \qquad (272)$$

where $\mu = 0, 1, 2, 3$ and $\nu = 1, 2, 3, 4$ and the summation convention applies. We note that the imaginary unit i can also be introduced onto the spatial coordinates (instead of the temporal coordinate) for the same purpose.[179]

Based on the above facts, the Minkowski spacetime and metric are characterized by the following properties:

1. The Minkowski spacetime, which is the "space" of Lorentz mechanics, is a 4D Euclidean[180] flat space since it can be represented by a diagonal metric tensor with all the diagonal elements being ± 1.
2. The "coordinate system" of the Minkowski spacetime (with the above given Cartesian system for the spatial part) is orthogonal because the metric is diagonal, and hence the coordinates of the spacetime are mutually orthogonal.

---

[177] We note that the value of the ordinary finite form σ is obtained from the differential infinitesimal form $d\sigma$ by integration.

[178] This will lead to sign reversal of the quadratic form from the commonly used definition.

[179] Hence, the sign of the quadratic form will follow the common definition that we use in this book.

[180] More rigorously, it should be described as pseudo (or quasi) Euclidean because technically it is not Euclidean. However, we do not go through these irrelevant details and mathematical technicalities which can be found in many textbooks (see for example Spain in the References) since they are of little value to the objectives of Lorentz mechanics as a physical theory. Hence, for simplicity and to avoid unnecessary distraction we describe it in this book as Euclidean.

3. The metric of the spacetime is like the metric of ordinary spaces (e.g. the 3D "spatial" space of classical mechanics) and hence it follows the same rules and possesses the same properties as those of ordinary spaces. However, unlike the metrics of ordinary spaces, the metric (or rather the quadratic form which is based on the metric and represents the square of the line element) of the Minkowski spacetime is not positive definite, and hence the quadratic form can be zero or negative as well as positive. Accordingly, the line element can be zero or imaginary as well as positive.[181]

4. The metric of the 4D Minkowski spacetime is a tensor (i.e. it possesses the property of being invariant under certain coordinate transformations) under the Lorentz transformations of spacetime coordinates but not under the Galilean transformations. The invariance property of the quadratic form (and hence the metric tensor on which the quadratic form is based) has been demonstrated in the main text of § 3.9.4, while the non-invariance of the quadratic form under the Galilean transformations of space coordinates and time has been demonstrated in the solved problems of that subsection.

5. As indicated before, the forms of the metric tensor that are given above belong to a spacetime with an underlying orthonormal Cartesian coordinate system for the 3D spatial part. Other types of metric tensor corresponding to different types of coordinate system for the 3D spatial part, such as cylindrical and spherical, can also be obtained (refer to the exercises of this section).

6. As we know, for a more general type of curvilinear coordinate system (coordinating a 3D spatial space instead of the rectangular Cartesian system) the metric of the space with spatial coordinates x^i is incorporated in the definition of the spatial interval (or infinitesimal line element) ds as given by the following quadratic form:

$$(ds)^2 = g_{ij} dx^i dx^j \tag{273}$$

where g_{ij} are the components of the covariant metric tensor of the space, x^i and x^j are general curvilinear spatial coordinates, and $i, j = 1, 2, 3$. This definition can be extended to the 4D spacetime of Lorentz mechanics by adding the temporal component to obtain the metric of the spacetime and its line element $d\sigma$, that is:

$$(d\sigma)^2 = g_{\mu\nu} dx^\mu dx^\nu \tag{274}$$

where $g_{\mu\nu}$ are the components of the covariant metric tensor of the spacetime, x^μ and x^ν are spacetime coordinates with an underlying general coordinate system for the spatial part, and $\mu, \nu = 0, 1, 2, 3$ (or $\mu, \nu = 1, 2, 3, 4$).

Solved Problems

1. What we mean by the term "quadratic form" of the spacetime?
   Answer: The quadratic form of the spacetime (or rather the quadratic form of the line element of the spacetime) is the mathematical expression that represents the square of the spacetime interval, i.e. $(d\sigma)^2$.

---

[181] For more details about these issues with compatible concepts, definitions, symbols and terminology to those used in the present book, the reader is referred to the other books of the author which are listed in the References in the back of the book.

6.3 Minkowski Metric Tensor

2. How can the form of the metric tensor be judged from the quadratic form of the line element of the Minkowski spacetime (assuming a rectangular Cartesian spatial system)?
   Answer: The numerical factors that multiply the spacetime coordinates in the expression of the quadratic form represent the diagonal elements of the metric tensor. For example, the following quadratic form of the line element:
   $$(d\sigma)^2 = +(dx^0)^2 - (dx^1)^2 - (dx^2)^2 - (dx^3)^2$$
   corresponds to the following form of the metric tensor:
   $$[g_{\mu\nu}] = \text{diag}\,[1, -1, -1, -1]$$
   where diag $[\cdots]$ stands for diagonal matrix with the given diagonal elements $[1, -1, -1, -1]$.
3. Why the Minkowski spacetime is regarded as flat space?
   Answer: The Minkowski spacetime is a flat space because it can be represented by a diagonal metric tensor with all the diagonal elements being ± 1.
4. Comment on the terminology that is used by some authors where they label $(ds)^2$ and $(d\sigma)^2$ as the metric of the 3D and 4D manifolds.
   Answer: It is rather common to label the square of the infinitesimal line element (or the quadratic form), i.e. $(ds)^2$ and $(d\sigma)^2$, as the metric of the space. The justification is obvious, that is the elements of the metric tensor are used as coefficients for the terms of the quadratic form, as can be seen from Eqs. 273 and 274, and hence the metric tensor is incorporated in the quadratic form. However, this terminology should not be understood to mean that $(ds)^2$ or $(d\sigma)^2$ is the "metric tensor" in its technical sense although it is still correct to call it "metric" in its generic sense. It is more appropriate to label $(ds)^2$ and $(d\sigma)^2$ as the quadratic forms of the line element of the particular space.
5. If the covariant metric tensor is given in matrix form by: $[g_{\mu\nu}] = \text{diag}\,[1, -1, -1, -1]$, show that $g_{\mu\nu} = g^{\mu\nu}$.
   Answer: According to the tensor identity: $g_{\mu\omega}g^{\omega\nu} = \delta_\mu{}^\nu$ (with $\delta_\mu{}^\nu$ being the Kronecker delta tensor in 4D), we should have (using matrix notation):[182]
   $$[g_{\mu\omega}][g^{\omega\nu}] = \text{diag}\,[1, -1, -1, -1][g^{\omega\nu}] = \text{diag}\,[1, 1, 1, 1]$$
   On solving this equation for $[g^{\omega\nu}]$, we obtain $[g^{\omega\nu}] = \text{diag}\,[1, -1, -1, -1]$, and hence $g_{\mu\nu} = g^{\mu\nu}$ (where the indexed g represent the components of these tensors).

Exercises

1. Why the Minkowski spacetime is the appropriate "space" for Lorentz mechanics?
2. What is the relation between the spacetime interval and the line element of the Minkowski spacetime?
3. What "homogeneous coordinate system" means? How can we homogenize the coordinates of the Minkowski spacetime?

---

[182] We can equally use the identity: $g^{\mu\omega}g_{\omega\nu} = \delta^\mu{}_\nu$.

4. What is the significance of the fact that a free particle in the Minkowski spacetime follows a geodesic trajectory?
5. Compare the "space trajectory" of a free particle in the 3D ordinary space with its "spacetime trajectory" in the 4D spacetime manifold.
6. Find the metric tensor and the quadratic form of the Minkowski spacetime with an underlying spatial cylindrical coordinate system.
7. Repeat the previous question with an underlying spatial spherical coordinate system.
8. Express the invariance of the quadratic form (and hence the invariance of the spacetime interval) using tensor notation.
9. Compare the Minkowski spacetime with an ordinary 4D Euclidean space.

6.4 Lorentz Transformations in Matrix and Tensor Form

The Lorentz spacetime coordinate transformations from an inertial frame O to another inertial frame O', where O and O' are in a state of standard setting, can be written in the following matrix form:

$$\mathbf{x}' \equiv \begin{bmatrix} x'^0 \\ x'^1 \\ x'^2 \\ x'^3 \end{bmatrix} = \begin{bmatrix} \gamma & -\beta\gamma & 0 & 0 \\ -\beta\gamma & \gamma & 0 & 0 \\ 0 & 0 & 1 & 0 \\ 0 & 0 & 0 & 1 \end{bmatrix} \begin{bmatrix} x^0 \\ x^1 \\ x^2 \\ x^3 \end{bmatrix} \equiv \mathbf{L}\mathbf{x} \qquad (275)$$

where $\mathbf{L}$ is the Lorentz matrix. Similarly, the opposite transformations from O' to O can be written in the following form:

$$\mathbf{x} \equiv \begin{bmatrix} x^0 \\ x^1 \\ x^2 \\ x^3 \end{bmatrix} = \begin{bmatrix} \gamma & \beta\gamma & 0 & 0 \\ \beta\gamma & \gamma & 0 & 0 \\ 0 & 0 & 1 & 0 \\ 0 & 0 & 0 & 1 \end{bmatrix} \begin{bmatrix} x'^0 \\ x'^1 \\ x'^2 \\ x'^3 \end{bmatrix} \equiv \mathbf{L}^{-1}\mathbf{x}' \qquad (276)$$

where $\mathbf{L}^{-1}$ is the inverse of the Lorentz matrix. As we see, Eqs. 275 and 276 can be obtained from each other by exchanging the primes and reversing the sign of β, which is the standard procedure for obtaining the opposite transformations. In tensor form the Lorentz matrix may be expressed as a mixed type tensor L^μ_ν and hence the transformation relations can be expressed as:

$$x'^\mu = L^\mu_\nu x^\nu \qquad (277)$$

As we see, both $\mathbf{L}$ and $\mathbf{L}^{-1}$ are symmetric matrices; moreover $\mathbf{L}\mathbf{L}^{-1} = \mathbf{L}^{-1}\mathbf{L} = \mathbf{I}$ where $\mathbf{I}$ is the 4×4 identity matrix (refer to the exercises).

We note that a contravariant 4-vector A^ν (i.e. a rank-1 tensor in the 4D Minkowski spacetime) is Lorentz transformed as:

$$A'^\mu = L^\mu_\nu A^\nu \qquad (278)$$

while a contravariant rank-2 4-tensor $B^{\mu\nu}$ is Lorentz transformed as:

$$B'^{\psi\omega} = L^\psi_\mu L^\omega_\nu B^{\mu\nu} \qquad (279)$$

6.4 Lorentz Transformations in Matrix and Tensor Form

More generally, a contravariant rank-n 4-tensor $C^{\mu_1,\ldots,\mu_n}$ is Lorentz transformed as:

$$C'^{\omega_1,\ldots,\omega_n} = L^{\omega_1}_{\mu_1} \cdots L^{\omega_n}_{\mu_n} C^{\mu_1,\ldots,\mu_n} \tag{280}$$

Similarly, a covariant 4-vector A_ν is Lorentz transformed as:

$$A'_\mu = M^\nu_\mu A_\nu \tag{281}$$

where M^ν_μ is the inverse Lorentz tensor (i.e. $(L^{-1})^\nu_\mu$) as given in matrix form by Eq. 276. In the same way, a covariant rank-2 4-tensor $B_{\mu\nu}$ is Lorentz transformed as:

$$B'_{\psi\omega} = M^\mu_\psi M^\nu_\omega B_{\mu\nu} \tag{282}$$

while a covariant rank-m 4-tensor $C_{\mu_1,\ldots,\mu_m}$ is Lorentz transformed as:

$$C'_{\psi_1,\ldots,\psi_m} = M^{\mu_1}_{\psi_1} \cdots M^{\mu_m}_{\psi_m} C_{\mu_1,\ldots,\mu_m} \tag{283}$$

Following the above pattern, a mixed type 4-tensor of m covariant indices and n contravariant indices $C^{\nu_1,\ldots,\nu_n}_{\mu_1,\ldots,\mu_m}$ is Lorentz transformed as:

$$C'^{\omega_1,\ldots,\omega_n}_{\psi_1,\ldots,\psi_m} = M^{\mu_1}_{\psi_1} \cdots M^{\mu_m}_{\psi_m} L^{\omega_1}_{\nu_1} \cdots L^{\omega_n}_{\nu_n} C^{\nu_1,\ldots,\nu_n}_{\mu_1,\ldots,\mu_m} \tag{284}$$

We should remark that the above rules are valid in general and not restricted to standard setting although the Lorentz matrix (and hence its inverse) will take a different more general form. In fact, if we note that the components of the Lorentz matrix in standard setting are given by:

$$L^\mu_\nu = \frac{\partial x'^\mu}{\partial x^\nu} \tag{285}$$

while the components of its inverse are given by:

$$M^\nu_\mu = \frac{\partial x^\nu}{\partial x'^\mu} \tag{286}$$

then we can infer that the components of Lorentz matrix in general (not necessarily in standard setting) should be so and hence we can apply the above transformation rules for transforming 4-tnsors in general where the Lorentz matrix and its inverse are obtained from Eqs. 285 and 286.

We finally remark that examples of 4-vectors include spacetime (position and displacement) 4-vector (see § 6.5.1), velocity 4-vector (see § 6.5.3) and momentum 4-vector (see § 6.5.5), while examples of rank-2 4-tensors include the Minkowski metric tensor (see § 6.3) and the field strength tensor (see § 6.5.7).

Solved Problems

1. How 4-vectors are Lorentz transformed between inertial frames in standard setting?

    Answer: A contravariant 4-vector is Lorentz transformed from frame O to frame O' as:

    $$A'^\mu = L^\mu_\nu A^\nu$$

where L^μ_ν is the Lorentz tensor as given in matrix form by Eq. 275. Similarly, a covariant 4-vector is Lorentz transformed from frame O to frame O' as:

$$A'_\mu = M^\nu_\mu A_\nu$$

where M^ν_μ is the inverse Lorentz tensor (i.e. $(L^{-1})^\nu_\mu$) as given in matrix form by Eq. 276.

2. How the covariant and contravariant forms of a 4-vector are obtained from each other?
   Answer: The covariant form of a contravariant 4-vector is obtained by lowering the index using the metric tensor, that is:

   $$A_\mu = g_{\mu\nu} A^\nu$$

   where $g_{\mu\nu}$ is the covariant metric tensor as given by Eq. 261. Similarly, the contravariant form of a covariant 4-vector is obtained by raising the index using the metric tensor, that is:

   $$A^\mu = g^{\mu\nu} A_\nu$$

   where $g^{\mu\nu}$ is the contravariant metric tensor as given by Eq. 261.

Exercises

1. Show that $\mathbf{L}$ and $\mathbf{L}^{-1}$ are inverses of each other by verifying the relations: $\mathbf{LL}^{-1} = \mathbf{L}^{-1}\mathbf{L} = \mathbf{I}$.
2. Express the relation $\mathbf{LL}^{-1} = \mathbf{L}^{-1}\mathbf{L} = \mathbf{I}$ in tensor form.

6.5 Vector, Tensor and Matrix Formulation

We summarize in this section the main elements of Lorentz mechanics in their tensor form. The formulation is largely based on a state of standard setting with a few exceptions. We also index the 3D entities with Latin letters ranging over $1, 2, 3$ (which correspond to the three dimensions of the ordinary space) and index the 4D entities with Greek letters ranging over $0, 1, 2, 3$ where 0 refers to the temporal variables. Also, for notational convenience and to show the correspondence between the Lorentzian and classical formulations in a more obvious fashion, we use the old convention about mass and hence we have proper mass symbolized with m_0 and improper mass (or Lorentzian mass) symbolized with m.

6.5.1 Spacetime Position and Displacement 4-Vector

The spacetime coordinates are represented by $\mathbf{x} = (ct, x, y, z)$ where $ct \equiv x^0$. Hence, we define the contravariant and covariant forms of the spacetime *position* 4-vector as:

$$[x^\mu] \equiv [x^0, x^1, x^2, x^3] = [ct, +x, +y, +z] \qquad (287)$$
$$[x_\mu] \equiv [x_0, x_1, x_2, x_3] = [ct, -x, -y, -z] \qquad (288)$$

Accordingly, the contravariant and covariant forms of the spacetime *displacement* 4-vector are given in their finite form by:

$$[\Delta x^\mu] \equiv [\Delta x^0, \Delta x^1, \Delta x^2, \Delta x^3] = [c\Delta t, +\Delta x, +\Delta y, +\Delta z] \qquad (289)$$

6.5.2 Quadratic Form of Spacetime Interval

$$[\Delta x_\mu] \equiv [\Delta x_0, \Delta x_1, \Delta x_2, \Delta x_3] = [c\Delta t, -\Delta x, -\Delta y, -\Delta z] \qquad (290)$$

This equally applies to the infinitesimal form of the displacement 4-vector (i.e. dx^μ and dx_μ) by replacing Δ with d.

6.5.2 Quadratic Form of Spacetime Interval

Based on what is given in § 6.5.1, the quadratic form of the spacetime interval in its finite differential form can be given by the following inner product between the covariant and contravariant forms of the displacement 4-vector:[183]

$$(\Delta \sigma)^2 = \Delta x_\mu \Delta x^\mu \qquad (291)$$

where summation over $\mu = 0, 1, 2, 3$ is implied. The invariance of the spacetime interval (and hence the length of the spacetime 4-vector) across various inertial frames can then be expressed in a compact form as:

$$(\Delta \sigma)^2 = \Delta x_\mu \Delta x^\mu = \Delta x'_\mu \Delta x'^\mu = (\Delta \sigma')^2 \qquad (292)$$

where the unprimed/primed symbols correspond to the unprimed/primed frames. As indicated earlier (see § 6.1), this invariance is justified by being an inner product of the covariant and contravariant forms of a 4-vector.

Exercises
1. Justify the invariance of the quadratic form of the spacetime interval by a simple reason.

6.5.3 Velocity

The velocity 3-vector **u** of an object is given by:

$$u^i = \frac{dx^i}{dt} \qquad (i = 1, 2, 3) \qquad (293)$$

where x^i are the spatial coordinates of the object and t is time. These space coordinates and time (and hence velocity) belong to a particular inertial frame and hence the velocity as defined above is not a tensor under the Lorentz transformations. To make it tensor, the 3D space coordinates are replaced by the 4D spacetime coordinates and the t-derivative is replaced by the τ-derivative. Accordingly, a velocity 4-vector **U** is defined as the τ-derivative of the spacetime coordinates and hence it is given by:

$$U^\mu = \frac{dx^\mu}{d\tau} \qquad (\mu = 0, 1, 2, 3) \qquad (294)$$

---

[183] In fact, this is no more than a simplification of the general relation of Eq. 274: $(d\sigma)^2 = g_{\mu\nu} dx^\mu dx^\nu$ where the infinitesimal form is replaced with a finite form and $g_{\mu\nu} \Delta x^\nu$ is replaced with Δx_μ where $g_{\mu\nu}$ is given by Eq. 261. Accordingly, the quadratic form may also be given as $(d\sigma)^2 = g^{\mu\nu} dx_\mu dx_\nu$ (or equivalently as $(\Delta \sigma)^2 = \Delta x^\nu \Delta x_\nu$) where $g^{\mu\nu}$ is also given by Eq. 261.

where x^μ are the spacetime coordinates of the observed object and τ is the proper time parameter as defined earlier (see § 6.1).

Exercises
1. What are the components of the contravariant velocity 4-vector?
2. What is the modulus of the velocity 4-vector? Comment on the result.

6.5.4 Acceleration

The 3D acceleration as defined previously (i.e. the second time derivative of position) is not a tensor under the Lorentz transformations. To make it tensor, the 3D spatial velocity **u** should be replaced by the 4D spacetime velocity **U** and the t-derivative should be replaced by the τ-derivative. Accordingly, an acceleration 4-vector **A** is defined as the τ-derivative of the velocity 4-vector and hence it is given by:

$$A^\mu = \frac{dU^\mu}{d\tau} = \frac{d^2 x^\mu}{d\tau^2} \qquad (\mu = 0, 1, 2, 3) \qquad (295)$$

Exercises
1. Define the acceleration 4-vector assuming a general curvilinear coordinate system.
2. Referring to the previous question, what is the equation of geodesics in the Minkowski spacetime?
3. Assuming that the Minkowski spacetime is coordinated by a rectangular Cartesian system, how the equation of geodesics in the Minkowski spacetime will simplify?

6.5.5 Momentum

Following the style of the previous formulations, a momentum 4-vector **P** is similarly defined as:[184]

$$P^\mu = m_0 \frac{dx^\mu}{d\tau} \qquad (\mu = 0, 1, 2, 3) \qquad (296)$$

which is identical in form to the classical definition.

Solved Problems
1. What are the temporal and spatial components of the momentum 4-vector? Comment on the results.
   Answer: The temporal component of the momentum 4-vector is:[185]

$$P^0 = m_0 \frac{dx^0}{d\tau} = m_0 c \frac{dt}{d\tau} = m_0 c \gamma = \frac{m_0 c^2 \gamma}{c} = \frac{E}{c}$$

---

[184] We follow in the label "momentum 4-vector" the common terminology although we think it is more appropriate to call it "energy-momentum 4-vector".

[185] We note that 0 in m_0 is a label and not an index and hence it has no connection with the index 0 in P^0 or x^0.

where E is the total energy (i.e. rest plus kinetic energy) and the identity $\frac{dt}{d\tau} = \gamma$ is used (see § 6.2). The three spatial components of the momentum 4-vector are:

$$P^i = m_0 \frac{dx^i}{d\tau} = m_0 \gamma u^i \qquad (i = 1, 2, 3)$$

where the identity $\frac{dx^i}{d\tau} = \gamma u^i$ is used in the last step (see § 6.2).[186] This momentum 4-vector is a generalization of the momentum 3-vector of classical and Lorentz mechanics where an energy temporal component is added.

Comment: we note that the energy of a physical system represents its temporal dynamic attribute while the momentum represents its spatial dynamic attribute. We also note that based on this combination of the energy and momentum into a 4-vector, an energy-momentum conservation law will emerge.[187] This can be easily formulated by noting that for a system of n massive objects we should have:[188]

$$\left(\sum_{j=1}^{n} P_j^\mu \right)_{in} = \left(\sum_{j=1}^{n} P_j^\mu \right)_{fi}$$

where $\mu = 0$ represents the conservation of energy while $\mu = 1, 2, 3$ represents the conservation of momentum and in and fi represent initial and final states of the system.

Exercises

1. Show that the rest energy is proportional to the length of the momentum 4-vector. What you conclude?
2. Using the result of the previous exercise and the definition of the momentum 4-vector, as well as other previously-given standard definitions, derive the momentum-energy relation.
3. How is the momentum 4-vector transformed between inertial frames in standard setting?

6.5.6 Force and Newton's Second Law

To have a tensorial force, the momentum 3-vector is replaced by the momentum 4-vector and the t-derivative is replaced by the τ-derivative in the previously given definition of force. Accordingly, the force 4-vector is defined as the τ-derivative of the momentum 4-vector and hence it is given by:

$$F^\mu = \frac{dP^\mu}{d\tau} \qquad (297)$$

---

[186] These components are identical to the components of the 3-momentum as defined before.

[187] The conservation of energy and momentum may be justified by the homogeneity (or symmetry) of time and space respectively, and hence the conservation of energy-momentum in Lorentz mechanics will be naturally explained as demonstration of the homogeneity of spacetime which is the space of Lorentz mechanics.

[188] We note that for simplicity we assume that the number of objects is conserved (i.e. the number in the initial state equals the number in the final state) which is not necessary for the general case. However, this generalization can be easily done.

6.5.6 Force and Newton's Second Law

$$\begin{aligned}
&= \frac{d}{d\tau}\left(m_0 \frac{dx^\mu}{d\tau}\right) \\
&= \frac{d}{dt}\left(m_0 \frac{dx^\mu}{dt}\frac{dt}{d\tau}\right)\frac{dt}{d\tau} \\
&= \frac{d}{dt}\left(m_0 \frac{dx^\mu}{dt}\gamma\right)\gamma \\
&= \gamma\frac{d}{dt}\left(\gamma m_0 \frac{dx^\mu}{dt}\right) \\
&= \gamma\frac{d}{dt}\left(m \frac{dx^\mu}{dt}\right)
\end{aligned}$$

which is the tensor form of Newton's second law.

Solved Problems

1. What are the temporal and spatial components of the force 4-vector?
   Answer: The temporal component of the force 4-vector is given by:[189]

$$F^0 = \gamma\frac{d}{dt}\left(m\frac{dx^0}{dt}\right) = c\gamma\frac{d}{dt}\left(m\frac{dt}{dt}\right) = c\gamma\frac{dm}{dt} = \frac{\gamma}{c}\frac{dE}{dt} = \frac{1}{c}\frac{dE}{dt}\frac{dt}{d\tau} = \frac{1}{c}\frac{dE}{d\tau}$$

while the three spatial components of the force 4-vector are given by:

$$F^i = \gamma\frac{d}{dt}\left(m\frac{dx^i}{dt}\right) \qquad (i = 1, 2, 3)$$

which, apart from the γ factor, takes the same form as in classical mechanics (i.e. force equals time derivative of momentum), considering the difference in the meaning of m.[190]

2. How is the force 4-vector transformed between inertial frames?
   Answer: Using Eq. 250, with L^μ_ν being given by Eq. 275, we obtain:

$$\begin{bmatrix} F'^0 \\ F'^1 \\ F'^2 \\ F'^3 \end{bmatrix} = \begin{bmatrix} \gamma & -\beta\gamma & 0 & 0 \\ -\beta\gamma & \gamma & 0 & 0 \\ 0 & 0 & 1 & 0 \\ 0 & 0 & 0 & 1 \end{bmatrix}\begin{bmatrix} F^0 \\ F^1 \\ F^2 \\ F^3 \end{bmatrix} = \begin{bmatrix} \gamma F^0 - \beta\gamma F^1 \\ -\beta\gamma F^0 + \gamma F^1 \\ F^2 \\ F^3 \end{bmatrix}$$

We note that, like momentum, the spatial components in the perpendicular directions to the direction of motion (i.e. y and z according to the standard setting) are Lorentz invariant across inertial frames.

Exercises

---

[189] In fact, this can be obtained directly from $P^0 = E/c$ and hence $F^0 = \frac{dP^0}{d\tau} = \frac{1}{c}\frac{dE}{d\tau}$.
[190] Also, apart from the γ factor it is the same as the non-tensorial form which was given earlier (also, see the exercises).

1. Using the given tensor formulations, show that the world line of a free massive particle is a straight line in the 4D Minkowski spacetime.
2. Give a tensor form of Newton's second law assuming a curvilinear coordinate system.
3. Show that a uniformly accelerated one dimensional motion is equivalent to a constant Lorentz force.
4. Find the relation between the spatial components of the tensorial force (i.e. $F^i = \frac{dP^i}{d\tau}$) and the non-tensorial force as defined earlier (i.e. $f^i = \frac{dp^i}{dt}$).

6.5.7 Electromagnetism and Maxwell's Equations

Referring to § 5.1.8, the equations of electric charge density ρ and electric current density **j** which for frame O are given by:[191]

$$\rho = \gamma \rho_0 \qquad \text{and} \qquad \mathbf{j} = \gamma \rho_0 \mathbf{u} \qquad (298)$$

have the same velocity dependence as the equations of mass and momentum in mechanics where $\rho_0, \rho, \mathbf{j}$ correspond to $m_0, m, \mathbf{p}$.[192] Accordingly, a current density 4-vector **J** (that corresponds to the momentum 4-vector **P**) is defined as:

$$J^\mu \equiv \left(J^0, J^1, J^2, J^3\right) = (c\rho, j_x, j_y, j_z) \qquad (299)$$

Now, the Poisson equation of electrostatics in 3D space is given by:

$$\nabla^2 \phi = -\frac{\rho}{\varepsilon_0} \qquad (300)$$

where ϕ is the electromagnetic scalar potential, ρ is the charge density and ε_0 is the permittivity of free space. This equation is not invariant under the Lorentz transformations and hence to be compatible with the formalism of Lorentz mechanics it requires amendment by replacing the Laplacian with the d'Alembertian, that is:[193]

$$\Box^2 \phi = -\frac{\rho}{\varepsilon_0} \qquad \text{or equivalently} \qquad \Box^2 \left(\frac{\phi}{c}\right) = -\mu_0 c \rho \qquad (301)$$

Now, since $c\rho$ is the first component of the current density 4-vector **J**, then ϕ/c should also be the first component of a 4-vector potential $\mathbf{A} \equiv (\phi/c, a_x, a_y, a_z)$ such that:[194]

$$\Box^2 \mathbf{A} = -\mu_0 \mathbf{J} \qquad \text{or} \qquad \Box^2 (\phi/c, a_x, a_y, a_z) = -\mu_0 (c\rho, j_x, j_y, j_z) \qquad (302)$$

---

[191] We should remind the reader that all the Lorentzian formulations (and electromagnetism in particular) belong to free space, and hence the electromagnetic phenomena in material media are not considered although some of free space formulations may apply.

[192] The reader is referred to § 4.4 about the conservation of charge and its invariant tensorial nature.

[193] We have: $c = (\mu_0 \varepsilon_0)^{-1/2}$ and hence the equivalence is justified. We also note that $\Box^2$ should be given by Eq. 255 for the sign in Eq. 300 to be consistent (this is to follow the commonly used form of this equation).

[194] To keep consistency of our notation (i.e. lower case symbols for 3-vectors and upper case symbols for 4-vectors) we are forced to abandon the common symbol of **A** 3-vector which is upper case and replace it with our lower case symbol **a**. Both these symbols should not be confused with the acceleration 3- and 4-vectors which are also symbolized with **a** and **A**. It is noteworthy that the 4-vector potential **A** is a combination of the two ordinary potentials, i.e. the scalar potential ϕ and the 3-vector potential **a**.

6.5.7 Electromagnetism and Maxwell's Equations

In brief, to make the laws of electromagnetism invariant under the Lorentz transformations we need to define a potential 4-vector that corresponds to the current density 4-vector such that:

$$\Box^2 A^\nu = -\mu_0 J^\nu \tag{303}$$

where $\nu = 0, 1, 2, 3$. Accordingly, the Poisson electrostatic equation as given by Eq. 300 is a special case of the zeroth component of Eq. 303 that corresponds to a frame in which the scalar potential ϕ is time independent or linear in time and hence the temporal term in Eq. 301 vanishes identically.

Now, the electric and magnetic field 3-vectors, **E** and **B**, are defined as:

$$\mathbf{E} = -\frac{\partial \mathbf{a}}{\partial t} - \nabla \phi \tag{304}$$

$$\mathbf{B} = \nabla \times \mathbf{a} \tag{305}$$

The six components of these 3-vectors are the independent non-zero elements of an anti-symmetric rank-2 tensor **S** in the 4D spacetime. This 4-tensor, which is called the field strength tensor, is given in matrix notation by:

$$\mathbf{S} = [S^{\mu\nu}] = \begin{bmatrix} 0 & E_x/c & E_y/c & E_z/c \\ -E_x/c & 0 & B_z & -B_y \\ -E_y/c & -B_z & 0 & B_x \\ -E_z/c & B_y & -B_x & 0 \end{bmatrix} \tag{306}$$

Now, since $S^{\mu\nu}$ is a rank-2 tensor it transforms (from unprimed frame to primed frame) as:

$$S'^{\psi\omega} = L^\psi_\mu L^\omega_\nu S^{\mu\nu} \tag{307}$$

which in matrix form becomes:

$$[S'^{\psi\omega}] = \begin{bmatrix} 0 & \frac{E_x}{c} & \gamma\left(\frac{E_y}{c} - \beta B_z\right) & \gamma\left(\frac{E_z}{c} + \beta B_y\right) \\ -\frac{E_x}{c} & 0 & \gamma\left(B_z - \frac{\beta E_y}{c}\right) & -\gamma\left(B_y + \frac{\beta E_z}{c}\right) \\ -\gamma\left(\frac{E_y}{c} - \beta B_z\right) & -\gamma\left(B_z - \frac{\beta E_y}{c}\right) & 0 & B_x \\ -\gamma\left(\frac{E_z}{c} + \beta B_y\right) & \gamma\left(B_y + \frac{\beta E_z}{c}\right) & -B_x & 0 \end{bmatrix} \tag{308}$$

This matrix form can be easily obtained by conducting the matrix multiplication as represented in tensor notation by Eq. 307 where **L** and **S** are given by Eqs. 275 and 306 (refer to the exercises).

On comparing Eq. 306 with Eq. 308, we see that the components of the electric and magnetic field 3-vectors are transformed between frames O and O' by the following transformation relations (see the solved problems):

$$E'_x = E_x \qquad\qquad cB'_x = cB_x \tag{309}$$
$$E'_y = \gamma\left(E_y - c\beta B_z\right) \qquad\qquad cB'_y = \gamma\left(cB_y + \beta E_z\right) \tag{310}$$
$$E'_z = \gamma\left(E_z + c\beta B_y\right) \qquad\qquad cB'_z = \gamma\left(cB_z - \beta E_y\right) \tag{311}$$

6.5.7 Electromagnetism and Maxwell's Equations

On making the following substitutions: $E/c \to B$ and $B \to -E/c$ in the field strength tensor **S** we obtain a new rank-2 4-tensor **T**,[195] that is:

$$\mathbf{T} = [T^{\mu\nu}] = \begin{bmatrix} 0 & B_x & B_y & B_z \\ -B_x & 0 & -E_z/c & E_y/c \\ -B_y & E_z/c & 0 & -E_x/c \\ -B_z & -E_y/c & E_x/c & 0 \end{bmatrix} \quad (312)$$

The transformation relations of Eqs. 309-311 remain valid following this transformation. This should be obvious when we note that the magnetic equations on the right of Eqs. 309-311 can be obtained from the corresponding electric equations on the left of Eqs. 309-311 and vice versa by the substitutions: $E \to cB$ and $-cB \to E$ which is identical to the above substitution (refer to the solved problems).

Referring to the rank-2 4-tensors $S^{\mu\nu}$ and $T^{\mu\nu}$ of Eqs. 306 and 312, Maxwell's equations can be given in tensor form by the following equations:

$$\frac{\partial S^{\mu\nu}}{\partial x^\nu} = \mu_0 J^\mu \quad (313)$$

$$\frac{\partial T^{\mu\nu}}{\partial x^\nu} = 0 \quad (314)$$

where $\mu, \nu = 0, 1, 2, 3$ and summation over ν is implied. The four equations of Maxwell correspond to the following four cases related to the differential equations of $S^{\mu\nu}$ and $T^{\mu\nu}$ (as given by Eqs. 313 and 314) and the value of the index μ, that is:

A. The first equation with $\mu = 0$: this corresponds to the first row of **S** matrix as given by Eq. 306 and we have:

$$\begin{aligned} \frac{\partial S^{0\nu}}{\partial x^\nu} &= \frac{\partial S^{00}}{\partial x^0} + \frac{\partial S^{01}}{\partial x^1} + \frac{\partial S^{02}}{\partial x^2} + \frac{\partial S^{03}}{\partial x^3} \\ &= 0 + \frac{1}{c}\left(\frac{\partial E_x}{\partial x^1} + \frac{\partial E_y}{\partial x^2} + \frac{\partial E_z}{\partial x^3}\right) \\ &= \frac{1}{c}\left(\frac{\partial E_x}{\partial x} + \frac{\partial E_y}{\partial y} + \frac{\partial E_z}{\partial z}\right) \\ &= \frac{1}{c}(\nabla \cdot \mathbf{E}) \\ &= \mu_0 J^0 \\ &= \mu_0 c \rho \end{aligned} \quad (315)$$

that is:

$$\nabla \cdot \mathbf{E} = \frac{\rho}{\varepsilon_0} \quad (316)$$

which is the first of Maxwell's equations (see Eq. 344 in § 12.1), i.e. Gauss law of electrostatics.

---

[195] This may be called dual field strength tensor.

6.5.7 Electromagnetism and Maxwell's Equations

B. The first equation with $\mu = 1, 2, 3$: this corresponds to the last three rows of **S** matrix as given by Eq. 306.

For $\mu = 1$ (i.e. the second row of **S**) we have:

$$\begin{aligned}\frac{\partial S^{1\nu}}{\partial x^\nu} &= \frac{\partial S^{10}}{\partial x^0} + \frac{\partial S^{11}}{\partial x^1} + \frac{\partial S^{12}}{\partial x^2} + \frac{\partial S^{13}}{\partial x^3} \\ &= -\frac{1}{c^2}\frac{\partial E_x}{\partial t} + 0 + \frac{\partial B_z}{\partial x^2} - \frac{\partial B_y}{\partial x^3} \\ &= -\frac{1}{c^2}\frac{\partial E_x}{\partial t} + \left(\frac{\partial B_z}{\partial y} - \frac{\partial B_y}{\partial z}\right) \\ &= \left[-\frac{1}{c^2}\frac{\partial \mathbf{E}}{\partial t} + (\nabla \times \mathbf{B})\right]_x \\ &= \mu_0 J^1 = \mu_0 j_x \end{aligned} \qquad (317)$$

For $\mu = 2$ (i.e. the third row of **S**) we have:

$$\begin{aligned}\frac{\partial S^{2\nu}}{\partial x^\nu} &= \frac{\partial S^{20}}{\partial x^0} + \frac{\partial S^{21}}{\partial x^1} + \frac{\partial S^{22}}{\partial x^2} + \frac{\partial S^{23}}{\partial x^3} \\ &= -\frac{1}{c^2}\frac{\partial E_y}{\partial t} - \frac{\partial B_z}{\partial x^1} + 0 + \frac{\partial B_x}{\partial x^3} \\ &= -\frac{1}{c^2}\frac{\partial E_y}{\partial t} + \left(\frac{\partial B_x}{\partial z} - \frac{\partial B_z}{\partial x}\right) \\ &= \left[-\frac{1}{c^2}\frac{\partial \mathbf{E}}{\partial t} + (\nabla \times \mathbf{B})\right]_y \\ &= \mu_0 J^2 = \mu_0 j_y \end{aligned} \qquad (318)$$

For $\mu = 3$ (i.e. the fourth row of **S**) we have:

$$\begin{aligned}\frac{\partial S^{3\nu}}{\partial x^\nu} &= \frac{\partial S^{30}}{\partial x^0} + \frac{\partial S^{31}}{\partial x^1} + \frac{\partial S^{32}}{\partial x^2} + \frac{\partial S^{33}}{\partial x^3} \\ &= -\frac{1}{c^2}\frac{\partial E_z}{\partial t} + \frac{\partial B_y}{\partial x^1} - \frac{\partial B_x}{\partial x^2} + 0 \\ &= -\frac{1}{c^2}\frac{\partial E_z}{\partial t} + \left(\frac{\partial B_y}{\partial x} - \frac{\partial B_x}{\partial y}\right) \\ &= \left[-\frac{1}{c^2}\frac{\partial \mathbf{E}}{\partial t} + (\nabla \times \mathbf{B})\right]_z \\ &= \mu_0 J^3 = \mu_0 j_z \end{aligned} \qquad (319)$$

On combining these three components we obtain (noting that **j** is the current density 3-vector):

$$\nabla \times \mathbf{B} = \frac{1}{c^2}\frac{\partial \mathbf{E}}{\partial t} + \mu_0 \mathbf{j} \qquad (320)$$

which is the fourth of Maxwell's equations (see Eq. 347 in § 12.1), i.e. Ampere's circuital law with the Maxwell correction term.

6.5.7 Electromagnetism and Maxwell's Equations

C. The second equation with $\mu = 0$: this corresponds to the first row of **T** matrix as given by Eq. 312 and we have:

$$\begin{aligned}
\frac{\partial T^{0\nu}}{\partial x^{\nu}} &= \frac{\partial T^{00}}{\partial x^0} + \frac{\partial T^{01}}{\partial x^1} + \frac{\partial T^{02}}{\partial x^2} + \frac{\partial T^{03}}{\partial x^3} \\
&= 0 + \frac{\partial B_x}{\partial x^1} + \frac{\partial B_y}{\partial x^2} + \frac{\partial B_z}{\partial x^3} \\
&= \frac{\partial B_x}{\partial x} + \frac{\partial B_y}{\partial y} + \frac{\partial B_z}{\partial z} \\
&= \nabla \cdot \mathbf{B} = 0
\end{aligned} \qquad (321)$$

which is the second of Maxwell's equations (see Eq. 345 in § 12.1), i.e. Gauss law of magnetism.

D. The second equation with $\mu = 1, 2, 3$: this corresponds to the last three rows of **T** matrix as given by Eq. 312.

For $\mu = 1$ (i.e. the second row of **T**) we have:

$$\begin{aligned}
\frac{\partial T^{1\nu}}{\partial x^{\nu}} &= \frac{\partial T^{10}}{\partial x^0} + \frac{\partial T^{11}}{\partial x^1} + \frac{\partial T^{12}}{\partial x^2} + \frac{\partial T^{13}}{\partial x^3} \\
&= -\frac{\partial B_x}{\partial x^0} + 0 - \frac{1}{c}\frac{\partial E_z}{\partial x^2} + \frac{1}{c}\frac{\partial E_y}{\partial x^3} \\
&= -\frac{1}{c}\left[\frac{\partial B_x}{\partial t} + \frac{\partial E_z}{\partial y} - \frac{\partial E_y}{\partial z}\right] \\
&= -\frac{1}{c}\left[\frac{\partial B_x}{\partial t} + \left(\frac{\partial E_z}{\partial y} - \frac{\partial E_y}{\partial z}\right)\right] \\
&= -\frac{1}{c}\left[\frac{\partial \mathbf{B}}{\partial t} + \nabla \times \mathbf{E}\right]_x \\
&= 0
\end{aligned} \qquad (322)$$

i.e.

$$\left[\frac{\partial \mathbf{B}}{\partial t} + \nabla \times \mathbf{E}\right]_x = 0 \qquad (323)$$

For $\mu = 2$ (i.e. the third row of **T**) we have:

$$\begin{aligned}
\frac{\partial T^{2\nu}}{\partial x^{\nu}} &= \frac{\partial T^{20}}{\partial x^0} + \frac{\partial T^{21}}{\partial x^1} + \frac{\partial T^{22}}{\partial x^2} + \frac{\partial T^{23}}{\partial x^3} \\
&= -\frac{\partial B_y}{\partial x^0} + \frac{1}{c}\frac{\partial E_z}{\partial x^1} + 0 - \frac{1}{c}\frac{\partial E_x}{\partial x^3} \\
&= -\frac{1}{c}\left[\frac{\partial B_y}{\partial t} - \frac{\partial E_z}{\partial x} + \frac{\partial E_x}{\partial z}\right] \\
&= -\frac{1}{c}\left[\frac{\partial B_y}{\partial t} + \left(\frac{\partial E_x}{\partial z} - \frac{\partial E_z}{\partial x}\right)\right] \\
&= -\frac{1}{c}\left[\frac{\partial \mathbf{B}}{\partial t} + \nabla \times \mathbf{E}\right]_y
\end{aligned} \qquad (324)$$

6.5.7 Electromagnetism and Maxwell's Equations

$$= 0$$

i.e.

$$\left[\frac{\partial \mathbf{B}}{\partial t} + \nabla \times \mathbf{E}\right]_y = 0 \tag{325}$$

For $\mu = 3$ (i.e. the fourth row of $\mathbf{T}$) we have:

$$\begin{aligned}
\frac{\partial T^{3\nu}}{\partial x^\nu} &= \frac{\partial T^{30}}{\partial x^0} + \frac{\partial T^{31}}{\partial x^1} + \frac{\partial T^{32}}{\partial x^2} + \frac{\partial T^{33}}{\partial x^3} \\
&= -\frac{\partial B_z}{\partial x^0} - \frac{1}{c}\frac{\partial E_y}{\partial x^1} + \frac{1}{c}\frac{\partial E_x}{\partial x^2} + 0 \\
&= -\frac{1}{c}\left[\frac{\partial B_z}{\partial t} + \frac{\partial E_y}{\partial x} - \frac{\partial E_x}{\partial y}\right] \\
&= -\frac{1}{c}\left[\frac{\partial B_z}{\partial t} + \left(\frac{\partial E_y}{\partial x} - \frac{\partial E_x}{\partial y}\right)\right] \\
&= -\frac{1}{c}\left[\frac{\partial \mathbf{B}}{\partial t} + \nabla \times \mathbf{E}\right]_z \\
&= 0
\end{aligned} \tag{326}$$

i.e.

$$\left[\frac{\partial \mathbf{B}}{\partial t} + \nabla \times \mathbf{E}\right]_z = 0 \tag{327}$$

On combining these three components we obtain:

$$\frac{\partial \mathbf{B}}{\partial t} + \nabla \times \mathbf{E} = \mathbf{0} \tag{328}$$

which is the third of Maxwell's equations (see Eq. 346 in § 12.1), i.e. Faraday's law of induction.

Solved Problems

1. Show that the components of the electric and magnetic field 3-vectors are transformed between frames O and O' by Eqs. 309-311.

   Answer: From Eq. 306, the field strength tensor is given in frame O' by:

   $$\mathbf{S}' = \begin{bmatrix} 0 & E'_x/c & E'_y/c & E'_z/c \\ -E'_x/c & 0 & B'_z & -B'_y \\ -E'_y/c & -B'_z & 0 & B'_x \\ -E'_z/c & B'_y & -B'_x & 0 \end{bmatrix}$$

   and from Eq. 308 the tensor is also given in frame O' (in terms of the components of frame O) by:

   $$\mathbf{S}' = \begin{bmatrix} 0 & \frac{E_x}{c} & \gamma\left(\frac{E_y}{c} - \beta B_z\right) & \gamma\left(\frac{E_z}{c} + \beta B_y\right) \\ -\frac{E_x}{c} & 0 & \gamma\left(B_z - \frac{\beta E_y}{c}\right) & -\gamma\left(B_y + \frac{\beta E_z}{c}\right) \\ -\gamma\left(\frac{E_y}{c} - \beta B_z\right) & -\gamma\left(B_z - \frac{\beta E_y}{c}\right) & 0 & B_x \\ -\gamma\left(\frac{E_z}{c} + \beta B_y\right) & \gamma\left(B_y + \frac{\beta E_z}{c}\right) & -B_x & 0 \end{bmatrix}$$

6.5.7 Electromagnetism and Maxwell's Equations

On comparing the corresponding elements of these matrices, the transformation relations of Eqs. 309-311 can be obtained right away. It should be noted that since the field strength tensor is an anti-symmetric rank-2 tensor in a 4D space, it has only 6 non-vanishing independent elements, which correspond to the 6 components of the electric and magnetic field 3-vectors, and hence we need only to compare 6 corresponding elements, i.e. either those in the upper right triangle of the matrix or those in the lower left triangle.

2. Show that the magnetic equations on the right of Eqs. 309-311 can be obtained from the corresponding electric equations on the left of Eqs. 309-311 and vice versa by the substitutions: $E \to cB$ and $-cB \to E$.
   Answer: These substitutions are equivalent to the substitutions: $E \to cB$ and $B \to -E/c$.
   Obtaining the magnetic equations from their electric counterparts:

$$E'_x = E_x \qquad \to (cB)'_x = (cB)_x \to \qquad cB'_x = cB_x$$
$$E'_y = \gamma \left(E_y - c\beta B_z \right) \to (cB)'_y = \gamma \left[(cB)_y - c\beta \left(-E/c \right)_z \right] \to cB'_y = \gamma \left(cB_y + \beta E_z \right)$$
$$E'_z = \gamma \left(E_z + c\beta B_y \right) \to (cB)'_z = \gamma \left[(cB)_z + c\beta \left(-E/c \right)_y \right] \to cB'_z = \gamma \left(cB_z - \beta E_y \right)$$

Obtaining the electric equations from their magnetic counterparts (by reversing the direction):

$$E'_x = E_x \qquad \leftarrow c\left(-E/c\right)'_x = c\left(-E/c\right)_x \leftarrow \qquad cB'_x = cB_x$$
$$E'_y = \gamma \left(E_y - c\beta B_z \right) \leftarrow c\left(-E/c\right)'_y = \gamma \left[c\left(-E/c\right)_y + \beta (cB)_z \right] \leftarrow cB'_y = \gamma \left(cB_y + \beta E_z \right)$$
$$E'_z = \gamma \left(E_z + c\beta B_y \right) \leftarrow c\left(-E/c\right)'_z = \gamma \left[c\left(-E/c\right)_z - \beta (cB)_y \right] \leftarrow cB'_z = \gamma \left(cB_z - \beta E_y \right)$$

Exercises

1. Show that:
$$S'^{\psi\omega} = L^\psi_\mu L^\omega_\nu S^{\mu\nu}$$
where the tensors involved are given in the text.

Chapter 7
Consequences and Predictions of Lorentz Mechanics

The main consequences and predictions of Lorentz mechanics are outlined in the following sections. Since all these consequences and predictions have been discussed previously within other contexts, the discussion in this chapter is rather brief to avoid repetition.

7.1 Merging of Space and Time into Spacetime

As seen earlier, space and time according to Lorentz mechanics are mixed in a spacetime manifold (or Minkowski space). This is vividly seen in particular in the main Lorentz spacetime coordinate transformations (Eqs. 90-93 and 95-98) where the transformations of both x-x' and t-t' involve both spatial and temporal factors. This is also reflected in the fact that the invariant line element of the underlying space of Lorentz mechanics is not the line element representing space or time but the line element representing spacetime (see § 3.9.4). Another example of this mix is seen in § 6 where 3-objects are replaced by 4-objects (e.g. 4-velocity instead of 3-velocity and d'Alembertian instead of Laplacian) to represent the actual dimensionality of the underlying manifold and its invariance properties.

In reference to the Lorentz transformations of Eqs. 90-93 and 95-98, we note that there is no involvement of a temporal factor in the transformations of the y-y' and z-z' coordinates because, due to the choice of standard setting, the motion is essentially one dimensional since it is only along the x-x' orientation. We also note that there is no involvement of a γ factor in the transformations of the y-y' and z-z' coordinates because the distortion of the spatial coordinates of spacetime is restricted to the direction of motion and hence a γ factor is needed to account for the spatial distortion in that direction. Accordingly, in a more general 3D motion all the spatial coordinates will have a time factor and a γ factor in their transformations so that the time factor compensates for the change of space coordinates due to the relative motion while the γ factor accounts for the spatial distortion.

Exercises
1. List some examples from Lorentz mechanics that demonstrate the merge of space and time into spacetime.

7.2 Length Contraction

Length contraction, which may also be called FitzGerald-Lorentz contraction, is either a consequence of the Lorentz space transformations or a cause of these transformations.

As discussed earlier, the proper length is the length of the object as measured in its rest frame, while the improper length is the length of the object as measured by an observer who is in a state of relative motion with respect to the rest frame of the object. According to Lorentz mechanics, an object is shortened when observed from an inertial frame which is in a state of relative motion with respect to the object. Hence, the length of an object as measured in any frame other than its rest frame is shorter than the length measured in its rest frame. This means that the length of the object contracts by the motion. As seen earlier, length contraction is expressed by the following relation:

$$L = \frac{L_0}{\gamma} \tag{329}$$

where L_0 is the proper length and L is the improper length. For $v > 0$, $\gamma > 1$ and hence it is always the case that $L < L_0$ which is the essence of the Lorentzian length contraction.[196]

Exercises

1. Summarize the main features of length contraction.

7.3 Time Dilation

This is a consequence of the Lorentz time transformation or a cause of it. As seen earlier, the proper time interval Δt_0 between two events is the length of time period as measured in the rest frame of the time measuring equipment, while the improper time interval Δt is the length of time period as measured in a frame that moves relative to that rest frame. According to the time dilation effect, the proper time interval is shorter than the improper time interval. Hence, the time interval as measured in any frame other than the rest frame will be longer than the time interval as measured in the rest frame. From another perspective, this means that time interval is subject to a contraction by the movement similar to the contraction of length. As seen earlier, time dilation is expressed by the following relation:

$$\Delta t = \gamma \Delta t_0 \tag{330}$$

where Δt_0 is the proper time interval and Δt is the improper time interval.[197]

Solved Problems

---

[196] The above statements (which follow the style of similar statements that are commonly found in the literature of Lorentz mechanics) should be subject to the right interpretation to rationalize the formalism. As we saw and will see (refer for example to § 1.6 and § 11), these statements may require modification to be consistent with the existence of an absolute frame and its implication on standardizing the observed spacetime coordinates.

[197] Again, the above statements (which resemble similar statements in the literature) should be subject to the right interpretation to rationalize the formalism, and hence they should not be seen as acceptance to the special relativistic interpretation which dominates the literature of Lorentz mechanics.

1. By considering three inertial frames which are in a state of motion relative to each other, show that time dilation and length contraction cannot be real effects if we have no absolute frame.
   Answer: The fact that these effects cannot be real if we have no absolute frame is obvious when we have three inertial frames which are in relative motion with respect to each other where time dilation and length contraction effects will be different in each frame as observed from the other two frames and hence by the principles of physical reality and truth (see § 1.6) these effects cannot be real because we cannot have two contradicting real effects. The change of frame[198] and similar arguments will not help to remove this real contradiction. Yes, if we have a single absolute frame then all the contractions of spacetime coordinates (in the form of length contraction and time dilation) will be referred to this unique absolute frame and hence even if two frames disagree on these contractions the effects are still real (with no contradiction) since they are ultimately referred to this real absolute frame, i.e. when these disagreeing frames perform the right transformations to the absolute frame to standardize their measurements they should agree on these real effects (see for example § 1.6).

Exercises
1. How is time dilation effect commonly stated in informal terms?
2. What is the essence of length contraction and time dilation?
3. Outline some features of time dilation effect.

7.4 Relativity of Simultaneity

This is a natural consequence of having frame dependent time as expressed in the Lorentz spacetime coordinate transformations. Accordingly, the simultaneity of events is not an invariant property across inertial frames under the Lorentz transformations and hence we can have two events that are simultaneous in one frame but not simultaneous in another frame. However, this does not necessarily imply the abolishment of absolute time unless we abolish the existence of absolute frame by adopting the special relativistic interpretation of the relativity principle. In this context, we remark that we should distinguish between the case of two events taking place at the same point in space and the case of two events taking place at two different points in space, i.e. points separated by a distance. For two events taking place at the same position in space, simultaneity is well defined since in any inertial reference frame time is uniquely defined for any particular point in space.[199] Hence, if two such events are simultaneous with respect to a particular inertial observer, they should be simultaneous with respect to all inertial observers although the time assigned to these events is generally different for these different observers. However, for two events taking place at two different locations in space, simultaneity is not a universal and well defined

---
[198] This may give the impression of moving between two parallel worlds when moving from one frame to another which is inconsistent with the existence of a unique physical reality.
[199] The meaning of this is that when two events are identical (i.e. being co-positional and simultaneous) in one frame then they are identical in all other frames.

7.4 Relativity of Simultaneity

concept since it belongs to the observer who in his reference frame the two events are taking place at the same instant of his time.

We should also remark that we must differentiate between the simultaneity of occurrence of two events and the simultaneity of observation of two events. The first means that the two events occur at the same time in a particular reference frame while the second means that the two events are observed at the same time by a particular observer who is at a particular location in a particular frame.[200] The simultaneity of observation is related to the relative distance between the observer and the two events and hence it has nothing to do with the simultaneity of occurrence. For example, if event A occurs at a position that is 1 light year away from the origin of coordinates of a particular frame O and event B occurs at a position that is 2 light years away from the origin of coordinates of O, then if the occurrence of B is 1 year earlier than the occurrence of A then the two events will be seen simultaneously by an observer who is at the origin of coordinates of O although they did not occur simultaneously in that frame. The significance of the relativity of simultaneity is related to the simultaneity of occurrence and not to the simultaneity of observation. This means that the relativity of simultaneity of observation is a trivial matter that does exist even in classical mechanics and hence it does not require any fundamental change of the paradigms of space and time and if they are absolute or not and if they are merged or not, or more technically if they are transformed by Lorentz transformations or by any other transformations like the Galilean transformations. The literature of special relativity is full of examples that demonstrate the confusion by some between these two concepts, i.e. the simultaneity of occurrence and the simultaneity of observation.[201]

Solved Problems

1. Define and discuss the concepts of simultaneity and relativity of simultaneity.

    Answer: Two events are described as simultaneous if they take place at the same instant of time. In classical physics, simultaneity of events is absolute because time is absolute and hence it is the same for all frames. In fact, all inertial observers according to classical physics can unify their time by shifting their origin of time by a translational transformation and unifying their unit of measurement by a linear constant scaling. But in Lorentz mechanics time is a frame dependent variable and hence each frame has its own time which cannot only differ with the time of other frames by a translational shift of origin and uniform constant scaling that affects the size of the unit of time but can also differ by a motion-dependent scaling through the Lorentz γ factor as well as by the dependency on the spatial coordinates. In fact, the ultimate cause of the relativity of simultaneity is the distortion of spacetime that is caused by motion through spacetime and is demonstrated by the effects of length contraction and time dilation. Accordingly, the relativity of simultaneity of occurrence is proprietary to Lorentz mechanics.

---

[200] In fact, reference frame represents a single global observer who is present everywhere at any time.

[201] We note that the relativity of simultaneity (which is supposed to be one of the fundamental principles of special relativity) may not be consistent with the claim that time dilation is an apparent and not real effect. This should also apply to the relativity of co-positionality in its relation to length contraction.

7.4 Relativity of Simultaneity

2. Does the relativity of simultaneity necessarily mean the abolishment of absolute time?
   Answer: In our view it does not and hence the existence of absolute time can still be justified despite the relativity of simultaneity that is implied by the formalism of Lorentz mechanics. In fact, the ultimate meaning of the concept of relativity of simultaneity (which is a factor in determining the fate of absolute time) is highly dependent on the philosophical and epistemological interpretation of the formalism of Lorentz mechanics. So, if we accept the special relativity view about the relativity principle which is based on the denial of the existence of absolute frame, then we should abolish absolute time. But if we adopt a different interpretation for the relativity principle (e.g. the classical interpretation) then there is still place for absolute time in Lorentz mechanics and hence the existence of absolute time can be compliant with the formalism of Lorentz mechanics and its implication of the relativity of simultaneity.

3. Show by a formal argument that simultaneity is frame dependent. On what parameter simultaneity depends?
   Answer: Let have two inertial frames, O and O', in a state of standard setting. The Lorentz transformation of the time interval that separates two events, A and B, between these two frames is given by:

$$t_B - t_A = \gamma \left[(t'_B - t'_A) + \frac{v}{c^2} (x'_B - x'_A) \right]$$

Now, if A and B are simultaneous in frame O' then we have $t'_B = t'_A$ and hence $t'_B - t'_A = 0$. The above transformation then becomes:

$$t_B - t_A = \gamma \frac{v}{c^2} (x'_B - x'_A)$$

As we see, the events A and B are simultaneous in frame O (i.e. $t_B = t_A$ and hence $t_B - t_A = 0$) only if we have $x'_B - x'_A = 0$, i.e. A and B are also co-positional in O' and hence they are identical. So, if $x'_B \ne x'_A$ then the two events will be simultaneous in frame O' but not simultaneous in frame O which means that the simultaneity of occurrence is frame dependent. Accordingly, we conclude that the simultaneity of two events in one frame (i.e. O) depends on the positional separation of these events in the other frame (i.e. O') although these events are simultaneous in O'.

Exercises

1. Find the condition for two events which are not simultaneous in frame O' to be simultaneous in frame O where O and O' are in a state of standard setting. Repeat the question assuming this time that the two events are simultaneous in frame O' but not in frame O.
2. Make a clear distinction between the simultaneity of occurrence and the simultaneity of observation in the context of relativity of simultaneity.
3. Show that the relativity of simultaneity of occurrence is proprietary to Lorentz mechanics and hence it does not exist in classical physics.
4. O and O' are inertial observers in a state of standard setting. In O frame, events A and B are simultaneous and they are spatially separated by a distance $x_B - x_A = 10^4$

m. Find the time interval between A and B in O' frame if their spatial separation in this frame is $x'_B - x'_A = 10^5$ m.

5. O and O' are inertial observers in a state of standard setting. In O frame, events A and B are observed to be temporally separated by $t_B - t_A = 0.5$ s and they are spatially separated by a distance $x_B - x_A = 10^{10}$ m while in O' frame they are observed to be simultaneous. What is the relative speed between O and O'?

6. Compare simultaneity in classical mechanics and in Lorentz mechanics according to the special relativistic interpretation. Is the denial of absolute time necessary to make sense of the relativity of simultaneity?

7.5 Relativity of Co-positionality

Like the relativity of simultaneity, the relativity of co-positionality is a natural consequence of the Lorentz space coordinate transformations since these transformations do not only depend on the space coordinates and time in the other frame but also on the relative motion which is embedded in the Lorentz γ factor. Accordingly, two events may be co-positional in one frame but not co-positional in another frame.[202] Again, the relativity of co-positionality does not necessarily imply the abolishment of absolute space unless we abolish the existence of absolute frame by adopting the special relativistic interpretation of the relativity principle. The relativity of simultaneity and the relativity of co-positionality are further investigated in the following solved problems where we compare the temporal and spatial coordinates of two events using the Lorentz spacetime coordinate transformations.

Solved Problems

1. Make thorough comparisons between two events, A and B, in spacetime as seen from two inertial frames, O and O', which are in a state of standard setting using the Lorentz spacetime coordinate transformations. Draw a conclusion.
   Answer: There are different ways for making such comparisons. However, we choose to base our comparisons on the difference in t and x coordinates between A and B as seen from O and O' using the basic Lorentz transformations. For conciseness, we label the differences as t_{BA} and x_{BA}. Accordingly, we have:

$$t_{BA} = \gamma\left(t'_B + \frac{vx'_B}{c^2}\right) - \gamma\left(t'_A + \frac{vx'_A}{c^2}\right) = \gamma\left(t'_{BA} + \frac{v}{c^2}x'_{BA}\right)$$

$$x_{BA} = \gamma\left(x'_B + vt'_B\right) - \gamma\left(x'_A + vt'_A\right) = \gamma\left(x'_{BA} + vt'_{BA}\right)$$

Now, we have three main cases of simultaneity or/and co-positionality that correspond to these equations:
   • A and B are simultaneous in O: hence, $t_{BA} = 0$ and since $\gamma \neq 0$ then from the first equation we have:

$$t'_{BA} + \frac{v}{c^2}x'_{BA} = 0$$

---

[202] Like the relativity of simultaneity, the relativity of co-positionality does not apply to identical events because being identical is a universal qualification and hence if two events are identical in one frame then they are identical in all frames.

7.5 Relativity of Co-positionality

So, if A and B are simultaneous in O' (i.e. $t'_{BA} = 0$) then $x'_{BA} = 0$, i.e. A and B are also co-positional in O' (i.e. identical). Similarly, if A and B are co-positional in O' (i.e. $x'_{BA} = 0$) then $t'_{BA} = 0$, i.e. A and B are also simultaneous in O' (i.e. identical).[203] Accordingly, if A and B are not simultaneous/co-positional in O' then they are not co-positional/simultaneous in O' (i.e. they are anti-identical in O'). This means that being simultaneous is frame dependent (which is known as the relativity of simultaneity).

• A and B are co-positional in O: hence, $x_{BA} = 0$ and since $\gamma \neq 0$ then from the second equation we have:

$$x'_{BA} + vt'_{BA} = 0$$

So, if A and B are simultaneous in O' (i.e. $t'_{BA} = 0$) then $x'_{BA} = 0$, i.e. A and B are also co-positional in O' (i.e. identical). Similarly, if A and B are co-positional in O' (i.e. $x'_{BA} = 0$) then $t'_{BA} = 0$, i.e. A and B are also simultaneous in O' (i.e. identical).[204] Accordingly, if A and B are not co-positional/simultaneous in O' then they are not simultaneous/co-positional in O' (i.e. they are anti-identical in O'). This means that being co-positional is frame dependent (which we call the relativity of co-positionality to treat space and time equally).

• A and B are identical in O: hence, $t_{BA} = 0$ and $x_{BA} = 0$ and since $\gamma \neq 0$ then from the above two equations we have:

$$t'_{BA} + \frac{v}{c^2} x'_{BA} = 0$$
$$x'_{BA} + vt'_{BA} = 0$$

On substituting from the second into the first we obtain:

$$t'_{BA} + \frac{v}{c^2}(-vt'_{BA}) = t'_{BA}\left(1 - \frac{v^2}{c^2}\right) = \frac{t'_{BA}}{\gamma^2} = 0$$

and hence $t'_{BA} = 0$ (because for $v < c$, $\frac{1}{\gamma^2} \neq 0$), i.e. A and B are simultaneous in O'. Also, on substituting from the first into the second we get:

$$x'_{BA} + v\left(-\frac{v}{c^2}x'_{BA}\right) = x'_{BA}\left(1 - \frac{v^2}{c^2}\right) = \frac{x'_{BA}}{\gamma^2} = 0$$

and hence $x'_{BA} = 0$ (because for $v < c$, $\frac{1}{\gamma^2} \neq 0$), i.e. A and B are co-positional in O'. This means that if A and B are identical in O then they are also identical in O'.
Conclusion: the relativity of simultaneity can only occur for events that are spatially separate (this can also be obtained from the time dilation formula). Equally, the relativity of co-positionality can only occur for events that are temporally separate (this can also be obtained from the length contraction formula). Hence, if two events are

---

[203] We note that since we consider these cases (i.e. simultaneous, co-positional, identical and anti-identical) mutually exclusive, then the case of being identical in O' should be excluded (refer to the third part of this exercise and the next exercise).

[204] Again, since we consider these cases (i.e. simultaneous, co-positional, identical and anti-identical) mutually exclusive, then the case of being identical in O' should be excluded.

7.5 Relativity of Co-positionality

identical in one frame then they are identical in all frames, i.e. being identical is a universal frame independent attribute.

2. Is it possible that two events are simultaneous but not co-positional in one frame, and they are co-positional but not simultaneous in another frame?
Answer: From the answer of the previous question, it is not. To show this formally,[205] let use the equations of the previous question, that is:

$$t_{BA} = \gamma\left(t'_{BA} + \frac{v}{c^2}x'_{BA}\right)$$
$$x_{BA} = \gamma\left(x'_{BA} + vt'_{BA}\right)$$

Now, let assume that the two events are simultaneous but not co-positional in frame O (i.e. $t_{BA} = 0$ and $x_{BA} \neq 0$), and they are co-positional but not simultaneous in frame O' (i.e. $x'_{BA} = 0$ and $t'_{BA} \neq 0$). Hence, we have:

$$0 = \gamma\left(t'_{BA} + 0\right)$$
$$x_{BA} = \gamma\left(0 + vt'_{BA}\right)$$

which is contradictory because from the first equation we should have $t'_{BA} = 0$ (since $\gamma \neq 0$), and hence from the second equation we should have $x_{BA} = 0$ (since $t'_{BA} = 0$ according to the first equation) which contradicts the assumption that $x_{BA} \neq 0$.[206]
In more details:[207]
(a) If two events are identical (i.e. simultaneous and co-positional) in one frame, then they should be identical in all frames, i.e. being identical is invariant across inertial frames.
(b) If two events are simultaneous (but not co-positional) in one frame, then they should be anti-identical in the other frames (assuming the other frames are in relative motion to exclude being simultaneous).
(c) If two events are co-positional (but not simultaneous) in one frame, then they should be anti-identical in the other frames (assuming the other frames are in relative motion to exclude being co-positional).
(d) If two events are anti-identical in one frame, then they can be simultaneous, or co-positional or anti-identical in the other frames (depending on the frame).
All these cases can be simply shown formally by using the two equations in the start of this answer, as we did above.[208] For example, if the two events are identical in frame O' (i.e. $t'_{BA} = 0$ and $x'_{BA} = 0$)then we should have:

$$t_{BA} = \gamma\left(0 + \frac{v}{c^2} \times 0\right) = 0$$

---

[205] This may also by shown graphically by using spacetime diagrams.
[206] Alternatively, we can say: because from the first equation we should have $t'_{BA} = 0$ (since $\gamma \neq 0$), and from the second equation we should have $t'_{BA} \neq 0$ (since $x_{BA} \neq 0$).
[207] We note that "simultaneous" and "co-positional" are sometimes used in their technical sense (and hence they are mutually exclusive with each other and with other classes) and sometimes used in their generic sense (i.e. "same time" and "same place"). This is also the case in some other places. The context should be consulted to remove any potential confusion.
[208] They may also by shown graphically by using spacetime diagrams.

7.5 Relativity of Co-positionality

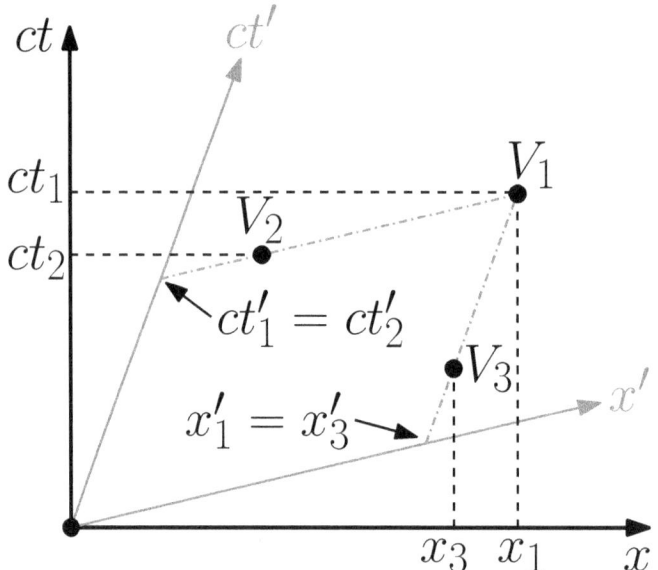

Figure 17: Demonstrating the relativity of simultaneity and the relativity of co-positionality on a 2D spacetime diagram representing two inertial frames.

$$x_{BA} = \gamma(0 + v \times 0) = 0$$

i.e. they are also identical in frame O. Also see the exercises of § 3.9.3.

3. Demonstrate on a figure like Figure 7 the relativity of simultaneity and the relativity of co-positionality.

    Answer: Referring to Figure 17, we see that V_1 and V_2 are simultaneous in O' but not simultaneous in O. Similarly, V_1 and V_3 are co-positional in O' but not co-positional in O.

Exercises

1. O and O' are inertial observers in a state of standard setting. In O frame, events A and B are co-positional and they are temporally separated by a time interval $t_B - t_A = 0.01$ s. Assuming that the y and z coordinates of the two events are identical, find the spatial separation between A and B in O' frame if their time separation in this frame is $t'_B - t'_A = 0.02$ s.
2. O and O' are inertial observers in a state of standard setting. In O frame, events A and B are observed to be temporally separated by $t_B - t_A = 0.005$ s and they are spatially separated by a distance $x_B - x_A = 10^4$ m while in O' frame they are observed to be co-positional. What is the relative speed between O and O'?
3. How do you compare the relativity of simultaneity and the relativity of co-positionality to the absolute simultaneity and absolute co-positionality in reference to the absolute frame? Try to link this to the speed of light as an invariant across all inertial frames.

7.6 Equivalence of Mass and Energy

According to Lorentz mechanics, mass and energy are equivalent and hence there is an amount of energy associated with the mass of an object even when it is at rest with respect to the observer.[209] The energy associated with the mass of an object is called its rest energy or mass energy. This energy is given by the Poincare mass-energy relation:

$$E_0 = mc^2 \qquad (331)$$

where E_0 is the rest energy, m is the mass[210] and c is the characteristic speed of light in vacuum. Accordingly, it is possible to convert mass to energy and vice versa where the mass-energy combination is conserved. As a result of the mass-energy equivalence, the units and measures of mass and energy can be used interchangeably where $1/c^2$ and c^2 are used as conversion factors. Hence, the mass of an object can be stated in energy units (e.g. joule or eV) while the energy of an object can be stated in mass units (e.g. gram or amu) with the above conversion factors. The equivalence between mass and energy in Lorentz mechanics is general and it includes all forms of energy[211] and hence an equivalent mass can be assigned to any type of energy like heat (see for example § 4.3.2).[212]

As discussed earlier, the mass-energy equivalence may also be established by purely classical arguments. If so, then this equivalence will not be a proprietary consequence of Lorentz mechanics, unlike time dilation or length contraction for instance which are entirely Lorentzian effects with no possible presence in classical mechanics. We should also remark that for the mass-energy equivalence to be a consequence of Lorentz mechanics we should have a valid argument that is completely based on the principles and framework of Lorentz mechanics so that this equivalence can be theoretically regarded as a consequence of this mechanics. This similarly applies to classical mechanics in determining if the mass-energy equivalence can be regarded as a classical prediction (if we have a purely classical argument in support of this equivalence) or not. The reader is referred to § 5.3.2 for more

---

[209] As indicated earlier, to have a more general sense of the equivalence between mass and energy the rest mass should also include (and hence depends on) all forms of non-kinetic energy. This will be clarified further in the following.

[210] As stated before, we follow the modern convention about mass except where we state otherwise (as in § 6.5 for example).

[211] If this should include the kinetic energy that is associated with the motion of massive objects, then it is more appropriate to follow the old convention about mass (or use what we call Lorentzian mass).

[212] In this context, we have two main types of energy: kinetic energy that is associated with the motion of massive objects, and non-kinetic energy like heat. The generalized concept of rest mass to include all forms of non-kinetic energy is one step in this generalization while the concept of non-rest mass (or Lorentzian mass) according to the old convention will extend this generalization to include even the kinetic energy. We note that the kinetic energy in this context means the kinetic energy of the object as a whole (as may be represented by its center of mass) relative to the observer and hence it should not include the kinetic energy that is based on the motion of the parts of the object (relative to its center of mass) which may be labeled as heat and hence it is part of its rest energy. We may call the first type the external kinetic energy and the second type the internal kinetic energy. Accordingly, the internal kinetic energy is a frame independent attribute (and hence it is invariant across all frames) while the external kinetic energy is a frame dependent attribute (and hence it varies across frames).

details.

Exercises

1. Find a common factor between the above consequences and predictions of Lorentz mechanics which are discussed in the sections of this chapter.
2. Give examples of subsidiary consequences and predictions that can be added to the above list of main consequences and predictions (i.e. merging of space and time into spacetime, length contraction, etc.) of Lorentz mechanics.

Chapter 8
Evidence for Lorentz Mechanics

In this chapter, we investigate the evidence in support of Lorentz mechanics. As discussed before, there are two main parts for Lorentz mechanics: the formalism and the interpretation. We note that all the real and alleged evidence is in support to the formalism and not to any particular interpretation. This is similar to the situation in quantum mechanics where all the evidence in support of quantum mechanics is regarded as support to the formalism of quantum mechanics and not to any particular interpretation such as the Copenhagen school or many-worlds interpretation. However, there is a general tendency to direct the evidence of Lorentz mechanics to the interpretation of special relativity and hence this theory gets undeserved credit.

Also, because of this strong tendency to endorse special relativity, some of the evidence in support of Lorentz mechanics may not be free of intentional or non-intentional bias. In fact, there are several reasons that make some of the claimed evidence of Lorentz mechanics, which is almost universally framed within the theory of special relativity, suspicious or controversial. Some of these reasons are outlined in the following points:

1. There is a general tendency in the scientific circles to support Einstein as well as fear of opposing him and hence some of the alleged evidence may have been twisted or stretched if not fabricated.[213]
2. There is another reason to be cautious about the validity of any experimental evidence, that is the analysis and conclusions are generally conducted and drawn by indoctrinated scientists who have no doubt about the validity of special relativity and hence they do not only have a strong tendency to interpret these experiments in favor of special relativity but they also have no hesitation to dismiss any evidence against special relativity since they regard such evidence as fallacy due to their strong conviction and hence they would rather assume mistakes and errors in the experiments or the equipment or even in their own understanding to save this theory. Moreover, some of the alleged experimental and observational evidence is based in its design and analysis on the framework of relativistic mechanics and hence the scientists just see what is supposed to be seen. In fact, the objective of some experimental and observational projects in this field is not to test the relativistic mechanics but to endorse this mechanics by searching for supporting evidence although these projects are normally marketed with a cover of objectivity and neutrality.
3. Except for the very trivial experiments that are conducted at the level of primary and secondary schools, the design of experiments, taking measurements, recording data, and interpreting the results are not a straightforward business as may be imagined by

---

[213] We should note that fraud in science is as common as in anywhere else. In fact, there are many reasons to make it more common.

some. Hence, there are many sources of uncertainty and question marks that normally associate any experimental result. For example, there are many sources of human and non-human errors as well as deviations caused by the equipment and measuring devices; some of these sources may be impossible to foresee or predict let alone take account of or correct. There are also many subjective and judgmental factors that enter in the experiment and its overall conclusions as well as many objective factors that can go, intentionally or unintentionally, one way or the other. The complexity and the impact of these negative factors are escalated substantially with the rise of the complexity of the experiment itself. Since Lorentz mechanics usually deals with extreme cases of high speed and/or extreme size and requires highly sophisticated equipment and procedures, as well as complex theoretical framework, the uncertainty in the results of such experiments is significantly high.

All these, among other reasons, contribute to the fact that many experimental results cannot be taken as unequivocal evidence, and that is why there are many disputes and controversies about the claimed evidence for or against Lorentz mechanics and its relativistic interpretation. In brief, more caution should be applied when dealing with any claimed evidence in support of this mechanics. However, unlike some who believe that Lorentz mechanics is a completely false and fabricated theory, we think there is strong evidence in support of the formalism of Lorentz mechanics (not its relativistic interpretation) although this does not mean that all the reported details should be accepted.[214] So, we can summarize our stand by saying: the formalism of Lorentz mechanics is generally supported by experiment and observation although there are many suspicious details, interpretations and analyses that require further investigation and verification. These uncertainties should justify the call for a revision (which could lead to amendment or replacement) of the current formalism of Lorentz mechanics as well as its interpretations (whether relativistic or not). In the following sections, we briefly investigate the evidence in support of Lorentz mechanics.

Solved Problems

1. If any evidence should be accredited to the formalism and not to the interpretation, then why some interpretations should be accepted while others should be rejected if the evidence does not belong to the interpretation?
   Answer: The interpretation gets its endorsement from its rationality (i.e. being logical and self consistent) as well as from its consistency with the formalism by making full sense of the formalism. The interpretation should also express the spirit of the formalism

---

[214] Also, this should not mean that the theory of Lorentz mechanics is final. Our view is that Lorentz mechanics is likely to be a good approximation to a more general and more perfect theory that can emerge in the future as we witnessed this in the past with classical mechanics where it is replaced and extended by modern theories like Lorentz mechanics and quantum physics. In fact, as the search for truth continues, all theories (scientific or else) should improve and hence the truth in its perfect form cannot be reached; in contrast to the common feeling (or conviction) in many stages of history (including nowadays) that the final truth has been reached. This feeling (which originates from laziness or short sight or worship of ancestors or whatever reason) is one of the main obstacles to the progress of science and to the process of evolution in general.

8.1 Success of Lorentz Transformations 253

which is a matter related to intuition and intellectual sense that any scientist should have experienced.

8.1 Success of Lorentz Transformations

In fact, this should be regarded as the most important evidence in support of Lorentz mechanics since the Lorentz spacetime coordinate transformations enters directly or indirectly in the formulation of many branches of science (e.g. quantum mechanics and its many applications like atomic and molecular physics) under the name of relativistic mechanics. The success of these Lorentzian (or relativistic) formulations is in our view the main and the most strong evidence in support of Lorentz mechanics although not all the claimed details should be accepted and taken for granted. Accordingly, no specific evidence (representing a particular instance or application of Lorentz mechanics like the ones that we will present in the following sections) is needed any more.

8.2 Mass-Energy Equivalence

This should be one of the strongest evidence in support of Lorentz mechanics due to the accumulation of many experimental and observational phenomena that are based on the conversion of mass to energy and vice versa. However, this depends on the establishment of the mass-energy equivalence relation solely and purely by Lorentzian arguments. So, if the mass-energy equivalence cannot be established by a valid Lorentzian argument then this equivalence will not belong theoretically to Lorentz mechanics and hence it cannot be considered as evidence in support of Lorentz mechanics.[215] Similarly, if the mass-energy equivalence can also be established by purely classical arguments (as well as by purely Lorentzian arguments) then this is not a decisive evidence in support of Lorentz mechanics since this equivalence, and all its consequences, can be explained classically. In this context, we should remark that there are several classical or semi-classical derivations of the mass-energy equivalence relation; some of which are given in § 5.3.2.[216]

8.3 Prolongation of Lifetime of Elementary Particles

This observational test is usually explained by the effect of time dilation although it can also be explained by length contraction since these two effects are complementary according to our previous and forthcoming investigation (see for example § 5.1.5) due to their common origin which is the contraction of spacetime coordinates by motion. The lifetime of unstable subatomic elementary particles,[217] like muons, which decay after a

---

[215] Although it may be a support in a sense, e.g. by belonging to a theory to which Lorentz mechanics is an approximation (refer to § 5.3.2 for more details).

[216] In this context, the reader is also referred to the derivations of the mass-energy equivalence relation by Larmor, Pauli, Lenard and Simhony (among others) which are reported in the literature.

[217] The lifetime of such particles is usually quantified in terms of half lifetime. However, for simplicity we present this in terms of lifetime (or average lifetime) to avoid unnecessary distraction from the main objective.

8.3 Prolongation of Lifetime of Elementary Particles

period of time following their creation or liberation from a bound system, are observed to be longer when they are observed in motion than their estimated proper lifetime in their rest frame and hence they travel longer distances during their lifetime in accord with the quantitative predictions of Lorentz mechanics. This means that if the predictions of classical mechanics are followed, then these particles should travel shorter distances than they actually do. For example, if an unstable subatomic particle has a proper lifetime of 2 seconds and it travels at a speed of 1 m/s in the laboratory frame, then according to classical mechanics it will travel 2 meters during its lifetime because its proper lifetime is its lifetime in any frame since lifetime is a frame-independent property of the particle. However, if its proper lifetime is dilated by a factor of 10 due to its motion relative to the observer (as it is the case according to Lorentz mechanics), then although its proper lifetime is still 2 seconds, its lifetime in the laboratory frame (relative to which the particle is traveling at a speed of 1 m/s) will be 20 seconds and hence it will travel in the laboratory frame a distance of 20 meters instead of 2 meters.

Solved Problems

1. Comment on the interpretation of prolongation of lifetime of elementary particles by length contraction instead of time dilation.
   Answer: It is possible to interpret this prolongation of lifetime in terms of length contraction where the distance seen by the elementary particle contracts by the γ Lorentz factor so that it travels a contracted distance during its proper lifetime with the same effect as time dilation. The reader is referred to one of the upcoming solved problems and some of the exercises of this section. He is also referred to the solved problems and exercises of § 4.1.2 for similar examples.

2. How is the prolongation of lifetime of elementary particles detected and measured?
   Answer: The prolongation of lifetime can be detected and measured by measuring the intensity of these particles in the cosmic ray, for example, at two different altitudes (e.g. at the top of a mountain and at sea level) where the difference between the two intensities, associated with the known proper lifetime of these particles and the data about their speed and traveled distance, will lead to the deduction and measurement of the prolonged lifetime.

3. A particle traveled a distance $d = 33$ m during its lifetime. If the particle was traveling at a speed $v = 0.6c$, find its improper lifetime t and its proper lifetime t_0.
   Answer: Its improper lifetime in the laboratory frame is:
   $$t = \frac{d}{v} \simeq \frac{33}{0.6 \times 3 \times 10^8} \simeq 1.83 \times 10^{-7}\,\text{s}$$
   Also, from the time dilation formula its proper lifetime is:
   $$t_0 = \frac{t}{\gamma} = \frac{d}{\gamma v} \simeq \frac{33\sqrt{1 - 0.6^2}}{0.6 \times 3 \times 10^8} \simeq 1.47 \times 10^{-7}\,\text{s}$$

4. Show, analytically, the equivalence between time dilation and length contraction as two ways for solving particle decay problems and any similar problems.

8.3 Prolongation of Lifetime of Elementary Particles

Answer: In this type of problems we have proper and improper lifetime, t_0 and t, where the former is the lifetime in the frame of the particle while the latter is the lifetime in the laboratory frame. We also have proper and improper distance, d_0 and d, where the former is the distance as seen from the particle frame while the latter is the distance as seen from the laboratory frame. Now, from the time dilation and length contraction formulae[218] we have:
$$t = \gamma t_0 \qquad \text{and} \qquad d = \gamma d_0$$
Moreover, since the relative speed v is the same in both frames then we have:
$$v = \frac{d}{t} = \frac{d_0}{t_0}$$
where the first equality represents the laboratory frame while the second equality represents the particle frame. So, from the laboratory frame we have:
$$d = vt = v\gamma t_0$$
where we used the time dilation formula. Similarly, from the particle frame we have:
$$\begin{aligned} d_0 &= vt_0 \\ \frac{d}{\gamma} &= vt_0 \\ d &= v\gamma t_0 \end{aligned}$$
where we used the length contraction formula. As we see, the two results are identical. We should remark that in the above method of tackling this problem and its alike we unified the proper/improper perspective in conceptualizing time dilation and length contraction effects and hence we have $t = \gamma t_0$ and $d = \gamma d_0$ (instead of $t = \gamma t_0$ and $d_0 = \gamma d$), i.e. we ascribed the proper value of both time and length to the same frame which is the frame of the particle. The common approach is to discriminate in the proper/improper perspective and hence the proper time is ascribed to the moving frame (i.e. particle frame) while the proper length is ascribed to the stationary frame (i.e. laboratory frame) and this justifies the difference in the "dilation" and "contraction" labels. Since we unified our proper/improper perspective then we should label both effects as dilation or both as contraction instead of attributing dilation to time and contraction to length. In essence, our method is based on the view that all the coordinates of spacetime (whether temporal or spatial) of the moving frame (particle) are seen from the stationary frame (laboratory) to contract by motion. Similarly, we can say that all the coordinates of spacetime of the stationary frame is seen to dilate by motion when compared to the coordinates in the moving frame.[219]

---

[218] The reader is referred to the remark in the end of this answer for justification of these formulae and symbolism.

[219] In more practical terms, the laboratory observer will see a contraction in both the temporal and spatial coordinates of the particle frame, which is equivalent to a dilation in both the temporal and spatial coordinates of the laboratory frame. The reader is referred to § 1.6 and § 11 for more details about the interpretation.

Exercises

1. An unstable subatomic particle has a proper lifetime $t_0 = 1$ microsecond and a speed $v = 0.99c$. What distance will this particle travel during its lifetime as observed in the laboratory frame?
2. An unstable subatomic particle is observed in the laboratory to travel 300 m at a speed of $v = 0.95c$. What is its proper lifetime? How far the particle will travel according to classical mechanics?
3. A subatomic particle with a proper lifetime $t_0 = 10^{-7}$ s is created in a particle collider and it is moving at a constant velocity $v = 0.7c$ at the instant of its creation. How far the particle will travel according to classical mechanics and according to Lorentz mechanics? Solve the second part once as a time dilation problem and once as a length contraction problem.
4. An elementary particle with a proper lifetime of 10^{-6} s is observed to travel a distance of 100 m during its lifetime. What is the speed of this particle?

8.4 Atomic Clock Experiment

This test is based on the effect of time dilation where precise atomic clocks placed on board airplanes are used to detect time dilation by performing a round trip around the world and comparing their time with the time recorded by identical clocks that stayed stationary on the Earth. It is claimed that the traveling clocks were observed to run slower as predicted by the time dilation effect of Lorentz mechanics. However, this sort of experiments may be criticized for being conducted in non-inertial frames and in a space that is not free of gravitational fields and hence significant errors may be introduced (let alone that the experiments may be wrong in principle). Moreover, the preparations and settings of such experiments usually introduce many sources of error. The result and analysis of this experiment may also be linked to the twin paradox challenge of special relativity (see § 10.1). Anyway, evidence like this cannot establish a big theory like the theory of Lorentz mechanics although they may, if reliably verified, endorse the theory somewhat and increase its credibility.

Exercises

1. A clock was placed on board a missile that moves with a constant velocity $v = 1000$ m/s. How long the missile should move (as seen from the frame of a stationary launch pad) for the clock to be 1 millisecond behind an identical clock that stayed stationary on the launch pad?

8.5 Stellar Aberration

This is regarded as another evidence in support of Lorentz mechanics. However, there are many controversial issues about this alleged evidence and its possible explanation by classical mechanics using the ether hypothesis. Hence, we do not go through the details of this alleged evidence. In fact, this alleged evidence and its alike are not needed at all if we believe in the existence of more clear and decisive evidence like the success of the Lorentz transformations (refer to § 8.1).

Chapter 9
Special Relativity

The special theory of relativity, which is originated by Poincare and elaborated by Einstein, is largely a philosophical and epistemological view to interpret and explain the Lorentz transformations and their direct and indirect consequences as represented by the formalism of Lorentz mechanics which was investigated in the previous chapters. The theory is based on the two famous postulates (i.e. the principle of relativity and the constancy of the observed speed of light) from which the Lorentz transformations are derived. In the following sections we investigate this theory which is the dominant theory in interpreting, explaining and structuring Lorentz mechanics and its formalism. In fact, it is considered by the scientific community as the only viable explanation of the Lorentz transformations and their consequences. In brief, in the eye of the overwhelming majority of physicists of modern times it is the same as what we call the mechanics of Lorentz transformations or Lorentz mechanics.

9.1 Characteristic Features of Special Relativity

In his attempt to explain the Lorentz transformations, Einstein adopted the main ideas of Poincare and elaborated on them to present what will be known as the special theory of relativity or special relativity. Accordingly, this theory, which became the dominant and even the only theory representing Lorentz mechanics, is based on two pillars: the formalism of Lorentz and the interpretation of Poincare. The theory also includes some additives and attachments like considering c as restricted and ultimate speed. In broad terms, special relativity is characterized by the following fundamental ideas:
1. The whole formalism and interpretation of Lorentz mechanics are based on two postulates which are the postulate of relativity and the postulate of invariance of the observed speed of light for all inertial observers. These postulates are used to derive the Lorentz spacetime coordinate transformations (see § 12.4.1) and from these transformations all other transformations and consequences of Lorentz mechanics are eventually derived.
2. The abolishment of the existence of any absolute frame of reference, and hence the abolishment of absolute space and absolute time.
3. The theory also includes some attachments like considering c as restricted speed to light and as ultimate speed to any massive and massless physical object.

We should remark that the above features are what characterize special relativity as an interpretative theory and hence they do not include the features that characterize Lorentz mechanics itself, such as the merge of space and time into spacetime or the contraction of spacetime coordinates by motion (refer for example to § 7), since these features belong to the main body of Lorentz mechanics and its formalism and hence they are not characteristic of any particular theory or interpretation of this mechanics like special relativity.

Exercises

1. Outline the main features of special relativity.
2. Can we consider the abolishment of absolute frame (and hence absolute space and absolute time) as an implication of the Lorentz spacetime coordinate transformations and hence as an endorsement to the special relativistic view?

9.2 Postulates of Special Relativity

As stated earlier, the special theory of relativity is based on two main postulates which are used to derive the Lorentz transformations of spacetime coordinates (refer to § 12.4.1) and subsequently the other mathematical formulations of Lorentz mechanics. These two postulates are:

1. The equivalence of all inertial frames in their validity for formulating the laws of physics. Accordingly, the laws of physics take the same form in all inertial frames and hence the laws of physics should be form invariant under the correct set of transformations of space coordinates and time between these frames. Although this postulate as stated here is plausible and it sounds like the principle of invariance of physical laws (refer to § 1.8), the distinctive feature of the relativity postulate in special relativity is the attached denial of the existence of any absolute frame. As we will see, this attachment is the main source of several valid challenges to this theory.[220]

2. The invariance of the observed speed of light in free space for all inertial observers and hence this speed takes the characteristic value c in all inertial frames regardless of the state of motion of the light source and the observer. Because special relativity is based on denying the existence of any absolute frame or any luminiferous medium for the propagation of light, this condition simply implies that the speed of light is independent of the relative motion between the light source and the observer with no need for any particular qualification, i.e. the motion of the source or the motion of the observer as such. As we saw in our previous investigation (see for example § 4.2.2), the formalism of Lorentz mechanics, and the velocity transformations in particular, suggests such an invariance of this speed due to the contraction of both the temporal and spatial coordinates by the same factor as a result of the motion through spacetime. However, the confusion in the literature of special relativity casts a shadow on the special relativistic meaning of this invariance as we will see later.

We note that some authors may add another postulate to special relativity that is: an object in uniform motion with respect to an inertial frame is in uniform motion with respect to all inertial frames. However, this added postulate (which seems redundant) is

---

[220] We should remark that the relativity principle of special relativity may be claimed to be more general than the classical principle of relativity in a sense that the former applies to all laws of physics while the latter is restricted to the laws of classical mechanics. However, this claim is not firmly established. In fact, the emergence of Lorentz mechanics as a necessity for the invariance of Maxwell's equations suggests that such a generalization is assumed even in classical physics (or at least by some classical physicists) before the emergence of Lorentz mechanics and special relativity. Yes, this restriction may tentatively apply to the Galilean formulation of the relativity principle but this is just a trivial historical factor since the Galilean principle (assuming it is restricted to the laws of mechanics) has been developed and enhanced over a long period of time since its appearance.

rarely considered in the literature of special relativity and hence we do not discuss it any more.[221] We should also note that special relativity, like other physical theories, also relies on other postulates, in the form of implicit hypotheses, such as the homogeneity and isotropy of space, the uniformity of time flow in a strict unique direction, and the time order of causal relations. However, since these postulates are generally accepted they are not stated explicitly although they are implicitly incorporated in many theorems and arguments.[222]

Solved Problems

1. The first postulate of special relativity may be stated in the literature in a number of different forms. State some of these forms and show their equivalence.
   Answer: Some of these variant forms are:
   • It is impossible to detect uniform translational motion through space.[223]
   • The physical laws take the same form in all inertial frames.
   • All inertial frames are valid for formulating the laws of physics.
   Regarding the first and second forms, since the physical laws take the same form in all inertial frames then it is impossible to detect uniform motion because there is no distinctive mark in these laws that characterizes some frames from other frames due to their state of motion to distinguish which frame is moving and which frame is not moving. The equivalence between the second and third forms is obvious (noting that "the laws of physics" in the third form means the laws in their known forms), and hence the equivalence between the first and third forms should also be established. Accordingly, all these forms are equivalent.

2. In the previous investigations (see for example § 1.13 and § 1.14), we distinguished between the characteristic speed of light and the observed speed of light. Discuss this issue with regard to the second postulate of special relativity.
   Answer: The characteristic speed of light is the constant c which is the same for all observers regardless of any theory (special relativity or else), while the observed speed of light is the speed that is actually observed by individual observers and hence in principle this observed speed is variable and can be equal to c or not. The second postulate of special relativity claims that the observed speed of light in any inertial frame is always equal to the characteristic speed of light c.

Exercises

1. What are the two postulates of special relativity?
2. Show by a simple non-rigorous argument that the equivalence of all inertial frames of

---

[221] The purpose of this postulate seems to ensure that the transformations are linear (refer to § 12.4.1). However, this condition should be guaranteed by the validity of Newton's first law, since the frames are assumed to be inertial, and hence it is redundant.

[222] Some of these assumptions may be lifted in other theories like general relativity which is based on a curved spacetime manifold. However, this is of no interest to us since the spacetime of Lorentz mechanics is unanimously flat.

[223] This sounds more like the Galilean relativity although it may be stated by some as the first postulate of special relativity.

reference in their validity for formulating the laws of physics implies that it is impossible to detect the state of any inertial frame as being at rest or uniform motion in space in an absolute sense by conducting any experiment in that frame.
3. According to special relativity, the speed of light is independent of the speed of its source and the speed of its observer. Express this premise more efficiently and compactly.

9.3 Assessing the Postulates of Special Relativity

9.3.1 Relativity Principle

As discussed earlier, there are two principles of relativity: a restricted principle which is part of classical mechanics, and unrestricted (or absolute) principle which is part of special relativity. The essence of the restricted principle of relativity is that there is no distinction between inertial frames but without the denial of the existence of absolute frame. Hence, this restricted principle is limited to inertial frames and does not extend to non-inertial frames whose distinction from inertial frames can be based on the existence of absolute frame. In contrast, the unrestricted relativity principle is based on the denial of the existence of absolute frame. There are two main challenges to the relativity principle in its unrestricted or special relativistic sense that denies the existence of absolute frame:
1. In the absence of absolute frame, how the real physical difference between inertial and accelerating frames can be explained? Alternatively, in the absence of absolute frame, being inertial and accelerating is arbitrary and conventional and hence it cannot result in any real physical effect. In contrast, the relativity principle in its classical sense that is based on (or at least compliant with) the existence of absolute frame can cope with this challenge since being inertial or accelerating is a real physical qualification thanks to the existence of a real absolute frame. The equivalence of all inertial frames can then be explained by the indifference of the absolute frame with regard to uniform translational motion (refer to § 2.2).[224]
2. In the absence of absolute frame, the reality of the characteristic Lorentzian effects like time dilation and length contraction (or rather the contraction of spacetime coordinates by motion) is questionable. Since these effects form the main physical basis that underlies the whole of Lorentz mechanics, as they are the basis for the Lorentz spacetime coordinate transformations, then the reality of Lorentz mechanics as a whole is questionable. For example, if two inertial frames are in a state of relative uniform translational motion and we have no absolute frame then the effect of time dilation should equally apply to both frames because they are both in a state of relative uniform motion and hence each frame will see the time of the other frame dilating. Accordingly, time dilation cannot be a real effect because this contradicts the principles of physical reality and truth (see § 1.6) since we will have two conflicting realities, i.e. the time of the first frame is behind the time of the second frame and the time of the second frame is behind

---
[224] As indicated earlier, this indifference may originate from the infinity of the space which makes all uniform translational motions equivalent in the absence of real beginning and real end of space. However, a more fundamental reason for this indifference will be discussed later.

the time of the first frame. The change of reference frame, or any justification like this, does not help to solve this fundamental conflict with the physical reality. Also, labeling these effects as apparent to hide this problem is no more than a verbal solution with no real substance. In brief, if "apparent" means it is not real then this is a denial of the reality of these effects, and if it means it is "real but apparent" then first this requires clarification to have a sensible and logical meaning and second it will not solve the problem of conflict with the principles of physical reality and truth as we still have two real conflicting realities although they are given the status of being "apparent". Similarly, if it means something else then it should be explained unambiguously to decide if it is sensible and logical or not.

In this regard, we should remark that there is unshaken scientific evidence in support of the validity of the restricted form of the relativity principle where the whole of classical mechanics, and even Lorentz mechanics, is based on this principle since according to both mechanics all the laws of physics equally apply in all inertial frames regardless of their state of rest or motion. On the other hand, there is no evidence in support of the unrestricted relativity principle because Lorentz mechanics is completely consistent with the restricted form of the relativity principle and can be fully explained by this form of the relativity principle according to interpretations other than special relativity. This means that the unrestricted relativity principle (and hence the denial of the existence of absolute frame as embedded in the first postulate of special relativity) is not needed to explain and justify the formalism of Lorentz mechanics because this mechanics can be explained by the restricted relativity principle. Accordingly, the unrestricted form, which is characterized by the denial of the existence of absolute frame, is redundant and hence it is at least unsupported by any scientific evidence if there is no such scientific evidence against it. Yes, if the unrestricted relativity principle is required for the explanation and justification of Lorentz mechanics, then Lorentz mechanics will be a supporting evidence to the unrestricted relativity principle.

Solved Problems

1. Why the restricted relativity principle should be accepted while the unrestricted relativity principle should be rejected?
   Answer: In brief, we need to assume the existence of an absolute frame for the above two reasons (i.e. to explain the difference between inertial and non-inertial frames and to justify the reality of the characteristic Lorentzian effects like time dilation) and since the restricted relativity principle is compliant with this assumption and it is sufficient to explain and justify Lorentz mechanics then the restricted relativity principle is acceptable. On the other hand, the unrestricted relativity principle faces the above two challenges so even if it is valid for the explanation and justification of Lorentz mechanics it cannot be accepted.

Exercises

1. What are the restricted and unrestricted forms of the relativity principle?
2. Contemplate on the sufficiency of the restricted form of the relativity principle as a viable explanation for both classical mechanics and Lorentz mechanics.

3. Let have two inertial observers, O and O', who are in a state of relative uniform motion. Show that according to the principles of special relativity, each observer will see the clock of the other observer run slow. What you conclude from this?
4. Investigate all the possibilities for the reality of length contraction and time dilation effects with the relativity postulate of special relativity.
5. What "real" and "apparent" mean in the context of the interpretation of length contraction and time dilation?

9.3.2 Invariance of Observed Speed of Light

Based on our previous investigation, the formalism of Lorentz mechanics suggests the constancy of the observed speed (but not the velocity) of light in all inertial frames due to the contraction of the temporal and spatial coordinates by the same Lorentz γ factor where this contraction is caused by the motion through spacetime. In fact, this constancy of speed is equivalent to the invariance of the spacetime interval across all inertial frames, as we demonstrated earlier (see for example § 3.9.3, § 4.1.2 and § 5.1.5; also see the exercises of the present subsection). More fundamentally, this is a consequence of adopting the speed of light for calibrating the measurements of space and time, as indicated earlier and will be detailed later. However, what is the interpretation of this constancy? While our investigation, which is based on analyzing the formalism of Lorentz mechanics, indicates that this constancy is in an apparent sense[225] that is consistent with the existence of absolute frame, the special relativity interpretation of this constancy suggests otherwise although some special relativists also state that this constancy is apparent.[226] The confusion and potential error in the special relativistic interpretation of this constancy originate from the fact that in special relativity there is no absolute frame and hence if we embrace a real interpretation of this constancy it is difficult to imagine and rationalize, while if we embrace an apparent interpretation it is difficult to justify in the absence of an absolute frame relative to which this constancy is apparent unless we interpret "apparent" as being equivalent to "illusory" or "unreal" which no one should accept since this constancy has real physical effects.

We should remark that the universality of the observed speed of light in its apparent sense according to our interpretation intuitively indicates the existence of an absolute frame relative to which this uniform contraction in spacetime coordinates, and hence the constancy of this speed, is observed. Otherwise, the constant c will be completely arbitrary and cannot be justified.

Solved Problems

1. Examine and illustrate the constancy of the observed speed of light across all inertial frames according to the formalism of Lorentz mechanics where the temporal and spatial coordinates of the spacetime are assumed to contract under the influence of motion by

---

[225] As explained before and will be clarified further, this "apparent sense" should not be understood to mean it is like illusion but it is based on the reality of the absolute frame which is the ultimate reference for all spacetime measurements.

[226] But our "apparent" seems not the same as their "apparent".

9.3.2 Invariance of Observed Speed of Light

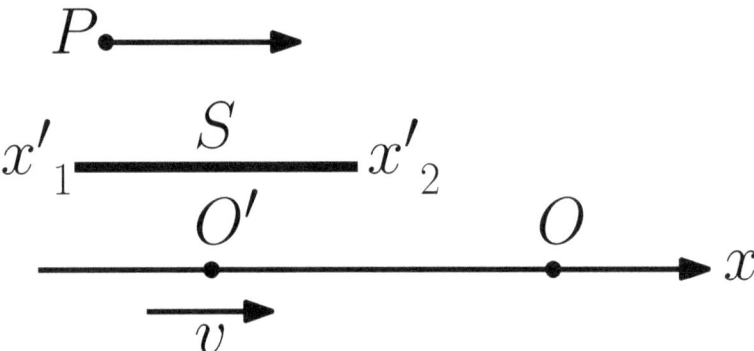

Figure 18: Schematic illustration representing the speed of a light photon P as measured in two inertial frames O and O' which are in a state of standard setting with a relative velocity v where a stick S, which is oriented along the common x direction, is at rest in frame O'.

the same Lorentz γ factor.

Answer: First, we unify our proper/improper perspective and hence we do not have time dilation and length contraction but we have time contraction and length contraction (or dilation of both) to express the fact that the temporal and spatial coordinates of the spacetime contract under the influence of motion by the same factor. So, let assume that we have a light photon P traveling in the positive x direction. We also have two inertial frames, O and O', which are in a state of standard setting with a relative speed v. Let also have a stick S oriented along the common x direction where S is at rest in frame O' and hence according to O it is traveling in the positive x direction with speed v. The situation is depicted schematically in Figure 18.

Now according to O, P arrives to x_1 at t_1 and arrives to x_2 at t_2. Since the length of the stick according to O is: $L = x_2 - x_1$ while the time interval between the two events of arrival is: $\Delta t = t_2 - t_1$, the speed of light will be measured by O to be:

$$s = \frac{L}{\Delta t} \tag{332}$$

By a similar argument, the speed of light will be measured by O' to be:

$$s' = \frac{L'}{\Delta t'} \tag{333}$$

where $L' = x'_2 - x'_1$ and $\Delta t' = t'_2 - t'_1$.

Now, let assume that the spatial and temporal coordinates of spacetime of frame O' are contracted by the motion relative to frame O and hence the length of objects and duration of time intervals will have a certain size in O frame and a contracted size in O' frame.[227] Now, if we have a photon that is observed by O to have speed c then

---

[227] This contraction of coordinates should be attributed to the contraction of objects and time (under the effect of motion) which results in contraction of the units of measurement. In other words, the contraction should be attributed to the spacetime itself.

9.3.2 Invariance of Observed Speed of Light

this photon will traverse a contracted distance during a contracted time interval in O' frame and because the characteristic speed of light is the calibration measure for the coordinates of space and time then O' should also see this photon to have speed c.
More technically, let assume that the spatial and temporal coordinates of spacetime of frame O' are contracted by the motion relative to frame O by the same γ factor and hence we have:

$$L' = L/\gamma \qquad (334)$$
$$\Delta t' = \Delta t/\gamma \qquad (335)$$

So, if the speed of light is observed to be c in frame O then we have:

$$\begin{aligned}
c &= s \\
&= \frac{L}{\Delta t} \\
&= \frac{\gamma L'}{\gamma \Delta t'} \\
&= \frac{L'}{\Delta t'} \\
&= s'
\end{aligned}$$

and hence the speed of light is also observed to be c in frame O'.
Similarly, if the speed of light is observed to be c in frame O' then we have:

$$\begin{aligned}
c &= s' \\
&= \frac{L'}{\Delta t'} \\
&= \frac{L/\gamma}{\Delta t/\gamma} \\
&= \frac{L}{\Delta t} \\
&= s
\end{aligned}$$

and hence the speed of light is also observed to be c in frame O.
The same analysis equally applies if we assume that the spatial and temporal coordinates of spacetime are contracted in frame O relative to the coordinates of frame O'. Also read the next problem and follow the continued analysis of this issue in § 9.7.1.

2. Assess and criticize the argument in the previous problem.
   Answer: There are three main criticisms to the argument in the previous problem:
   • The first is that since the arrival of the photon to the two ends of the stick is neither simultaneous or co-positional in any frame then the full Lorentz spacetime transformations should apply rather than the length contraction and time dilation formulae. The answer is that this is based on a misconception. What we need in calculating the speed of the photon is the traversed length and the required time interval that separate these

9.3.2 Invariance of Observed Speed of Light

events (i.e. arrival to the two ends) and both of these (i.e. length and time interval) can be obtained separately from the length contraction and time dilation and hence the speed can be estimated.[228] Anyway, let assume that the argument in the previous problem is incorrect or suspicious and hence we should use the full Lorentz spacetime coordinate transformations. So, if the speed of light is observed to be c in frame O then we have:

$$c = s$$
$$c = \frac{L}{\Delta t}$$
$$c = \frac{x_2 - x_1}{t_2 - t_1}$$
$$c = \frac{\Delta x}{\Delta t}$$
$$c = \frac{\gamma(\Delta x' + v\Delta t')}{\gamma\left(\Delta t' + \frac{v\Delta x'}{c^2}\right)}$$
$$c = \frac{\Delta x' + v\Delta t'}{\Delta t' + \frac{v\Delta x'}{c^2}}$$
$$c\Delta t' + \frac{v\Delta x'}{c} = \Delta x' + v\Delta t'$$
$$c\Delta t' - v\Delta t' = \Delta x' - \frac{v\Delta x'}{c}$$
$$c\Delta t'\left(1 - \frac{v}{c}\right) = \Delta x'\left(1 - \frac{v}{c}\right)$$
$$c\Delta t' = \Delta x'$$
$$c = \frac{\Delta x'}{\Delta t'}$$
$$c = \frac{L'}{\Delta t'}$$
$$c = s'$$

and hence the speed of light is also observed to be c in frame O'.

Similarly, if the speed of light is observed to be c in frame O' then we have:

$$c = s'$$
$$c = \frac{L'}{\Delta t'}$$
$$c = \frac{x'_2 - x'_1}{t'_2 - t'_1}$$
$$c = \frac{\Delta x'}{\Delta t'}$$
$$c = \frac{\gamma(\Delta x - v\Delta t)}{\gamma\left(\Delta t - \frac{v\Delta x}{c^2}\right)}$$

---
[228] Alternatively, we are using both time dilation and length contraction simultaneously.

9.3.2 Invariance of Observed Speed of Light

$$c = \frac{\Delta x - v\Delta t}{\Delta t - \frac{v\Delta x}{c^2}}$$

$$c\Delta t - \frac{v\Delta x}{c} = \Delta x - v\Delta t$$

$$c\Delta t + v\Delta t = \Delta x + \frac{v\Delta x}{c}$$

$$c\Delta t\left(1 + \frac{v}{c}\right) = \Delta x\left(1 + \frac{v}{c}\right)$$

$$c\Delta t = \Delta x$$

$$c = \frac{\Delta x}{\Delta t}$$

$$c = \frac{L}{\Delta t}$$

$$c = s$$

and hence the speed of light is also observed to be c in frame O.
In fact, we used the above argument (i.e. using time dilation and length contraction in the form of contraction of spacetime coordinates) instead of starting from the Lorentz transformations to highlight the origin of this invariance of the observed speed of light, i.e. the contraction of spacetime coordinates by motion.

• The second is that the displacement of the stick during the journey of the photon from one end to the other end should be added algebraically to the length of stick in calculating the speed of light in the frame relative to which the stick is moving. The answer is that we assumed that the speed of the photon is c in one frame and hence any velocity component of the stick due to the relative motion should be passed to the photon in the other frame as we found earlier. In other words, since the speed of the photon is assumed to be c in one frame then this frame will be like the source of light relative to the other frame because the photon of light takes its characteristic speed c in the former frame.[229] Anyway, even if this criticism is valid against the original argument (which is based on using the time dilation and length contraction), it is not valid against the derivation from the Lorentz transformations since any displacement will be taken into account by the transformations.

• The third is that this (i.e. scaling of c up or down due to the scaling of spacetime coordinates by the same factor) should equally apply to any speed. The reply is that the calibration of space and time measurement is based on the characteristic speed of light and not based on any other speed and hence this should apply only to light.

Exercises

1. Discuss and evaluate the argument that is given in the first solved problem.
2. O and O' are two inertial observers in a state of standard setting. Using the Lorentz spacetime coordinate transformations, show that if O observes the speed of light to be c then O' should also observe the speed of light to be c. Comment on this question.

---

[229] In fact, this is based on assuming the photon to be like a projectile since it has a velocity component from its source as we found earlier from analyzing the formalism of Lorentz mechanics.

3. O and O' are two inertial observers in a state of standard setting where a light signal that propagates in all directions is emitted at $t = t' = 0$ at the common origin of coordinates. Using the invariance of the spacetime interval (which was established in § 3.9.4), show that if a 3D light signal is observed by O to propagate in a spherical shape[230] centered at his origin of coordinates, then O' should also see this light signal to propagate in a spherical shape centered at his origin of coordinates. Can you generalize the result? Comment on this question.
4. Referring to the previous question, what about the source of light and if it should be in O frame or in O' frame for the concluded invariance to hold true?
5. Compare between the representation of light signal in the ordinary 3D space and in the Minkowski 4D spacetime.
6. Discuss the claim that the second postulate of special relativity is confirmed by experimental observation.

9.3.3 Overall Assessment of Special Relativity Postulates

Based on the previous discussions and questions, we can see that the postulates of special relativity should be adjusted as follows:
1. The relativity principle should be amended to comply with the existence of an absolute frame so that the difference between inertial and accelerating frames can be explained and the reality of the Lorentzian effects like time dilation and length contraction can be justified.
2. The constancy of the observed speed of light should be amended by interpreting this constancy in an apparent sense due to the spacetime contraction by the movement relative to the absolute frame. It should also be adjusted to include the fact that although the speed of light is independent of the speed of its source, its velocity is dependent on the velocity of its source since it has a velocity component from the velocity of its source (refer for example to the solved problems and exercises of § 4.2.2).

These adjustments, which are generally based on analyzing the formalism of Lorentz mechanics, are required to eradicate some causes of logical inconsistency and defects in this theory although these amendments may mean the abolishment of this theory altogether (at least the Einstein version).

Solved Problems
1. Assess the implication of the formalism of Lorentz mechanics on the issue of the invariance of the speed of light.
   Answer: From analyzing the formalism of Lorentz mechanics (e.g. by applying the Lorentzian spacetime or velocity transformations to the speed of light), we can conclude the following:
   • The observed speed of light is invariant across all inertial frames and it takes the characteristic constant value c although this should be interpreted in an apparent sense due to the spacetime contraction which arises from the motion relative to the absolute

---

[230] The spherical shape can be ascribed to the wave front of the signal.

frame.
- The velocity of light is frame dependent since it depends on the motion of the source of the light signal (refer to § 4.2.2 and § 4.2.3).

Exercises
1. Why we need to assume the existence of an absolute frame of reference?
2. Based on the results that were obtained from analyzing the formalism of Lorentz mechanics, discuss the issues of light propagation model and the invariance of the speed of light and try to link these to the issue of absolute frame.

9.4 Abolishment of Fundamental Concepts

The abolishment of fundamental concepts, like absolute space and absolute time, is a common claim in the literature of special relativity. These claims are normally based on Lorentzian concepts and consequences like the relativity of simultaneity. However, we have a number of reservations on these claims; some of these reservations are outlined in the following points:

1. The abolishment of absolute space and absolute time is a result of the special relativistic interpretation of Lorentz mechanics and the unrestricted relativity principle in particular which is based on the denial of the existence of absolute frame. Hence, this abolishment is not a necessary consequence of the formalism of Lorentz mechanics. In simple terms, concepts like relativity of simultaneity and relativity of co-positionality are compatible with the concepts of absolute time and absolute space in the presence of an absolute frame, which is the reference for all frames, whose time is regarded as the absolute (or reference) time and whose space is regarded as the absolute (or reference) space.

2. There is no sensible meaning to this abolishment within the wider domain of science even if we accept their "abolishment" within the framework of Lorentz mechanics or one of its interpretations. The reason is that these concepts can be regarded as conceptual tools that can be used in any physical theory if they proved to be theoretically or practically useful and can provide a better insight and sense of direction in our physical and conceptual adjustment to the physical world. So, even if these concepts proved to have no place in Lorentz mechanics or special relativity, this does not mean that they are wrong or redundant in science. In brief, as long as these concepts can be used sensibly and logically in a scientific theory they are legitimate and hence they cannot be abolished even if they proved to have no place in Lorentz mechanics or special relativity. In fact, even within the framework of Lorentz mechanics they can prove to be useful within a present or future interpretation or amendment.

3. Even if these concepts do not serve a purpose in science they cannot be abolished from the general knowledge of mankind like philosophy and hence the claim that special relativity abolished the classical view about absolute space and absolute time forever should be rejected.

4. As indicated earlier, there is an instinctive nature and biological roots to these concepts

9.4 Abolishment of Fundamental Concepts

which are obtained through a long history of evolution and conceptual development and adaptation. Hence, the claimed abolishment must be questioned and dismissed. If we have to abolish something we may need to abolish the special relativistic interpretation rather than these intuitive concepts and hence we should embrace or look for a more intuitive interpretation to Lorentz mechanics that accommodates these fundamental concepts.

We should remark that the above discussion about the abolishment of absolute space and absolute time also applies to other similar features of Lorentz mechanics and its interpretations even those features that characterize Lorentz mechanics itself such as the merge of space and time into spacetime. Hence, although these are legitimate and justified within the framework of Lorentz mechanics they should not be seen to represent universal and absolute facts that apply across all branches and theories of science let alone other forms of human knowledge like philosophy. This is mainly due to their interpretative and philosophical and epistemological nature.

Solved Problems

1. Contemplate on the abolishment of the independence of space and time because of their merge into spacetime.

   Answer: The claim about the abolishment of the independence of space from time and the merge of these two concepts into a spacetime according to the formalism of Lorentz mechanics should be treated like the claim of the abolishment of absolute space and absolute time according to special relativity. First, the merge of space and time to become spacetime does not mean the abolishment of these concepts as independent concepts because this merge is essentially a technicality of the Lorentzian formalism. In fact, this merge is not complete even in Lorentz mechanics where we still see the two have distinct features, e.g. we are free to move in space but not in time, or we can move in space in all directions but we can move in time only in one direction, i.e. forward. Accordingly, the claim of the abolishment of the independence of space and time should be considered with caution and conditionally like the abolishment of absolute space and absolute time. Anyway, even if such abolishment occurred within the framework of Lorentz mechanics, or one of its interpretations like special relativity, it should be limited to this branch of physics and within its domain of validity. Hence, it does not mean that such independence does not exist or it is illegitimate to exist in other interpretation, or other scientific branch or theory, or other forms of human knowledge like philosophy. We should also note that such a merge does exist even in classical mechanics, as represented by the Galilean x coordinate transformation, although it is much less significant.

Exercises

1. Contemplate on the abolishment of absolute space.
2. Discuss the following quote, which is attributed to Minkowski, and its significance in relation to the abolishment of space and time as separate entities: "From henceforth, space by itself, and time by itself, have vanished into the merest shadows and only a kind of blend of the two exists in its own right".

9.5 Controversies within Special Relativity

There are many examples of controversies and contradictions within the special relativity camp where some take one position while others take different or opposite position.[231] For example, time dilation and length contraction are seen by some as real effects while they are seen by others as apparent effects; moreover, the meaning of "apparent" seems to have different interpretations. A number of these controversies are discussed within other contexts in this book (see for example § 10) while the rest are not worthwhile to go through and investigate and hence we refer the reader to the literature of special relativity for details.

Exercises
1. List some of the difficult questions that face special relativity.

9.6 Thought Experiments in Special Relativity

Thought experiments play a focal role in special relativity and hence they enter in many of its arguments and in its logical structure. In this section, we present one example of thought experiments that are found in the literature of special relativity. Although it could be claimed that special relativity should not be assessed and judged ultimately by these thought experiments (most of which may be suitable for demonstrative and pedagogical purposes or even for popularizing science), they reflect the general attitude in special relativity and the type of logic that is used to establish this theory. Moreover, special relativity as an interpretative theory should be damaged and even collapse if its arguments and methods of presentation are imperfect because these aspects are essential part of interpretation unlike the formalism which is a rigorous mathematical structure that should not be affected if these rather marginal aspects are compromised as long as the formal aspects are correct. Anyway, we think the establishment of any theory should be based on more firm foundations, whether experimental or theoretical, and not on these thought experiments (see § 1.11).

9.6.1 Train Thought Experiment

In this thought experiment, which is related to the relativity of simultaneity and the abolishment of absolute time, a traveler P_1 inside a train sits in the middle of the train and another person P_2 is standing still at the midpoint of the platform. At the instant when the middle of the train passes by the midpoint of the platform, lightning strikes hit both ends of the train (version 1) or two points on the platform opposite to the ends of the train (version 2).[232] According to the claim of special relativity, while P_2 will observe

---
[231] In fact, there is hardly any scientific theory that is as controversial and contradictory among its followers (let alone non-followers) as special relativity. Since it is supposed to be a test bed for ingenuity, there is no shortage of proposals or lack of imagination in interpreting and defending this theory.

[232] In fact, there are many versions of this thought experiment where some of these versions come as an improvement to previous versions and hence they are designed to avoid potential traps and errors in

9.6.1 Train Thought Experiment

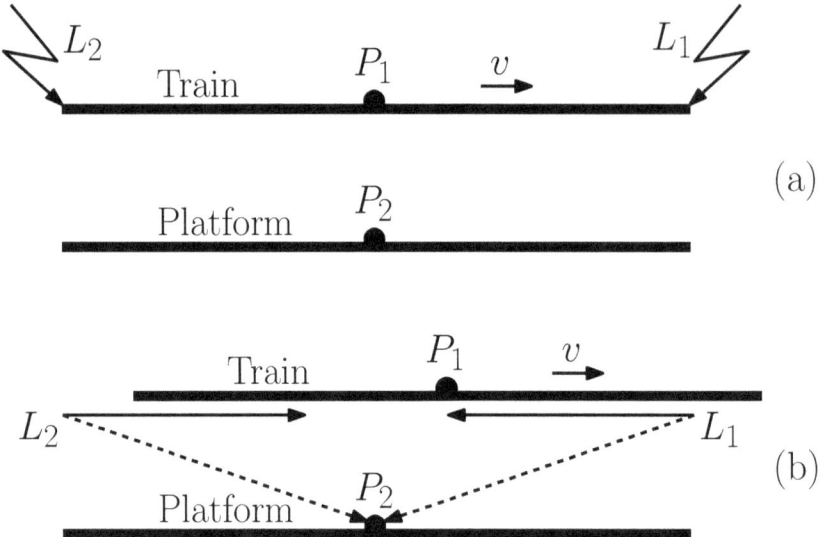

Figure 19: Train thought experiment where in (a) lightning strikes (L_1 and L_2) hit the front and back of the train (or platform) and in (b) the light signals from these strikes are seen at later times by P_2 simultaneously because he stays in the middle of the platform and by P_1 non-simultaneously because he is moving towards the front light signal and hence he sees it before the back light signal.

these events simultaneously (and hence to him they are simultaneous events) because he stays at rest in the middle between the two striking points, P_1 will observe the front lightning before the back lightning (and hence to him they are not simultaneous events) because during the time of propagation of the two light signals he has moved forward and hence the front signal needs less time to reach him than the back signal. The situation is depicted schematically in Figure 19. In brief, according to the logic of this thought experiment the same events are simultaneous to one person (i.e. P_2) but not simultaneous to another person (i.e. P_1) and therefore simultaneity (and hence time) is relative and not absolute.

In the following points we discuss this thought experiment highlighting some potential criticisms to its logic and analysis (or at least to its presentation which creates unduly confusion):[233]

1. Regarding version 1, let remove the platform and P_2 and repeat this thought experiment. Should P_1 then observe these events simultaneously or not? If he observes them simultaneously then the presence of the other frame and observer (i.e. platform and P_2) has this effect which no one can explain, while if he still observes them non-simultaneously

---

the older versions. We will give some examples of these versions later.

[233] This does not mean that we reject the conclusion of this experiment about the relativity of simultaneity (i.e. of occurrence although the thought experiment is more appropriate to be about the relativity of simultaneity of observation) which may be derived from the formalism of Lorentz mechanics (regardless of its significance and interpretation) but we want to highlight the potential vulnerability of special relativity (and indeed any theory) when it is presented in this tattered form.

9.6.1 Train Thought Experiment

then we should need an external, and may be absolute, rest frame relative to which the train is moving in the forward direction and relative to which the light propagates in both directions with a constant speed.[234] If we dig deep into this version of the train thought experiment and its logical roots, we see that the whole logic of this experiment is based on the existence of an absolute frame of reference in the background which is at rest, or even identical, with the frame of P_2 and that is why we see in this thought experiment the front light signal reaches P_1 earlier than the rear light signal. It seems that the thought experimenter unwittingly identified himself with the person on the platform and that is why he assumed that everything (except the train and the traveler) in this experiment belongs to his frame including the front and back striking points and even the speed of light. In fact, according to this version, it may be claimed that the events should be simultaneous for P_1 and non-simultaneous for P_2 because the striking points belong to the frame of P_1. In brief, according to the available special relativistic interpretations of this version, the whole theory could be demolished by this thought experiment because some of these interpretations are based, wittingly or unwittingly, on abolishing the postulate of unrestricted relativity due to the implicit presumption of the existence of an absolute frame in the background. In fact, some of these special relativistic interpretations should even lead to the disposal of the second postulate of special relativity about the constancy of the speed of light since this speed should have been presumed to be dependent on the speed of its observer to explain why the front light signal is faster than the back signal in reaching the middle of the train. If the speed of light is really independent of the motion of the train relative to the platform then both signals should arrive to the middle of the train at the same time unless we assume the presence of a frame (other than the train frame) relative to which the train is moving and in which the light is propagating with speed c. This implicitly-presumed frame should resemble the frame of a propagation medium of a classical wave model.

2. Regarding version 2, it faces similar criticisms to those directed to version 1. Moreover, we can assume a classical projectile model for the propagation of light and hence we can logically presume that the rest frame of the source of light is the frame of the platform and hence the velocity of light as observed by P_1 has a component from the velocity of its source, as we found earlier, and this may justify the logic of the special relativistic interpretation about the non-simultaneity. However, this justification is inconsistent with the framework of special relativity where the speed (and even the velocity) of light is supposed to be independent of the motion of its source. In fact, this justification should lead to the disposal of the second postulate of special relativity about the constancy of the speed of light since this speed should have been presumed dependent on the speed of its source (like a classical projectile model) to explain why the front light signal is faster than the back signal in reaching the middle of the train. If the speed of light is really independent of the motion of the train relative to the platform then both signals should have the same speed relative to P_1 and hence they should arrive to the middle of the train at the same time.[235] We should remark that,

---

[234] This is inline with a classical wave propagation model.
[235] It may be claimed that the arrival of the two signals non-simultaneously to P_1 is because the lightning

9.6.1 Train Thought Experiment

unlike the special relativistic interpretation, the presumed added velocity component should not affect the invariance of the speed of light according to our interpretation to this invariance and hence the speed of light is c in both frames despite the added velocity component in the P_1 frame (refer to the exercises and § 9.7.1).

3. This thought experiment should also be criticized for being based on a confusion between the simultaneity of observation and the simultaneity of occurrence (see § 7.4) because even if one of the observers (say P_1) does not observe these events simultaneously, he can (knowing his movement) easily conclude that they occurred simultaneously by accounting for the time difference in his observations.[236] So, even if we assume that the logic of the train thought experiment is fine, this just shows that the whole experiment and its outcome lead to a very trivial conclusion that is: some simultaneous occurrences are not simultaneous observations (and vice versa although this may not be concluded from this experiment), due to the difference in the distance between the local observer and the locations of the events and the finity of the speed of light, and hence the simultaneity of observation is not absolute. This conclusion should be very obvious to every one (or at least to the physicists and astronomers) and does not represent any revolution in the definition of time and simultaneity and the emergence of a new theory that abolishes the old concepts of time and simultaneity of classical mechanics as it is supposed to be claimed by the outcome of this thought experiment. In this regard, we refer to two examples from the history of classical physics about the awareness and acceptance of the relativity of simultaneity of observation[237] in classical physics long before the emergence of Lorentz mechanics and its precursors: (a) Galileo attempt to measure the speed of light which should be based (tentatively at least) on the assumption that this speed is finite and (b) Romer determination of the speed of light from his observation of the eclipses of one of the Jupiter moons. Both these examples indicate that the finity of the speed of light, and hence the time lag between the occurrence and observation which leads to the relativity of simultaneity of observation, were known long before the emergence of Lorentz mechanics and these thought experiments.

4. As discussed earlier, even if the logic and analysis of this thought experiment is perfect, it does not lead to the special relativistic conclusion about the abolishment of absolute

---

strikes are non-simultaneous relative to P_1 and not because of the non-invariance of the speed of light. However, this claim is meaningless because if we already assume that these events are non-simultaneous relative to P_1 and simultaneous to P_2 then we demolished this thought experiment and its objective, and hence this assumption will be completely arbitrary and requires justification.

[236] In fact, this is the major refute to the train thought experiment since the two observers (P_1 and P_2) are local and not global observers and hence this thought experiment (as presented in the literature of special relativity) may be useful for demonstrating the relativity of simultaneity of observation but not the relativity of simultaneity of occurrence. Accordingly, we believe that the train thought experiment (as presented in most sources of special relativity literature) is an instance of the former rather than the latter as it is supposed to be. However, we did not present this as the main refute to this thought experiment to allow for other potential variations and presentations of this thought experiment and to be thorough in our attack to the logic of this thought experiment (whether it is about the relativity of simultaneity of observation or about the relativity of simultaneity of occurrence).

[237] The relativity of simultaneity of observation is essentially based on the finity of the speed of signals (which are usually the light signals) that carry the information and facilitate the observation.

9.6.1 Train Thought Experiment

time, which is usually the purpose of this thought experiment, because the relativity of simultaneity of occurrence is compatible with the existence of absolute frame and hence absolute time, as discussed earlier (see for example § 7.4).

5. We should also note that the phrasing of this thought experiment (as well as other phrasings that are found in the literature) is based on the assumption of the existence of a universal frame in the background where the lightning strikes hit the train or the platform in simultaneity to the passing of the middle of the train by the midpoint of the platform as can be understood from the statement "At the instant when the middle of the train passes by the midpoint of the platform, lightning strikes hit both ends of the train ... etc.".[238] It is obvious that according to the relativity of simultaneity this simultaneity should be ascribed to a particular frame to be sensible. In fact, this background frame can be detected in many of the special relativity arguments and thought experiments despite the denial of the existence of such a frame.

We remark that some authors (according to another version) seem to suggest that lightning strikes hit two points on the platform opposite to the front and back end of the train and they caused clocks on the ground in front and back to stop simultaneously but they caused clocks at front and back of train to stop in different times. We think this claim is arbitrary and cannot be proved or justified. We should also remark that there are many other variants of this thought experiment where the designers of these modified versions tried to avoid some of the traps of the original forms of this thought experiment. However, most (if not all) these variants suffer from the same or similar challenges and hence they should not be considered as different thought experiments or deserve to be addressed independently.

To conclude, although legitimate consequences and predictions like the relativity of simultaneity (in its correct interpretation) can be established from analyzing the formalism of Lorentz mechanics, they cannot be established by this type of shaky logic and dodgy thought experiments which may lead to wrong conclusions such as the claim of the abolishment of absolute frame. In fact, questionable arguments and thought experiments like this one are behind the rejection of Lorentz mechanics or the skepticism about its sensibility and rationality by some who use these arguments and thought experiments to attack even the formalism despite the experimental evidence in its support. As we saw, the special relativity methods and logic made the relativity of simultaneity (as represented by the train thought experiment) and its alike a fertile ground for confusion and provided many causes, whether legitimate or illegitimate, for the rejection of Lorentz mechanics altogether. In brief, we believe that reliable science should not be based on thought experiments, especially like this one, although they may serve as an accessory to demonstrate and structure some scientific ideas and arguments if they are phrased and presented correctly and if the science of these experiments has already been established on a firm ground such as by analyzing the formalism.

Solved Problems

1. Make a brief assessment of the special relativistic interpretation of the train thought

---

[238] In fact, even the simultaneity of the lightning strikes should be questioned and not only the simultaneity between these strikes and the passing by the midpoint.

experiment.

Answer: We can say:

• Regarding version 1, to make sense of the special relativistic interpretation we need to assume a classical wave model for the propagation of light where P_2 is at rest relative to the propagation medium while P_1 is in motion relative to this medium. The characteristic speed of light c will then belong to the propagation medium and hence the observed speed of light should be dependent on the motion of the observer relative to the medium. These assumptions will lead to the collapse of both postulates of special relativity due to the existence of an absolute frame, represented by the medium of propagation, and the variance of the speed of light. Moreover, the thought experiment should be considered an example for the relativity of simultaneity of observation rather than the relativity of simultaneity of occurrence since the former is sufficient to explain the outcome of this experiment while the latter is not needed to explain this outcome. This could be endorsed by the fact that the simultaneity and non-simultaneity in this experiment belong to the observations of localized, rather than global, observers.

• Regarding version 2, we can say what we said about version 1, where we assume a wave propagation model. We may also consider a classical projectile model for the propagation of light where the characteristic speed of light belongs to the rest frame of the light source, represented in this version by the frame of P_2, and hence the front light signal is faster than the back light signal relative to P_1 and that is why the front signal is observed by P_1 before the back signal. This assumption should lead to the collapse of the second postulate of special relativity. Moreover, the whole experiment should be about the relativity of simultaneity of observation rather than occurrence. We note that if we consider a projectile model in version 1 then we should conclude that the events will be simultaneous to P_1 and non-simultaneous to P_2 because the source of light (i.e. the front and back of train) are in the frame of P_1 and hence P_2 should observe the back signal before the front signal.

Exercises

1. Describe another variant of the train thought experiment where a light signal is emitted at the center of the train.
2. Discuss the difference between the simultaneity of observation and the simultaneity of occurrence.
3. Make an argument in support of the claim that the train thought experiment is an example of the relativity of simultaneity of observation and not the relativity of simultaneity of occurrence.
4. Regarding version 2 of the train thought experiment, justify why the added velocity component is rejected as a possible rationale for the special relativistic interpretation of non-simultaneity although this added component can be concluded from analyzing the formalism and hence it is accepted according to our view.
5. Analyze the two versions of the train thought experiment (as given in the main text) using the Lorentz spacetime coordinate transformations and hence conclude the relativity of simultaneity. Comment on the results.

9.7 Light Clock   277

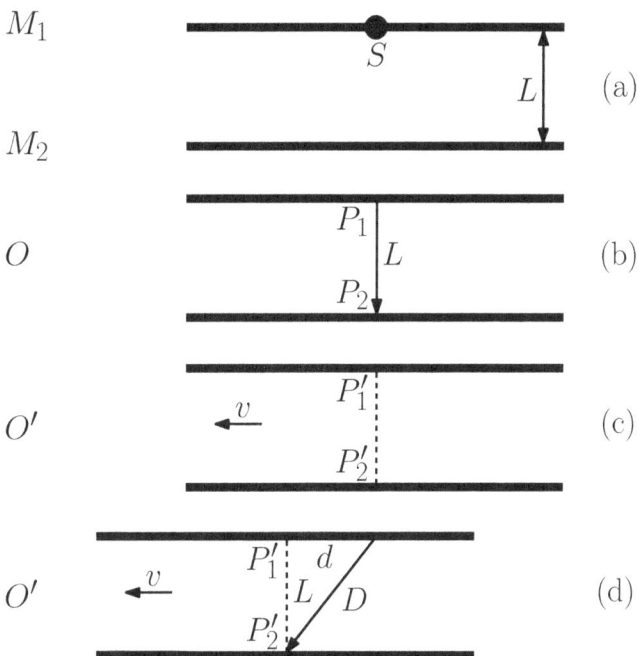

Figure 20: A schematic diagram that illustrates the construction and setting of light clock and its functionality as seen by two inertial observers who are in a state of relative uniform motion.

6. Analyze the other variant of the train thought experiment, which is given in exercise 1, using the principles of Lorentz mechanics as obtained from analyzing its formalism. Comment on the results.

9.7 Light Clock

Light clock is an abstract device that is commonly used in the literature of special relativity to demonstrate the effect of time dilation as a result of motion and even to derive the Lorentz transformations with the support of the postulates of special relativity. The essence of this device is a light signal that is used to measure time interval by its journey between two parallel mirrors where it is shown, according to the special relativity view, that this clock ticks at different rates for two inertial observers who are in a state of relative uniform motion. A schematic diagram of this device and its functionality as seen by the two observers is shown in Figure 20.

In the following points, we outline the construction and functionality of this device and the physical principles and assumptions that the device is based upon:
1. As seen in Figure 20 (a), the light clock is made of two parallel mirrors M_1 and M_2 separated by a distance L with a light source S positioned in the middle of M_1.
2. The time measuring mechanism of this clock is based on counting the ticks where each tick is made of the forward or backward journey of the light signal, which is emitted from S, between M_1 and M_2.

9.7 Light Clock

3. The light clock is in the rest frame of the inertial observer O (Figure 20 b) and is moving to the left with a constant speed v according to a second inertial observer O' (Figure 20 c).
4. When a light signal is emitted from S it will be seen by O to follow the route $P_1 P_2$ (Figure 20 b) and it will be seen by O' to follow the slant route (shown as solid line in Figure 20 d) that connects P_1' (as in Figure 20 c) to P_2' (as in Figure 20 d) due to the displacement of the clock by a distance d to the left during the time of the tick according to O'.
5. Now, since the light speed is the same (i.e. c) for both observers (at least according to the second postulate of special relativity), then the time interval of this tick according to O is equal to $\Delta t = L/c$ while the time interval of this tick according to O' is equal to $\Delta t' = D/c$.[239] Now, from the Pythagoras theorem we have:

$$D^2 = L^2 + d^2 \tag{336}$$
$$(c\Delta t')^2 = (c\Delta t)^2 + (v\Delta t')^2 \tag{337}$$
$$(\Delta t')^2 = (\Delta t)^2 + \beta^2 (\Delta t')^2 \tag{338}$$
$$(\Delta t')^2 - \beta^2 (\Delta t')^2 = (\Delta t)^2 \tag{339}$$
$$(\Delta t')^2 (1 - \beta^2) = (\Delta t)^2 \tag{340}$$
$$\Delta t' \sqrt{1 - \beta^2} = \Delta t \tag{341}$$
$$\frac{\Delta t'}{\gamma} = \Delta t \tag{342}$$

and hence the time interval of a tick as measured by the two observers will be related by:

$$\Delta t' = \gamma \Delta t \tag{343}$$

which is the formula of the time dilation effect according to Lorentz mechanics (see Eq. 330).

We should remark that $d = v\Delta t'$ is used instead of $d = v\Delta t$ because the displacement distance d of the light signal is seen in O' frame. This can be seen from Figure 20 (c) and (d). We should also remark that the demonstration of the light clock may also be based on counting a tick by the forward-backward journey. This will not introduce any fundamental change in the above explanation and derivation although it requires the assumption of isotropy of space with regard to the speed of light which is already assumed since the speed of light is supposed to be the same in all inertial frames regardless of location, direction and observer. In fact, the whole idea of light clock is based on the constancy of the speed of light and hence the forward and backward ticks should be equal in this sense. Anyway, if this clock is supposed to be used for actual time measurement then using the forward-backward journey as a single tick will be more safe since this tick

---

[239] We note that we are assuming here that time is measured by discrete rather than continuous units. This should motivate a deeper examination to the notion and definition of time, its quantification and calibration.

9.7 Light Clock

will be the same even if the times of the two journeys are not equal.[240] However, we preferred the one-way journey in our demonstration because it is sufficient for the purpose of derivation and is more tidy and simple.

Based on the above explanation of the light clock, we see that the essence of the time dilation effect, as demonstrated by the light clock, is that because the distance that the light signal should travel between the two mirrors during each tick is longer in the frame of O' than in the frame of O while the speed of this signal is identical in both frames, then the time of a tick as measured in the frame of O' should be longer than the time as measured in the frame of O. Accordingly, this light clock will be seen by O' to tick at a slower rate than its ticking rate in the frame of O (or than the ticking rate of an identical clock that is in the frame of O') since the interval of each tick in the frame of O is shorter than the interval in the frame of O'.[241]

As we will see in the exercises, light clock seems to follow the rules of classical mechanics in some aspects. Also, light clock may be criticized by the claim that the time measurement mechanism is based on the orientation of the clock. Accordingly, if the light clock is rotated 90° for example the time measurement mechanism will change (refer to the details of Michelson-Morley experiment in § 2.6 and § 12.2 as well as the questions of the present section). However, this should be linked to the length contraction effect in the direction of motion (which exists in Lorentz mechanics but not in classical mechanics) as discussed in the exercises of this section and in other parts of the book, although this may not be sufficient to address this issue as we will see. It is also linked to the displacement of the clock during the tick and the added velocity component of light from the velocity of its source, as will be discussed and analyzed in detail in § 9.7.1. We also note that "shorter" and "longer" in the above discussion requires the concept of a universal and possibly even absolute time to regulate and standardize the flow of time. However, this can be refuted by the principles of special relativity (and even Lorentz mechanics) where the definition of time is based on the speed of light[242] which is universal and absolute although this may also depend on the interpretation of this as being real or apparent for example. More details about these issues will be discussed in the solved problems and exercises as well as in § 9.7.1.

---

[240] Being "equal" or "not equal" here is based on using a continuous standard of measurement and calibration by an external agent which may be challenged. However, if the standard of spacetime is the actual speed of light then they should be equal, and this indicates the deep roots of the invariance of the speed of light in Lorentz mechanics. In fact, we may even say that the assumption of isotropy and homogeneity will necessarily imply this invariance. These philosophical and epistemological issues about the meaning of time and its measurement and calibration require deep inspection.

[241] This should be fully justified by adopting discrete rather than continuous units for time measurement, as indicated earlier.

[242] Or may be: the *actual* speed of light which may explain its universal and absolute nature since it is the most fundamental attribute in the measurement and calibration of spacetime with all other "relative" concepts and attributes (like time) being derived from it and based upon it. This is unlike classical mechanics where the fundamental attributes are space and time which may explain their invariance properties, as discussed earlier and will be clarified more later. This should highlight the deep philosophical and epistemological difference between classical and Lorentz mechanics and their theoretical frameworks.

Solved Problems

1. Assess light clock and its use as a basis for time measurement.

   Answer: There are several observations to note about light clock and its use; some of these observations are:

   • Light clock is based on considering the speed of light as the standard for time and space measurement. This means that if we use a physical phenomenon other than the speed of light as a standard for time and space measurements, we could have different transformations of spacetime coordinates from those of Lorentz. Moreover, the characteristic Lorentzian effects and consequences, like length contraction and relativity of simultaneity, may not be needed to explain the physical laws. Accordingly, we may obtain a completely consistent physical theory for spacetime coordinate transformations, which is different from the Lorentz transformations and Lorentz mechanics, if we change this standard (i.e. light speed) for spacetime calibration and measurement.

   • Basing the measurement of time on the speed of light may cast a shadow on the generality of the definition of spacetime and hence the time may not necessarily flow with the same rate if we use a physical phenomenon other than the speed of light to standardize space and time measurements. This issue is related to a number of important issues such as the issue of aging of travelers that we discussed earlier, i.e. the aging of traveling living beings may not follow the rate of physical time flow according to Lorentz mechanics since the biological clock may not be based on the speed of light as a standard for time flow (see for example § 3.4 and § 4.1.2).

   • Light clock is based on using a discrete measurement unit (i.e. tick)[243] and this should have an impact on the definition of time and its measurement. Accordingly, the nature of time and its measurement could change if we adopt a measurement unit of different nature.

Exercises

1. Derive the time dilation formula from the light clock using a simple plot to illustrate the underlying physical principles.
2. Show that light clock, as described in the text, follows the rules of classical mechanics in some aspects where the light signal behaves like a classical projectile.
3. Make another argument for the case that the light clock follows the rules of classical mechanics in some aspects.
4. Assess a possible criticism to the light clock that its functionality (and hence time and time dilation) is based on its orientation.[244]

---

[243] This means that the unit of time is represented by "the actual distance of a tick (representing a unit of time) divided by c" instead of "the actual distance of a tick (but not representing a unit of time) divided by c".

[244] We note that there is nothing wrong in principle with the dependence of functionality on orientation since the physical principles on which the light clock rests can depend on orientation. The purpose of this question and its alike is to give typical examples about the framework of special relativity and its potential vulnerability to criticism due to questionable arguments and interpretations even when the results may be correct. In brief, our focus is potential criticism to this particular interpretation

5. Make a formal argument for the case that if we accept the logic and argument of special relativity then the functionality and time measuring mechanism of light clock should depend on its orientation in a way that is inconsistent with the framework of this theory. Comment briefly on the result of your analysis.
6. Assess the consequences of the claim that the light clock is dependent in its functionality on its orientation within the framework of special relativity.
7. What is the point of the previous exercises where some results that have already been shown to be obtainable from the formalism of Lorentz mechanics were challenged and shown to be wrong?

9.7.1 Assessing Light Clock

There are several aspects to the light clock that require further scrutiny and analysis. In the following we outline some of these aspects. Before that, we should remark that the light clock is supposed to be based on the second postulate of special relativity whose essence is the invariance of the observed speed of light. Now, let analyze this abstract device further and examine the effect of these aspects on the concluded results. Let consider light as a corpuscular phenomenon (say photons) and hence what is observed by O and O' is a photon emitted from the light source at the top mirror and it hits the lower mirror at a later time. The same analysis equally applies to a wave of light where the wave front plays the role of photon. We refer the reader in this analysis to Figure 21 which is a simplified version of Figure 20 with minor modifications. We will see from the analysis that for the light clock to work it should be assumed that the velocity of light is dependent on the velocity of its source but its speed is not. In other words, while the speed of light is independent of the velocity and speed of its source, the velocity of light is dependent on the velocity of its source. This is because the light is supposed, according to the working principle of light clock, to have a sidewise velocity component from the sidewise motion of its source. Although such a conclusion can be obtained from analyzing the formalism of Lorentz mechanics, as we saw for example in § 4.2.2 and § 4.2.3, special relativity fails to provide convincing explanation according to its principles and postulates.

In Figure 21 we plot the light clock where we label two opposite points of shortest distance on the two opposite sides of the clock. The unprimed labels A and B belong to observer O while the primed labels A' and B' belong to observer O'. At a certain instant of time $t = t' = 0$ the two observers agree on the spatial coordinates of these points, and this is depicted in (a). At this instant of time (i.e. $t = t' = 0$) a photon is emitted from point A-A' in the direction of point B-B'. At a later time t_O the photon is seen by O to hit B, and this is depicted in (b). Because the clock is in the rest frame of O the photon is supposed to be seen by O to follow the straight line AB which is shown in (b). Regarding O', his situation in this experiment is represented by part (c) of Figure 21. In fact, there are two main possibilities for the path of the photon according to O': either the velocity of light is independent of the velocity of its source or it is not. In the following, we investigate both these possibilities as far as O' is concerned.

---

and its presentation.

9.7.1 Assessing Light Clock

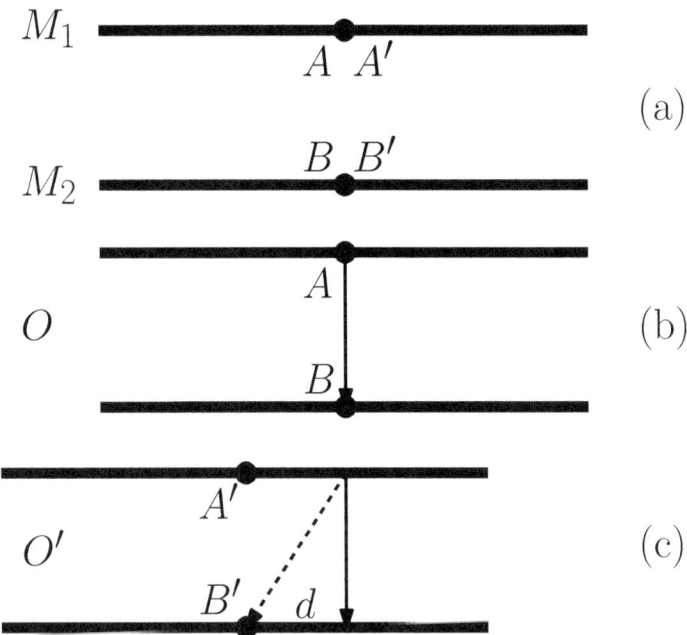

Figure 21: A schematic diagram that illustrates the setting of light clock and its functionality as seen by two inertial observers who are in a state of relative motion where the point of emission and the point of impact of a photon are labeled.

(i) The velocity of light is independent of the velocity of its source: this possibility means that as soon as the light photon is emitted, it will move with speed c and in the direction of emission regardless of the velocity of its source in the direction of emission or in the lateral direction (i.e. the perpendicular direction to the direction of emission). In fact, this possibility is based on assuming that light follows similar rules to those of waves in classical mechanics (see § 1.13.2). This possibility may seem to be consistent with the second postulate of special relativity in some aspects since the propagation of light following its emission from the source is independent of the motion of the source because it does not have any component from the velocity of its source. According to this possibility, the light may be thought to follow the solid path in Figure 21 (c). However, this possibility should be rejected for the following reasons:
• This possibility will destroy the whole logic of the light clock device according to special relativity and the derivation of time dilation formula from this abstract device as outlined in § 9.7. Hence, it should be rejected at least from a special relativistic view.
• There is no justification for the light signal to follow this route unless we adopt a wave propagation model and hence we believe first in the existence of a propagation medium (e.g. ether) and we believe second that O' is actually in the rest frame of this medium and hence O is moving to the left relative to the medium with speed v.
• Also this possibility is against the principles of reality and truth since the point of impact of the photon on the mirror according to O' will be different from the point of impact according to O unless we assume that the point of impact seen in (c) as the end

9.7.1 Assessing Light Clock

of the solid path is the same for O which is inline with the proposition that O is moving relative to the medium with speed v. Accordingly, the path as seen by O is not as depicted in (b).

(ii) The velocity of light is dependent on the velocity of its source: according to this possibility, the light will follow the dashed path in Figure 21 (c) because the velocity of light will have a component in the sidewise direction, which is obtained from the sidewise motion of its source, according to O'. In fact, this possibility is based on assuming that light follows similar rules to those of projectiles in classical mechanics (see § 1.13.1), and hence its velocity depends on the relative motion between the source and observer.[245] However, this possibility can also be challenged by the fact that it is inconsistent with the commonly held view (which is explicitly expressed in the literature of special relativity) that the velocity of light is independent of the velocity of its source. Yes, if we assume that "velocity" is used negligently to mean speed then this may be an acceptable excuse to rectify the situation. However, special relativity as an interpretative theory is still required to explain why the speed of light is independent of the speed of its source while the velocity of light is dependent on the velocity of its source.[246] Accordingly, we do not have a convincing special relativistic explanation to the supposed functionality of the light clock even according to this possibility (i.e. ii). Now, since the whole logic of the light clock and the derivation of the time dilation formula according to special relativity are based on this possibility, we conclude that the light clock device is inconsistent with the principles of special relativity and with the second postulate in particular (at least because special relativity fails to provide a convincing explanation that is based on its postulates and principles).

Based on the above analysis, we can conclude that the light clock, if it is really working according to the prescription of special relativity, is an evidence against special relativity as it indicates the fallacy of its second postulate or an inconsistency in the framework of special relativity because the functionality of light clock is either based on a violation of the second postulate or at least it is not interpreted correctly within the framework of special relativity which is an interpretative theory. Accordingly, time dilation as derived from light clock can be seen as an evidence against special relativity and in support of other interpretations of Lorentz mechanics. In brief, according to both possibilities (i) and (ii), the light clock does not follow the logic of special relativity and its interpretative framework.

Now, let analyze this thought experiment further, but this time according to our approach of postulating the Lorentz spacetime coordinate transformations, where we follow in the setting and performance of this clock what is stated in the literature of special relativity

---

[245] In § 1.13.1 only motions in one dimension were assumed (i.e. the motion of signal and the relative motion between source and observer are in the same orientation) while here we are assuming motions in two dimensions (i.e. the motion of signal and the relative motion between source and observer are in perpendicular orientations) and hence we need here the concept of velocity in multi-dimension.

[246] We suggested an explanation for this by the contraction of the spacetime coordinates and the added velocity component to the clock which compensates for any added velocity component to the photon from the velocity of the source, as will be detailed later.

9.7.1 Assessing Light Clock

regardless of the above possibilities, (i) and (ii), although they may still play a role in its interpretation later. So, let assume that light is made of photons (with no regard to the rules that it follows in its propagation and whether it behaves like wave or like projectile), and hence what is observed by O and O' is a photon emitted from the light source at the top mirror and it hits the lower mirror at a later time. We refer the reader again to Figure 21 in this analysis where (a) represents the mirror when the photon was emitted and (b) represents the mirror when the photon hits the lower mirror as seen by O and hence we are assuming that in O frame the speed of the photon is c and it follows the path AB. As we see, the mirror during the time of travel of this photon across the clock has moved to the left according to O', as seen in Figure 21 (c), and hence we need to find the photon path and its speed according to O'. In other words, we need to find the velocity of the photon in O' frame where the path indicates the direction and the speed represents the magnitude of this velocity.

To solve this problem we use the basic Lorentz spacetime coordinate transformations which are postulated according to our approach. So, to find the path and speed of light in O' frame we need to find the velocity of the photon in this frame and hence we use the Lorenz velocity transformations, which are derived directly from the Lorentz spacetime coordinate transformations as shown in § 5.1.2. On applying these velocity transformations to the photon we find that according to O' the photon has a velocity component v in the direction of motion of the clock and a velocity component $\frac{c}{\gamma} = c\sqrt{1-(v/c)^2}$ in the perpendicular direction to the motion (refer to the exercises of this section and § 4.2.2 and § 4.2.3). On adding these components vectorially, we conclude that the speed of the photon in O' frame is c and its velocity makes an angle $\theta' = \arctan\sqrt{(c^2/v^2)-1}$ to the direction of motion (i.e. $\frac{\pi}{2} - \theta'$ to the direction of photon velocity in O frame) and hence it follows the dashed path that is shown in Figure 21 (c). Now, by using the Pythagoras theorem (as done by special relativists in § 9.7) we obtain the time dilation formula (and potentially even length contraction according to one possibility) as obtained in special relativity. In fact, the derivation of time dilation (and subsequently length contraction) is redundant in our approach because according to our approach these effects are either obtained directly from the Lorentz spacetime coordinate transformations, which are postulated, or these effects are postulated as a basis for the derivation of the Lorentz spacetime coordinate transformations (see § 12.4.2). Anyway, the derivation of these effects is harmless and can serve as a consistency check.

In brief, from our analysis to the light clock, which is completely based on analyzing the Lorentzian formalism, we reach two important conclusions:
1. The functionality of the light clock (as described by special relativists) is completely based on the Lorentz spacetime coordinate transformations. Now, since these transformations originate from the contraction of the spacetime coordinates by the motion through spacetime where this contraction originates from (or reveals itself by) the effects of length contraction and time dilation, then the constancy of the speed of light is no more than a demonstration of these effects. More fundamentally, these effects are related to the use of the speed of light for calibrating the measurements of spacetime

9.7.1 Assessing Light Clock

coordinates and hence the whole story originates from our convention to adopt this calibration. So, if we adopt another calibration standard, then the speed of light could be frame dependent while another quantity (i.e. the standard quantity) will be invariant and hence we should obtain spacetime coordinate transformations that differ from the Lorentzian transformations and a physical theory that differ from Lorentz mechanics although both theories could be correct as they provide accurate description of the physical phenomena but from different aspects and perspectives.

2. Although the speed of light is independent of the speed (and velocity) of its source, its velocity is dependent on the velocity of its source. So, the constancy of the speed of light can be seen from a different angle as the result of combining two intriguing effects which balance each other to make the speed of light constant despite the relative motion between the frames. These two effects are the distortion of spacetime coordinates due to the motion and the dependence of the velocity of light on the velocity of its source since a velocity component is added to the velocity of light from the velocity of its source and this added component neutralizes the effect of the displacement of the objects (represented by the clock) in the moving frame. The result of combining these effects is the constancy of the speed of light and its invariance across all inertial frames. In brief, a photon behaves like a material projectile and hence it has a velocity component from the velocity of its source like any material object, so apart from its intrinsic speed which is relative to its source, it is in a state of rest relative to the objects in its source frame and hence its velocity in any frame (source frame or else) originates from its intrinsic velocity only. Now, the scaling of time and space by the same γ factor in any moving frame relative to the source frame will lead inevitably to having a constant speed across all frames because in any frame c is scaled by unity from the c of the source frame.[247] The logic of the last step is based on the fact that c in Lorentz mechanics is the standard for spacetime coordinate calibration.

Solved Problems

1. Using the Lorentzian velocity composition formulae (instead of special relativity principles), analyze the light clock in two orientations: (a) along the x direction (i.e. the mirror is oriented along the direction of its motion as in Figure 21), and (b) along the y direction (i.e. the mirror is oriented perpendicular to the direction of its motion and hence it is rotated 90° from its orientation in Figure 21). Hence, find the relation between the time in the two frames (i.e. derive the time dilation formula if possible) and assess how time is measured by this clock considering the speed of light as the standard for time measurement. Also, assess the results with regard to the orientation of the clock and how time is calculated in each case.
   Answer: In both cases we label O', O and the light signal with 1, 2 and 3.
   (a) Along the x direction: in this case the signal is propagating along the y direction in

---

[247] The language used here (e.g. "the scaling of time and space by the same γ factor in any moving frame...etc.") is meant to clarify the basic idea; otherwise this should be understood in accord with our interpretation where an absolute frame, which is the ultimate reference for any spacetime contraction and scaling, does exist.

9.7.1 Assessing Light Clock

O frame (i.e. the clock rest frame) and hence we have $u_{x21} = v$, $u_{x32} = 0$ and $u_{y32} = c$. Therefore, the velocity components of the light signal in O' frame are:

$$u_{x31} = \frac{u_{x32} + u_{x21}}{1 + \frac{u_{x32} u_{x21}}{c^2}} = \frac{0 + v}{1 + 0} = v$$

$$u_{y31} = \frac{u_{y32}}{\gamma \left(1 + \frac{u_{x21} u_{x32}}{c^2}\right)} = \frac{c}{\gamma(1+0)} = \frac{c}{\gamma} = c\sqrt{1 - (v/c)^2}$$

$$u_{z31} = 0$$

Accordingly, the speed of the light signal in O' frame is:

$$\begin{aligned} |\mathbf{u}_{31}| &= \sqrt{(u_{x31})^2 + (u_{y31})^2 + (u_{z31})^2} \\ &= \sqrt{v^2 + c^2\left(1 - (v/c)^2\right) + 0^2} \\ &= \sqrt{v^2 + c^2 - v^2} \\ &= c \end{aligned}$$

and it is restricted to the xy plane where in O' frame it makes an angle θ' with the x axis, that is:

$$\theta' = \arctan \frac{u_{y31}}{u_{x31}} = \arctan \frac{c\sqrt{1-(v/c)^2}}{v} = \arctan \sqrt{(c^2/v^2) - 1}$$

So, in both frames the speed of light is c although the velocity is different due to the difference in direction[248] where in frame O it is along the y axis while in frame O' it is making an angle θ' with the x axis (or $\frac{\pi}{2} - \theta'$ with the y axis).

Now, since we take the characteristic speed of light as the standard for time measurement then in O frame we have:

$$\Delta t = \frac{L}{c}$$

while in frame O' we have:

$$\Delta t' = \frac{D}{c} = \frac{\sqrt{d^2 + L^2}}{c} = \frac{\sqrt{(v\Delta t')^2 + (c\Delta t)^2}}{c}$$

Now, from the Pythagoras theorem and the last two equations we have:

$$\begin{aligned} D^2 &= d^2 + L^2 \\ (c\Delta t')^2 &= (v\Delta t')^2 + (c\Delta t)^2 \\ (c\Delta t')^2 - (v\Delta t')^2 &= (c\Delta t)^2 \\ (\Delta t')^2 (c^2 - v^2) &= (c\Delta t)^2 \end{aligned}$$

---

[248] In fact, this difference in direction is one aspect of the distortion of spacetime by the motion which the theory of Lorentz mechanics is based upon.

9.7.1 Assessing Light Clock

$$c\Delta t' \sqrt{1-(v/c)^2} = c\Delta t$$
$$\Delta t' = \gamma \Delta t$$

which is the time dilation formula.

(b) Along the y direction: we use in this case the u_{x31} formula (because the signal is propagating along the x direction in O frame) where $u_{x21} = v$ and $u_{x32} = c$, that is:

$$u_{x31} = \frac{u_{x32} + u_{x21}}{1 + \frac{u_{x32} u_{x21}}{c^2}} = \frac{c+v}{1+\frac{v}{c}} = \frac{c+v}{c+v} c = c$$

while the other two velocity components of the light signal in O' frame (i.e. u_{y31} and u_{z31}) are zero because u_{y32} and u_{z32} are zero. Now, because the propagation of the light signal is restricted to the x dimension then the speed of the light signal in O' frame is equal to $|u_{x31}|$, i.e. c. So, in both frames the speed of light is c and it is along the common x-x' axis.

Now, since we take the speed of light as the standard for time measurement then in O frame we have:

$$\Delta t = \frac{L}{c}$$

while in O' frame (where the coordinates of spacetime of O frame are seen to be contracted by the γ factor relative to the coordinates in O' frame) we have:[249]

$$\Delta t' = \frac{L'}{c} = \frac{\gamma L}{c} = \frac{\gamma c \Delta t}{c} = \gamma \Delta t$$

which is the time dilation formula as derived in case (a). Regarding the displacement of the light clock during the signal journey across the two mirrors, it will be dealt with in the explanation of the constancy of the speed of light that will follow in the next question.

Assessment: If we compare the analysis of case (a) and case (b) we see that we are not using the same method in our time measurements and hence the results are not consistent. While in case (a) we use in our time calculation the actual distance traversed by the signal, in case (b) we use the distance traversed by the signal across the clock (i.e. excluding the added distance $d = v \Delta t'$ due to the displacement of the clock during the signal transmission across the clock).[250] Hence, if we take case (a) as a prototype for calculating time then we should have:[251]

$$\Delta t' = \frac{L' + d}{c}$$
$$\Delta t' = \frac{\gamma L + d}{c}$$

---

[249] The length contraction formula that we use here is based on our unified proper/improper perspective that we explained earlier with O' being the frame of observation from which O is seen to be moving.

[250] We are considering here the forward or backward journey of the signal as a complete tick. This should be justified later when we consider the more general case of clock orientation.

[251] In fact, d should be considered as signed quantity to account for the difference in direction.

9.7.1 Assessing Light Clock

$$\Delta t' = \frac{\gamma c \Delta t + v \Delta t'}{c}$$

$$\Delta t' - \beta \Delta t' = \gamma \Delta t$$

$$\Delta t' = \frac{\gamma \Delta t}{1 - \beta}$$

and hence we will not get the time dilation formula. This means that the functionality of the light clock depends on its orientation and hence this clock is suitable for time measurement only in its orientation in case (a), i.e. along the direction of motion. In fact, if we do not distinguish between the time count of the clock and the flow of time (i.e. we define time by the time count mechanism of the clock regardless of its orientation) then this will lead to the conclusion that the time itself (according to our definitions and conventions) depends on the orientation of the light clock, which should be absurd.

2. Referring to the first question, explain in each case why the speed of light is the same in both frames.
   Answer: To explain the constancy of the speed of light in case a (i.e. when the clock is oriented along the orientation of motion), we note that the velocity of light in O' frame has two components: a component in the direction of motion (i.e. u_{x31}) which comes from the velocity of the source, and a component in the perpendicular direction to the motion (i.e. u_{y31}) which comes from the velocity of light in O frame. The first component is v which is obvious because it is the velocity of the source, while the second component is c/γ which requires an explanation. The reason that the second component is c/γ is that the length in the y direction is not subject to the length contraction while the time of transmission is subject to the time dilation. Hence, when we transform this velocity component from O frame (i.e. $c = \Delta y / \Delta t$) to O' (i.e. $\Delta y' / \Delta t'$) we have only a single γ factor in the denominator of the speed of light where this factor comes from the time dilation effect (or the contraction of temporal coordinate), that is:

$$\frac{\Delta y'}{\Delta t'} = \frac{\Delta y}{\Delta t'} = \frac{\Delta y}{\gamma \Delta t} = \frac{c}{\gamma}$$

So, when we add these two velocity components vectorially, by using the Pythagoras theorem, to obtain the speed of light in O' frame we get:

$$\sqrt{v^2 + \frac{c^2}{\gamma^2}} = \sqrt{v^2 + c^2(1 - v^2/c^2)} = \sqrt{c^2} = c$$

i.e. the speed of light is also c in O' frame.
To explain the constancy of the speed of light in case b (i.e. the clock is oriented perpendicular to the orientation of motion), we note again that the velocity of light in O' frame should have two components: a component in the direction of motion v which comes from the velocity of the source, and a component in the direction of light propagation c'_x which comes from the velocity of light in O frame. We may call v the

9.7.1 Assessing Light Clock 289

extrinsic component and call c'_x the intrinsic component because the former originates from the source motion while the second originates from the intrinsic velocity of light itself (i.e. relative to its source). Unlike the previous case, both these components are along the same orientation, which is the x orientation, and hence the components should be added algebraically. Now, since both of these components can be in the positive and negative x direction, we have 4 cases, i.e. the combinations of $\pm|v|$ with $\pm|c'_x|$. However, because v and c'_x are 1D velocities, all these cases are represented by the single expression: $v+c'_x$ where both v and c'_x can be positive or negative. Now, since the speed of light in O' frame (as obtained above from the u_{x31} velocity composition formula) is c, then we should have:[252]

$$c = v + c'_x \qquad \text{i.e.} \qquad c'_x = c - v$$

So, how to explain the equality $c'_x = c - v$ (i.e. the intrinsic velocity component of light in O' frame is equal to c minus the velocity of the source)? In fact, there are two factors that determine what c'_x should be: the contraction of spacetime coordinates of O frame by the motion and the displacement of the clock in O' frame during the transmission of light across the clock. Regarding the first factor, the signal is propagating along the orientation of motion (i.e. x) and hence it will be subject to both length contraction and time dilation effects unlike case (a) where the signal was subject to time dilation only. So, when we transform c from O frame to O' frame we obtain:

$$\frac{\Delta x'}{\Delta t'} = \frac{\gamma \Delta x}{\gamma \Delta t} = \frac{\Delta x}{\Delta t} = c$$

and hence this factor will not affect the speed of light in O' frame. Regarding the second factor, it has opposite effect to the added (or extrinsic) velocity component from the source and hence it neutralizes (or annihilates) this added velocity component. Now, since the added velocity component due to the motion of source is v then the added velocity component due to the displacement of the clock should be $-v$. In brief, although the photon has an added velocity component from the velocity of its source, the light clock also has this added velocity component from the velocity of O frame. Accordingly, if we exclude the inherent part (as represented by c) of the velocity of light then the photon and the clock are in a state of relative rest. The result is that the photon will also be seen to be moving with speed c in O' frame because although the photon has an extra speed from the motion of its source, the light clock is also has this extra speed from its frame and hence what remains is the inherent speed of light c which in O' frame is also c where the latter c can be regarded as a scaled version of c in O frame (i.e. $c = \frac{\gamma \Delta x}{\gamma \Delta t}$) with the scale factor being unity due to the contraction of spatial and temporal coordinates by the same γ factor. Therefore, the velocity of the photon in O'

---

[252] In fact, we should have written $c = |v + c'_x|$ or $\pm c = v + c'_x$ because c represents a speed and hence it is positive. However, to ease the notation we write $c = v + c'_x$ where c in this equality (and its alike) represents a 1D velocity.

9.7.1 Assessing Light Clock

frame can be expressed formally as:[253]

$$v + c'_x = v + c - v = c$$

where in the middle equality one v represents the added velocity component due to the motion of source while the other v represents the compensated velocity component due to the motion of the clock. These two velocities are opposite in sign because they have opposite effects on the calculation of the resultant velocity of the photon in its journey across the light clock since one belongs to the photon (and hence it should be included in this calculation) while the other belongs to the clock (and hence it should be excluded).

The reader is also referred to the following questions to culminate the analysis and conclusions.

3. Repeat the first question but this time the light clock is in an arbitrary orientation where the direction of the light beam of the clock in the rest frame of the clock makes an angle $0 \le \theta \le \pi$ with the common x-x' axis.

Answer: As in the first question, we label O', O and the light signal with 1, 2 and 3. The light signal in O frame has two velocity components: $u_{x32} = c \cos \theta$ and $u_{y32} = c \sin \theta$. We also have $u_{x21} = v$. Hence, the velocity components of the signal in O' frame are:

$$u_{x31} = \frac{u_{x32} + u_{x21}}{1 + \frac{u_{x32} u_{x21}}{c^2}} = \frac{c \cos \theta + v}{1 + \frac{v \cos \theta}{c}} = \frac{c^2 \cos \theta + vc}{c + v \cos \theta}$$

$$u_{y31} = \frac{u_{y32}}{\gamma \left(1 + \frac{u_{x21} u_{x32}}{c^2}\right)} = \frac{c \sin \theta \sqrt{1 - (v/c)^2}}{1 + \frac{v \cos \theta}{c}} = \frac{c^2 \sin \theta \sqrt{1 - (v/c)^2}}{c + v \cos \theta}$$

$$u_{z31} = 0$$

Accordingly, the speed of the light signal in O' frame is:

$$|\mathbf{u}_{31}| = \sqrt{(u_{x31})^2 + (u_{y31})^2 + (u_{z31})^2}$$

$$= \sqrt{\left(\frac{c^2 \cos \theta + vc}{c + v \cos \theta}\right)^2 + \left(\frac{c^2 \sin \theta \sqrt{1 - (v/c)^2}}{c + v \cos \theta}\right)^2 + 0^2}$$

$$= c \frac{\sqrt{(c \cos \theta + v)^2 + \left(c \sin \theta \sqrt{1 - (v/c)^2}\right)^2}}{c + v \cos \theta}$$

$$= c \frac{\sqrt{c^2 \cos^2 \theta + 2cv \cos \theta + v^2 + c^2 \sin^2 \theta - v^2 \sin^2 \theta}}{c + v \cos \theta}$$

---

[253] In fact, this approach is the same as the approach used in case (a) but because the two components in case (a) are orthogonal they are added vectorially by using the Pythagoras theorem, while in case (b) the two components are along the same orientation and hence they are added algebraically.

9.7.1 Assessing Light Clock

$$\begin{aligned}
&= c\frac{\sqrt{c^2\left(\cos^2\theta + \sin^2\theta\right) + 2cv\cos\theta + v^2\left(1 - \sin^2\theta\right)}}{c + v\cos\theta} \\
&= c\frac{\sqrt{c^2 + 2cv\cos\theta + v^2\cos^2\theta}}{c + v\cos\theta} \\
&= c\frac{\sqrt{(c + v\cos\theta)^2}}{c + v\cos\theta} \\
&= c\frac{c + v\cos\theta}{c + v\cos\theta} \\
&= c
\end{aligned}$$

and it is restricted to the xy plane where in O' frame it makes an angle θ' with the x axis, that is:

$$\theta' = \arctan\frac{u_{y31}}{u_{x31}} = \arctan\left(\frac{\sin\theta\sqrt{1 - \beta^2}}{\cos\theta + \beta}\right)$$

So, in both frames the speed of light is c although the velocity is different due to the difference in direction where in frame O it makes an angle θ $(= \arctan[\sin\theta/\cos\theta])$ with the x axis while in frame O' it makes an angle θ' with the x axis.

Now, since we take the speed of light as the standard for time measurement then in O frame we have:

$$\Delta t = \frac{L}{c}$$

Regarding O' frame, the distance traversed by the signal during a tick is D (see Figure 22). Now, in O frame L has an x component $L_x = L\cos\theta$ and a y component $L_y = L\sin\theta$. In O' frame, the x component of O frame is seen contracted by length contraction relative to its length in O' frame and hence we have: $L'_x = \gamma L\cos\theta$ while the y component is not affected and hence we have $L'_y = L\sin\theta$. Hence, in O' frame D is given, according to the Pythagoras theorem, by:

$$D^2 = (d + \gamma L\cos\theta)^2 + L^2\sin^2\theta = (v\Delta t' + \gamma c\Delta t\cos\theta)^2 + (c\Delta t)^2\sin^2\theta$$

Accordingly, the time $\Delta t'$ in O' frame is given by:

$$(\Delta t')^2 = \frac{D^2}{c^2} = (\beta\Delta t' + \gamma\Delta t\cos\theta)^2 + (\Delta t)^2\sin^2$$

To verify this formula, we test it for the two cases which we already derived in the first question, that is case (a) which corresponds to $\theta = \pi/2$ and case (b) which corresponds to $\theta = 0$ or $\theta = \pi$.

Now, for $\theta = \pi/2$ we have $\cos\theta = 0$ and $\sin\theta = 1$ and hence the derived expression becomes:

$$\begin{aligned}
(\Delta t')^2 &= (\beta\Delta t')^2 + (\Delta t)^2 \\
(\Delta t')^2\left(1 - \beta^2\right) &= (\Delta t)^2
\end{aligned}$$

9.7.1 Assessing Light Clock

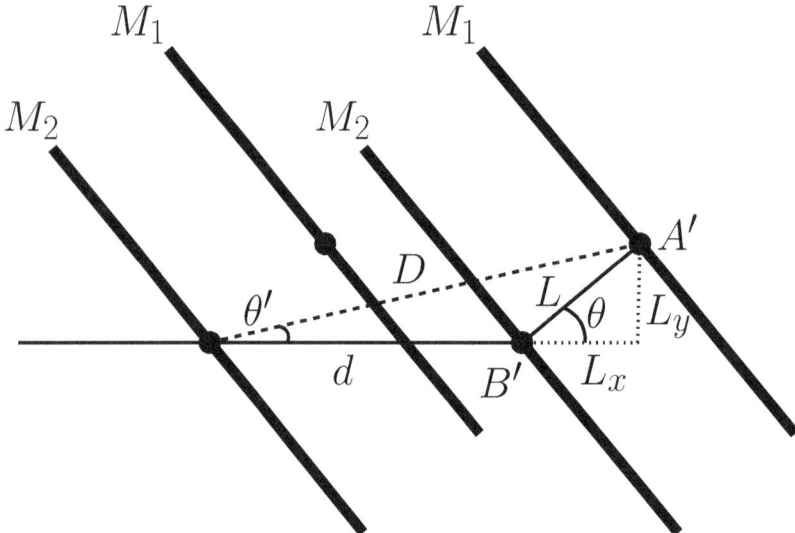

Figure 22: A schematic diagram that illustrates the setting of light clock in a slant state at the emission of the signal (right) and when the signal reached the opposite mirror (left). We note that the clock is not tilted by the same amount in both frames but this does not affect the given analysis. In fact, this diagram is inconsistent from some aspects (since it displays variables and parameters that belong to two different frames) and hence it has only demonstrative value.

$$(\Delta t')^2 = \gamma^2 (\Delta t)^2$$
$$\Delta t' = \gamma \Delta t$$

which is the time dilation formula as derived in the first question.

As for $\theta = 0$, we have $\cos\theta = 1$ and $\sin\theta = 0$ and hence the derived expression becomes:

$$(\Delta t')^2 = (\beta \Delta t' + \gamma \Delta t)^2$$
$$\Delta t' = \beta \Delta t' + \gamma \Delta t$$
$$\Delta t' = \frac{\gamma \Delta t}{1 - \beta}$$

which is the formula that we derived for case (b) in the assessment of the first question where we followed a derivation method that is consistent with the derivation method of case (a).

As for $\theta = \pi$, we have $\cos\theta = -1$ and $\sin\theta = 0$ and hence the derived expression becomes:

$$(\Delta t')^2 = (\beta \Delta t' - \gamma \Delta t)^2$$
$$\Delta t' = \beta \Delta t' - \gamma \Delta t$$
$$\Delta t' = -\frac{\gamma \Delta t}{1 - \beta}$$

Although this formula looks different from the previous formula because of the minus sign, it is essentially the same because they differ in sign only and hence it will not

9.7.1 Assessing Light Clock

affect the magnitude of time although it affects the sense due to the change of direction of motion. In fact, it is sufficient to reduce the θ range to $0 \le \theta \le \pi/2$ and hence avoid this minor problem without affecting the results.

4. Assess the results and analysis of the first and third questions.
Answer: We have two main cases: $\theta = \pi/2$ (i.e. the light signal is perpendicular to the direction of motion in frame O) and $\theta \ne \pi/2$ (i.e. the light signal has a velocity component in the direction of motion in frame O). For case (a) in the first question, the analysis should stand as it is because that is how light clock is defined. For case (b) in the first question (as well as the general case in the third question), we should revise our definition of time measurement and decide whether it should be based on the actual distance traversed by the photon across the clock (i.e. as in case a) or on the distance traversed by the photon excluding the distance traversed due to the displacement of the clock. The analysis in the third question (as well as the analysis of case b in the assessment of the first question) is based on the former approach. The conclusion is that if we want to obtain the time dilation formula in case (b) of the first question then we should use the latter approach and hence we will not be consistent since we use the first approach in case (a) of the first question. Accordingly, if we should be consistent in our measurement and calculation of time in both cases regardless of the clock orientation then the functionality of the clock will be dependent on its orientation. Regarding the third question, the time dilation formula will not be obtained in the general case (i.e. apart from the two special cases of being parallel and perpendicular) whether we follow the first or the second approach, which also leads to the conclusion that the functionality of the clock is dependent on its orientation.

We note that the above analysis and conclusion about the orientation dependency of light clock may be challenged by being based on a one-way tick and hence this orientation dependency should not occur if we use a two-way tick. However, the use of two-way tick will not address the conceptual aspect of the problem where the time in each one of the two parts has a different rate and hence it is conceptually different from the time in the original orientation even if they are made practically equal by averaging over the two parts. This conceptual difficulty will be felt when trying to use the above formulae in the calculation of the time of a two-way tick where we need to define $\Delta t'$ as the average (rather than the actual) time of half the two-way tick (as compared to Δt which is the actual time). Hence, in that sense the functionality of light clock (and even time) is still orientation dependent even if this dependency is removed at the practical level. In fact, the conceptual aspect (rather than the practical aspect) of light clock is the important problem because the purpose of light clock in special relativity is not the actual measurement of time but it is the conceptualization of time and its relation to the motion of the time measuring mechanism. Anyway, as long as we have right to define time by a one-way tick (in one way or the other) as we did in its original orientation, which no one can deny, this orientation dependency is still there and needs to be addressed at least from a conceptual perspective (or from the perspective of the relation between time and time counting mechanism).

5. Draw conclusions from the results of the previous questions which are based on the

9.7.1 Assessing Light Clock

assessment of light clock.

Answer: Based on the analysis of the previous questions we can reach the following conclusions:

• Although the observed speed of light is frame independent the velocity is not. This originates from the fact that the light has a velocity component from its source where this component will annul the effect of any velocity component due to the motion of the traversed object (i.e. light clock in our investigation) and hence what remains is the inherent velocity of light which keeps the invariance of its magnitude (or speed) across frames due to the scaling of space and time by the same γ factor. Accordingly, light according to the formalism of Lorentz mechanics essentially follows in its propagation similar velocity composition rules to those of classical mechanics (i.e. the Galilean composition formulae). The main difference between the Lorentzian and classical composition rules is the result of the contraction of spacetime coordinates (under the influence of motion through spacetime) which exists only in Lorentz mechanics, and that is why the two mechanics differ.[254]

• The analysis demonstrated that the whole theory of Lorentz transformations and Lorentz mechanics is based on adopting the speed of light for calibrating the space and time measurements and hence in principle a completely different theory (that can replace Lorentz mechanics as an alternative) may emerge if we adopt a different standard for this calibration.[255] Although both theories can be valid, there may be certain advantages/disadvantages in one theory or the other or even in each.

• If we have to be consistent in our time measurement method by the light clock then the functionality of the light clock will be dependent on its orientation and hence the time dilation formula can be obtained from the light clock only in one orientation in which the light signal is perpendicular to the direction of motion in the rest frame of the clock. Therefore, light clock may not be an ideal way for obtaining time dilation formula and conceptualizing and quantifying the relation between time and the motion of the time counting mechanism.

• We may also conclude that by following the formalism of Lorentz mechanics and hence drawing results from analyzing this formalism (instead of relying on special relativity or any other theory or using abstract devices like light clock), we can obtain more reliable results with much less effort and hassle.

Exercises

1. Repeat the analysis of the light clock for case (a) in the first question (refer to the solved problems) according to classical mechanics by using the Galilean velocity composition. Comment on the results.

---

[254] In fact, for a complete rationalization of this explanation the effect of this difference should be associated with what is coming in the next point about the adoption of the speed of light for calibrating the space and time measurements. This makes the speed of light in Lorentz mechanics invariant across inertial frames and the time is variant, in contrast to classical mechanics where the opposite is true (refer to the exercises of this subsection).

[255] Although this theory should not be classical, it could resemble classical (i.e. being amended classical) and this depends on the nature of the adopted standard.

9.7.1 Assessing Light Clock

2. Show that the functionality of light clock is based on the fact that while the observed velocity of light is frame dependent, since it has a component from its source (and accordingly the characteristic speed of light c is relative to its source like classical projectile but with the added assumption of the contraction of spacetime coordinates under the influence of motion), the observed speed of light is frame independent.

Chapter 10
Challenges, Criticisms and Controversies

In this chapter, we discuss some challenges and criticisms to Lorentz mechanics and its logical structure in particular. As we will see, most of these challenges are in fact directed to the special relativity interpretation and not to the formalism of Lorentz mechanics and hence if they proved to be valid they only affect the integrity and rationality of special relativity.

10.1 Twin Paradox

The essence of the twin paradox (which is also known by other names like the clock paradox) is that a twin traveled to another planet and returned to the Earth to meet his twin brother who stayed on the Earth. Now, since each twin was in a state of motion relative to the other twin then according to the time dilation effect he should age less and hence each twin should be younger than the other twin which is blatant contradiction. In fact, this is a valid challenge to special relativity because in the absence of absolute frame time dilation effect should equally apply to both twins since each twin is in a state of motion relative to the other twin. In other words, when we have no absolute reference frame relative to which time dilation takes place, the situation is symmetrical to both twins and hence in the eye of the first twin the second twin should experience time dilation because he is moving while in the eye of the second twin the first twin should experience time dilation for the same reason. Accordingly, this paradox is not a challenge to the formalism of Lorentz mechanics but to the special relativity interpretation. In fact, it is also not a valid challenge to any interpretation that is not based on the denial of the existence of absolute frame since time dilation will take place with reference to the absolute frame and hence the aging of each twin will be determined accordingly. In other words, the aging of each twin is based on his motion relative to the absolute frame and not on his motion relative to the frame of the other twin.

The twin paradox reveals a flaw in the special relativity interpretation which is the clash of this interpretation with the principles of physical reality and truth (see § 1.6). This may be demonstrated more vividly by another example in which three twins are involved where all these twins are in a state of uniform motion relative to each other and hence we have two different rates of aging for each twin due to his different speed relative to each one of the other two twins. So, if the speed of twin C relative to twin A is $0.1c$ and the speed of twin C relative to twin B is $0.2c$ then what is the real rate of aging of twin C if we believe in the rules of physical reality and truth? The change of frame, or indeed any similar explanation given by special relativists as will be discussed next, is not sufficient to remove this contradiction with the principles of physical reality. In fact, the reality of the rate of aging cannot be denied because in the absence of other twins (i.e. when we

10.1 Twin Paradox

have a single individual) we should have a "real" rate of aging, and hence we should accept the existence of physical reality which is based on the existence of a unique absolute frame relative to which the reality of other frames are defined and determined. Accordingly, this challenge is obviously not valid if there is a unique absolute frame relative to which the rate of aging is determined because the rate of aging of C is then determined by his motion relative to this unique absolute frame and not by his motion relative to any one of the other twins.

In the literature of special relativity, there are several answers that are proposed to revoke this challenge. In the following subsections, we investigate these answers and assess their validity.

Solved Problems

1. Imagine two inertial observers in a state of standard setting. Following the instant $t = t' = 0$, each one of these observers keeps watching and recording the time of the other observer. What will they see? Base your answer on the view of special relativity.
   Answer: If we follow the logic of special relativity then each one of these observers should see the time of the other observer running slow. However, because the rate of slowing is identical since the speed of the relative motion (and hence the Lorentz γ factor) is the same, then they should also see that their clocks are synchronized. So, if we follow the view that time dilation is real then we should accept the contradiction that while the time of one observer is really slow it remains synchronized and hence there is no real effect of this presumed real slowness. Similarly, if we follow the view that time dilation is apparent, then we face other contradictions since observations like prolongation of lifetime of elementary particles (refer to § 8.3) are based on having real (not apparent) time dilation effect. Consequently, challenges like twin paradox should be addressed and answered by special relativity whether time dilation is claimed to be real or apparent. As we will see in the following subsections, all the proposed answers to the twin paradox within the framework of special relativity are not convincing, at least for the stated case in this question where there is no accelerated motion or change of frame or justification for calling general relativity. In fact, this question highlights the fact that special relativity is not compatible with the principles of physical reality, at least according to this version of twin paradox, since we have two conflicting realities, i.e. the time of the first twin is slower and hence it is behind the time of the second twin while the time of the second twin is slower and hence it is behind the time of the first twin. Alternatively, it may be claimed that time dilation effect is apparent and this will lead to other non-realistic consequences which affect the reality of the whole Lorentz mechanics.

Exercises

1. Discuss the twin paradox and analyze its essence and implications.
2. Discuss why a similar "twin paradox" is not usually proposed with respect to length contraction as proposed for time dilation.
3. Discuss the logical inconsistency between the first postulate of special relativity and the twin paradox.

10.1.1 Time Dilation Effect is Apparent

According to this answer, time dilation effect is apparent and not real physical effect and hence there is no contradiction if each twin sees the other twin aging at a slower rate because this aging is not real and hence it has no physical reality to have a conflict. Hence, each twin will see the time of the other twin running slow with no physical consequences. If this is the case, then time dilation effect is just an illusion. This obviously contradicts the fact that time dilation is supposed to be real with real physical consequences like the prolongation of lifetime of decaying particles. We also note that if time dilation effect is based on an argument like light clock, which is common in the literature of special relativity, then it cannot be claimed that it is apparent and not real effect, because this type of argument inevitably implies the reality of this effect. In fact, the claim that physical effects like time dilation are apparent in this sense discredits the whole of Lorentz mechanics as a scientific theory and makes it a theory of illusions. Also, this answer obviously contradicts the commonly adopted special relativity narrative where the traveling twin is admitted to be the younger twin.

10.1.2 Traveling Twin is Distinguished by being non-Inertial

This answer is based on the claim that the traveling twin will age less because he is traveling (unlike his brother) since he is distinguished by being in a non-inertial frame, due to the acceleration and deceleration which are required for his frame to start the journey and come back, and hence time dilation applies to him selectively because he is the really traveling twin not to his twin brother who is not accelerating and decelerating and hence he is not traveling. In fact, the problem is addressed in this answer as if it is a linguistic matter about labeling rather than a physical problem that should be analyzed carefully where the different stages of motion (i.e. inertial and non-inertial) should be treated separately according to the appropriate physical principles. Anyway, this answer can be challenged by the following:
1. It can be challenged by imagining a completely symmetrical situation where both twins accelerate and decelerate and accelerate again to come back to their initial meeting point. So, who will be aging less?
2. In fact, we do not even need to accelerate and decelerate at all because the two twins during their perpetual uniform relative motion (where we may assume that they are in a state of standard setting)[256] can watch the frame of the other twin to see how time is progressing and hence if any one should observe time dilation effect then it should be symmetrical and hence the other should similarly observe this effect.
3. If the traveling twin is not in an inertial frame then why we should apply the time dilation formula which is based in its derivation on assuming inertial frames. In simple words, if acceleration destroys the state of the traveling frame as being inertial even

---
[256] The "standard setting" here is meant to refer to a potential challenge that "any uniform motion between two frames should have started with accelerated motion" because if this affects the inertiality of the frames then we should have no state of standard setting where the two frames are supposed to be inertial.

10.1.2 Traveling Twin is Distinguished by being non-Inertial 299

during its non-accelerating phase, then it should not be subject to the time dilation formula of inertial frames, and if acceleration did not destroy its state as inertial frame during its non-accelerating phase, then we can ask: the time of which frame is dilating during this phase? and hence we return to the starting point of this paradox.

4. A brief period of acceleration and deceleration should not have such a magical effect by changing the nature of the whole journey and affecting even the long periods of inertial state if we assume that the total time of the journey is the result of accumulation of the time spent during the inertial stages plus the time spent during the accelerating stages which is a logical assumption. Interestingly, some special relativists even claimed that most of the aging of the traveling twin occurs during his turn around to come back to the Earth.[257] This claim is arbitrary and baseless.[258]

5. Also, how the inertial phase of the journey will sense the effect of this acceleration and deceleration especially when this acceleration and deceleration occur after the inertial phase. For example, let have two inertial frames in a state of standard setting where the twins meet at the origin of coordinate systems of these frames at time $t = t' = 0$. They then continue their inertial journey for a while and one of these frames decelerates and reverses the direction. So, how the time during the inertial phase in the accelerated frame will sense this and hence apply the time dilation effect, because prior to this the inertial phase does not know about this future deceleration.

6. If acceleration should be credited for this change in aging then we may imagine two journeys with different periods of inertial state and with an identical acceleration phase, i.e. the period and magnitude of acceleration are identical in both journeys. Accordingly, we may ask: should this identical acceleration phase have the same effect on aging because the acceleration is identical, or the effect should be different due to the different length of the inertial phase in the two journeys. Both answers will lead to logical inconsistencies in the special relativity interpretation which we do not need to go through.

7. In the absence of an absolute frame according to special relativity, even acceleration should be relative and should identically apply to both frames (i.e. we can regard frame A accelerating relative to frame B, and we can regard frame B accelerating relative to frame A, or indeed we can regard each frame accelerating relative to the other frame) due to the absence of an absolute frame. Hence, how we identify which twin is accelerating/decelerating so that his time will be dilated. Yes, if we believe in the existence of a unique absolute frame relative to which acceleration is defined then

---

[257] If we accept this claim then if someone traveled inertially for a year (as seen from another inertial frame) and accelerated for a second then most of his aging will occur during one second.

[258] The "proof" of such claims is usually based on abstract devices like spacetime diagrams and light cone. However, these devices have only demonstrative value and hence they have no value in establishing scientific theories. Mathematical consistency, beauty and aesthetic factors do not guarantee the reality and correctness of any physical theory or give it any legitimacy if the theory is logically inconsistent or lack experimental support. In fact, this is the case with many elaborate theories of modern physics which are no more than "beautiful" theoretical and mathematical structures with no real physical substance as they do not represent any physical reality although the "beauty" of the theory is usually exploited to give it physical legitimacy.

10.1.2 Traveling Twin is Distinguished by being non-Inertial

this challenge will be addressed but then this answer will be redundant because we do not have any contradiction or paradox to answer. In fact, the existence of a unique absolute frame will demolish the entire special relativity interpretation. Interestingly, one of the proposed fixes to the problem of having symmetrical acceleration is the claim that "while the accelerations and decelerations between the two twins were reciprocal, the forces in the situation acted on the space twin alone". In fact, any symmetry in acceleration/deceleration should extend to force especially in the cases where the agent of force has no particular physical association with a particular frame.[259] Moreover, even if we accept such a distinction between acceleration and force this fix cannot solve the paradox when we have symmetrical accelerations caused by symmetrical forces and hence which twin should experience the time dilation effect in the absence of absolute frame. Anyway, this fix is totally unable to address the paradox in its purely inertial version or during the inertial stages in its accelerated version.

In fact, this answer (i.e. traveling twin is distinguished by being non-inertial) can be attacked by other challenges and revokes but the above are more than enough. Finally, we should point out to the fact that the claim "traveling twin is not inertial" is misleading because he is inertial in most parts of the journey and he is not inertial only in brief acceleration/deceleration periods. This misleading presentation and phrasing may be meant to provide the excuse for applying time dilation selectively to the traveling frame over the entire journey and hence justify the total rejection of the paradox instead of going through the detailed analysis of the different stages of the journey which will lead inevitably to the validity of this paradox in part of the journey at least.[260]

We should also remark that there is a claim in the relativity literature that the traveler twin should age more and not the one who stayed at home, and hence time dilation should apply to the inertial twin.[261] This claim can similarly be refuted by some of the above arguments, e.g. the justification of applying time dilation selectively during the inertial phases of the journey, the justification of ascribing the total effect to brief periods of acceleration and deceleration, answering the totally symmetric or non-accelerating version of the twin paradox, etc. In fact, this claim should also address the arguments in support of the above claim that the staying at home twin should age more since his brother is the really traveling twin (the opposite is also true). In fact, there are many other refutes to

---

[259] The distinction between the two frames by force is like their distinction by acceleration, and hence it has real meaning only if it is referred to an absolute frame which determines to which observer the force is really applied.

[260] Although some analyzed this problem in detail considering the different stages, their analysis is still based on selective aging, whether in the inertial stage or in the accelerated stage, and hence they failed to address the main issue in this paradox which is: in the absence of an absolute frame how can we identify which frame is traveling or accelerating so that we apply time dilation to it and not to the other frame.

[261] In fact, this claim contradicts the alleged results of the atomic clock experiments (see § 8.4) where the traveling clocks (which are the ones that are more appropriate to consider as non-inertial than the stationary clocks) are reported to be delayed (i.e. run slower). This should also apply to the prolongation of lifetime of elementary particles. Moreover, this claim contradicts the common narrative of this paradox in the literature of special relativity where the traveling twin is claimed to be the younger.

10.1.3 Traveling Twin has Two Inertial Frames

this claim but it is not worthwhile to go through.

Exercises
1. Analyze the traveling twin motion as partly inertial and partly accelerating, and hence derive the logical consequences of this analysis.

10.1.3 Traveling Twin has Two Inertial Frames

This answer is based on claiming that since the traveling twin needs to change his inertial frame on the return journey, he has two inertial frames. Accordingly, the difference that is based on his need to change the frame will distinguish him as the traveler because otherwise he will not need to change his frame. Hence, the "aging less" effect as a result of time dilation will apply to him only because he is the traveling twin not his twin brother who does not need to change his frame and hence he is not traveling. This answer can be challenged by the following:
1. This answer in essence is the same as the previous answer (i.e. traveling twin is distinguished by being non-inertial) which is discussed and challenged in § 10.1.2, and hence it can be similarly revoked. In brief, the change of frame is equivalent to the decelerated and accelerated motion in the middle of the journey.
2. In the absence of an absolute frame, each twin can be considered as the one who needs to change his frame, so the situation is still symmetrical. In other words, in the absence of a third frame (whether absolute or not) relative to which the traveler changed his frame while the other did not, the "staying at home" twin can be seen as the one who allegedly changed his frame. In brief, in the absence of a third reference frame relative to which the change of frame can be defined sensibly and unambiguously, the role of the two twins with regard to the change of frame can be exchanged and hence the "traveler" will become "staying at home" and vice versa.[262] Therefore, the situation is still symmetrical despite this arbitrarily-selected change of frame.
3. In fact, this answer (like the previous answer) addresses the problem as if it is a linguistic rather than a physical problem. Although, we have no objection to label the "traveler" twin as such, there is no fundamental physical reason to change our physics (as represented by the equivalence of applying the relativity principle in the absence of an absolute reference frame) because of this arbitrary label.

Exercises
1. Assess the concept of "change of frame" in the context of twin paradox.

10.1.4 Calling for General Relativity

This answer claims that this paradox can be addressed within the framework of general relativity rather than special relativity because it involves accelerated frames. In fact, this is the strangest of all answers because any theory that needs another theory to fix its logical

---
[262] In fact, in the absence of absolute frame we can also claim that both changed their frames symmetrically or non-symmetrically.

defects is an illogical theory and hence it should be rejected. Although a scientific theory can be incomplete because it is an approximation or a limiting case to another theory (like classical mechanics in its relation to Lorentz mechanics), the theory itself should be logically consistent although it may be partially in conflict with experiment or observation. For example, although classical mechanics is incomplete it is logically consistent. However, the twin paradox does not show such a clash with experiment or observation that can be fixed by calling another theory but it shows a conflict with logic[263] and hence the theory itself should be rejected if it needs the help of another theory to fix its logical inconsistency because this inconsistency is fixed (if it is really fixed) by the other theory not by the theory itself and hence the theory is still illogical. Moreover, this answer is based on the assumption that special relativity is a special or limiting case to general relativity which is a controversial issue even among the followers of these relativity theories (see § 3.1). As we stated previously, this assumption is not based on a solid foundation and hence it could be challenged. In fact, most of those who alleged this fix to this paradox did not produce a detailed solution[264] to this paradox according to the framework of general relativity; all they did is to refer this problem to general relativity where they claimed that general relativity is capable of addressing this paradox. In brief, all the alleged fixes to this challenge by calling general relativity are void and do not make sense and hence they are not worthwhile to investigate and refute. The interested reader should consult the literature of special relativity about this controversy. Anyway, even if general relativity is able to address the accelerated version of the twin paradox, we still need to address the inertial version of this paradox where special relativity, rather than general relativity, is definitely the right theory to address such a paradox since it is completely within the domain of validity of special relativity. This is also the case during the inertial stages of the accelerated version of this paradox because although general relativity may be the right theory to address the paradox during the accelerating stages, special relativity is definitely the right theory to address the paradox during the inertial stages and hence a special relativistic answer to this paradox is still required.

10.2 Barn-Pole Paradox

This may also be called the pole-barn paradox or the ladder paradox as well as other names. In fact, this paradox may be considered the length contraction version of the twin

---

[263] As shown above, the essence of the twin paradox challenge to special relativity is the logical incompatibility between the principle of relativity in its special relativistic sense and the selective aging of a particular inertial observer. In fact, this challenge applies to general relativity as much as it applies to special relativity and hence the two theories are the same in their need for an absolute frame to make sense of their logic whether in dealing with inertial frames or accelerating frames. In brief, our need for an absolute frame to have a sensible and realistic definition of accelerating motion is as much as our need for an absolute frame to have a sensible and realistic definition of uniform motion.

[264] Some who claimed to provide such detailed solutions based on calling general relativity did not produce more than irrelevant (and potentially wrong) technical details since they failed to address the main issue, i.e. justifying the selective aging (whether during inertial or accelerating motion) which is incompatible with the unrestricted principle of relativity (whether special or general).

10.2 Barn-Pole Paradox

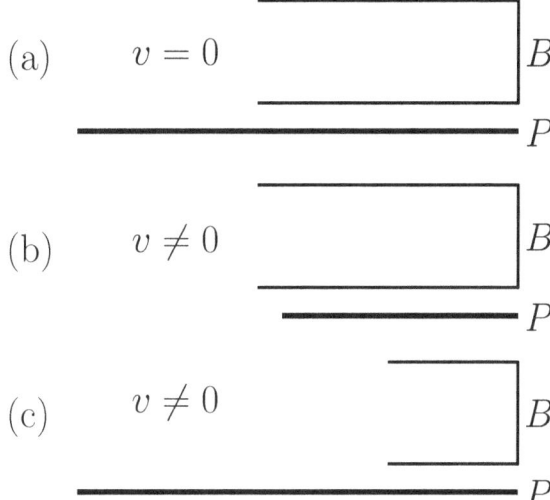

Figure 23: Schematic illustration of the barn-pole paradox where (a) represents the situation when the pole and the barn are in a state of relative rest, (b) represents the situation as seen from the barn frame when they are in a state of relative motion, and (c) represents the situation as seen from the pole frame when they are in a state of relative motion. In this illustration, B stands for barn, P for pole and v for the relative speed between the pole and the barn.

paradox which is the time dilation version. The essence of the barn-pole paradox is that a pole, whose proper length is bigger than the proper length of a barn and hence it is longer than to fit inside the barn when they are at relative rest, is moving uniformly at very high speed relative to the barn. As seen from the barn frame, the pole is shorter when it moves due to length contraction effect and hence it can fit inside the barn, while from the pole frame the barn is the one that is shortened due to this effect and hence the problem is aggravated by this relative motion. So, will the pole fit or not fit inside the barn? This paradox is schematically demonstrated in Figure 23.

Before discussing the special relativity answer to this paradox, we should address this paradox within the established formalism of Lorentz mechanics. As discussed earlier in the context of twin paradox and time dilation effect, the barn-pole paradox can be easily answered if we assume the existence of a unique absolute frame relative to which the length contraction effect takes place. The reason is that all lengths will then be referred to this unique and real absolute frame and hence the length of each object is uniquely and realistically defined. So, if the state of rest and motion relative to the absolute frame dictates the shortening of the pole relative to the barn by the right amount to fit then the pole should fit inside the barn; otherwise it will not fit. This situation does not exist in special relativity because each frame is referred to the other frame and hence there is no unique realistic sense of length. Accordingly, the pole will fit and will not fit at the same time from different perspectives depending on the frame of observation.

This paradox is usually addressed in special relativity by calling the relativity of simultaneity. The essence of the special relativistic answer is that fitting or not fitting depends

10.2 Barn-Pole Paradox

on the concept of simultaneity because "fitting" means that both ends of the pole are inside the barn at the same instant of time. Now, since simultaneity is a relative concept, the pole will fit in one frame and it will not fit in the other frame.[265] However, this answer is not convincing for the same reasons as the twin paradox answer if length contraction is real effect due to the conflict with the principles of physical reality. Again, the reason for this dilemma is the denial of the existence of absolute frame in special relativity and hence if we accept the existence of such a frame there should be no paradox because length contraction will take place due to the motion relative to the absolute frame and hence the order relation (i.e. shorter, equal, and longer) between the length of the barn and the length of the pole will be well defined in reference to the unique absolute frame. In the following points, we analyze this paradox further:

1. First, fitting does not depend on the concept of simultaneity alone but it also depends on the physical length which is a real physical attribute. Regardless of our definition of simultaneity, the pole can only fit if it is shorter than the barn when its two ends are observed simultaneously. Now, "being shorter" is a real physical attribute that depends on other two real physical attributes, i.e. length of pole and length of barn. So, no convention or definition of simultaneity can fix this paradox unless the underlying physics satisfies the condition of being shorter. Accordingly, if length contraction is apparent (as claimed by some special relativists) and not real, the pole will not fit regardless of any consideration of frames and simultaneity. Although the claim of being apparent is not inconsistent with the principles of physical reality, it will demolish the reality of Lorentz mechanics itself. In contrast, if length contraction is real (as claimed by other special relativists) then we have two conflicting realities: fitting according to the barn frame, and not fitting according to the pole frame. Again, this is based on the principles of physical reality and its unique nature, as discussed in § 1.6. So, accepting the special relativity view (either way) will force us to embrace a non-realism view.[266] We should also question the validity of the relativity of simultaneity in its special relativistic interpretation as a logical answer to this paradox since the relativity of simultaneity according to our interpretation is consistent with the existence of absolute frame and absolute time and hence there is an absolute simultaneity which is the ultimate reference to all simultaneities.

2. Interestingly, some special relativists tried to answer this paradox as if it is a problem

---

[265] We note that being simultaneous to a particular frame (i.e. barn frame to do the fit) and non-simultaneous to the other frame (but not the other way around) is an arbitrary choice that requires justification.

[266] We note that the above special relativity answer must be based on assuming that length contraction is real effect; otherwise the answer is meaningless. In fact, the whole paradox will not emerge if we assume that length contraction is apparent and not real because the pole will fit or not according to its real (or rest) length relative to the real (or rest) length of the barn. However, this seems to escape the notice of some special relativists who adopted these contradicting views, i.e. addressing the barn-pole paradox by the relativity of simultaneity and claiming length contraction is apparent. Anyway, although this paradox will not arise if length contraction is apparent, special relativity still needs to answer other challenges based on the non-realism of Lorentzian effects like length contraction and time dilation.

related to the speed of sending signals and information across the pole and subsequent effects to the fitting of the pole inside the barn. They simply assume that the fit has already been done and hence because the speed of sending information does not exceed c then as soon as the front end of the pole hits the front end of the barn while the rear end of the pole is inside the barn, then a signal will be sent from the front end to the back end. Because the speed of this signal is finite then there will be no problem in fitting the pole inside the barn. The problem, according to these special relativists, will then occur not because of the fit, which is already done, but if the barn will be able to resist the push of the rear end when the signal arrives and this is a very trivial problem. In fact, this answer is no more than a meaningless diversion by converting the problem from being a logical inconsistency problem to a physical problem about the consequences of entering the barn which can allegedly be fixed by limiting the speed of sending signals. This attempt is obviously based on assuming that the contraction occurs to the pole and not to the barn, and this is the real problem because the whole paradox is based on the symmetry of applying the relativity principle to both frames, and this answer does not address this issue. So, we can still ask why length contraction occurs to the pole but not to the barn if the relativity principle is symmetric and it equally applies to both frames due to the absence of absolute frame according to special relativity.

Exercises
1. Discuss the relation between the relativity principle and the barn-pole paradox.
2. A 2 m long pole is required to fit inside a 1 m long barn. What is the minimum speed required to do this fitting? In which frame (barn frame or pole frame) this fitting can happen? What about the other frame involved in this process? Base your answer on special relativity.
3. Discuss the special relativistic answer to the barn-pole paradox challenge that the fit will occur but because of the limited speed of sending signal from the front end of the pole to its back end the problem will occur later.
4. What is the significance of the barn-pole paradox according to special relativity?

10.3 Other Paradoxes

There are many other paradoxes that are synthesized to challenge Lorentz mechanics or rather its special relativistic interpretation. One of these paradoxes is the so-called "twirling pole paradox" where an extremely long pole is held from one of its ends and twirled so that its other end traces a circle. Now, if the pole is long enough and the angular speed of twirling is sufficiently large the other end could move faster than the speed of light in violation of special relativity. Another similar paradox is the so-called "superluminal scissors paradox" where a very long pair of scissors is supposed to be closed in an infinitesimal time interval so that the contact point of its two arms moves at the speed of light or faster. However, we do not see any need to go through these alleged paradoxes which are based on highly hypothetical thought experiments with virtually no realistic basis or perspective.[267] We think the logical structure of special relativity can be challenged by more profound and solid arguments than these trivial paradoxes and hence we see no need for discussing and assessing these alleged paradoxes. Similarly, these alleged paradoxes are less than capable to make a real challenge to Lorentz mechanics (if they are supposed to be so) and its theoretical framework which enjoys an overwhelming experimental support. Anyway, even if these paradoxes are worthy, they can be addressed simply by claiming that the speed restrictions (whether those imposed by special relativity or those imposed by the formalism of Lorentz mechanics) exclude such lateral or virtual speeds (also refer to § 10.4).

10.4 Speeds Exceeding c

As seen earlier, the lateral (or sweep) speed of light or the speed of other hypothetical (or virtual or imaginary) phenomena may exceed the characteristic speed of light in vacuum c (refer to § 3.8.1 and § 10.3). This fact is used to challenge the speed restrictions in Lorentz mechanics and special relativity in particular. However, these challenges can be easily addressed by excluding the lateral speeds and the speeds of hypothetical objects[268] from entering into the Lorentzian formalism due to the fact that these usually represent speeds of imaginary or subjective objects (like the point of contact of scissors arms) and not speeds of real physical objects. Anyway, we should remark that some of the speed restrictions in special relativity are questionable; moreover the speed restrictions of the current formalism of Lorentz mechanics could be subject to limitations and amendments, as explained before. Nevertheless, even if some speed restrictions of Lorentz mechanics are violated, this will not discredit this mechanics as a whole although it indicates that this mechanics is an approximate valid theory (thanks to its experimental support like classical mechanics) and hence it requires amendments or search for a more general and reliable theory.

10.5 Non-Local Reality of Quantum Mechanics

This may be considered as another challenge to Lorentz mechanics, and its special relativistic interpretation in particular, where a non-local reality that is supposed to underlie some quantum mechanical phenomena, like the instantaneous correlation of the spin of two separated photons, may be regarded as violation to the Lorentzian speed restrictions because of the presumed transmission of information by a speed that is greater than c or even infinite. However, these controversial issues require deep inspection and thorough investigation within the framework of quantum mechanics which is not the subject of the

---

[267] This type of paradoxes should remind us of the medieval contemplation about "how many angels can dance on the head of a pin?".

[268] We may define these hypothetical objects as objects that cannot be characterized and quantified by real physical quantities and parameters like mass, energy and momentum. For example, the point of contact of scissors arms cannot have mass or carry energy. These hypothetical objects may also be characterized by some as those which do not carry energy or information.

present book and hence we do not go further in this discussion. Nevertheless, as we advocated earlier the theory of Lorentz mechanics is not final and hence it can be subject to amendments in the future that lift some of the speed restrictions imposed by the current formalism of Lorentz mechanics. Moreover, some of the alleged restrictions (e.g. on massless objects) are questionable even under the current framework of Lorentz mechanics, and hence the violation of such restrictions may discredit or affect a particular interpretation (mainly special relativity) but not necessarily the formalism of Lorentz mechanics or other interpretations.

Chapter 11
Interpretation of Lorentz Mechanics

There are several interpretations to Lorentz mechanics and its formalism. The dominant interpretation is the theory of special relativity of Poincare and Einstein which we investigated in the previous chapters of the book and in § 9 in particular. Another interpretation is the Lorentz interpretation which is discussed and investigated in many textbooks and research papers on this subject. There are other interpretations that can be found in the research literature of Lorentz mechanics. However, instead of going through these interpretations (some of which lack sufficient details to be fully assessed and fairly judged) and repeat what is already available in the literature of this subject, we propose a number of criteria that should be satisfied and a number of elements that should be incorporated in any acceptable interpretation. The criteria are based on commonly accepted rules of logic and science while the elements are largely extracted from analyzing the formalism of Lorentz mechanics. Accordingly, there may be more than one acceptable interpretation to Lorentz mechanics and its formalism.

Solved Problems

1. Why we, as scientists, should bother about the interpretation and its validity or invalidity if we have the right formalism? In other words, what is the importance of having correct interpretation as well as correct formalism?[269]

    Answer: The interpretation of any scientific theory is very important and could be as important as the formalism itself. This importance does not arise only from the necessity of having deep insight in the physical laws (which is very important factor in itself), but also because the interpretation provides guidance and assistance for the future investigation and the expansion and modification of the existing theories as well as steering the scientific research in the correct direction. In brief, science is an adventure to understand the world and this understanding will not be complete if we do not have the right interpretation of the formalism because the formalism is a collection of blind rules while the interpretation is the insight that provides an understanding for the present and a compass and a sense of direction for the future. This issue may be illuminated by quantum mechanics where the successful formalism failed so far to provide complete understanding of the quantum phenomena due to the absence of a generally convincing interpretation. Another example is Lorentz mechanics itself where the interpretation of special relativity directed modern physics in certain directions which, at least, may not be ideal for the progress of science.

Exercises

---

[269] This question is based on the assumption that the interpretation of scientific theories belongs to the philosophy of science and hence it may not be relevant or important to science.

1. What is the relationship between the experimental evidence of a theory and its interpretation?
2. Refute the claim that the validity and invalidity of interpretation is of little practical value to science since in the presence of a correct formalism the interpretation will have little impact on the progress of science even if the interpretation was wrong.
3. Outline the main features of the Lorentz interpretation of Lorentz mechanics and compare it to the interpretation of special relativity.

11.1 Criteria for Acceptable Interpretation

Based on generally accepted epistemological and logical rules, any acceptable interpretation of Lorentz mechanics and its formalism should satisfy the following criteria:

1. It should be logically consistent and hence any interpretation that contains contradictions and inconsistencies should be rejected with no extra effort to prove its validity or invalidity.
2. It should be compliant with the principles of physical reality and truth (see § 1.6) so that it provides an honest picture of the unique reality that is supposed to exist for the proposed interpretation to be justifiable.
3. It should be consistent with other known facts and well-established theories and hence it should not contradict any known facts outside the theory. For example, if we have to provide a viable interpretation to Lorentz mechanics we cannot accept propositions that contradict the known laws of statistical mechanics or quantum mechanics.[270]
4. It should be thorough and hence it provides an explanation to the theory of Lorentz mechanics as a whole and not just parts of it. Some interpretations are proposed as fixes to certain problems and hence they should not be considered as legitimate interpretations unless they can provide consistent and thorough explanation to the main body of Lorentz mechanics.

Exercises

1. Try to justify the above criteria for accepting any proposed interpretation of Lorentz mechanics.
2. What is the relation between science and logic?

---

[270] In this regard, we should remark that although differences that violate the rules of logic or the principles of reality and truth are unacceptable, differences in theoretical formulation and conceptualization, especially those related to the philosophical and epistemological aspects of a physical theory, should be acceptable. Although, the reality and truth are unique, they can be expressed, described and theoretically structured in many different ways and that is why we see different branches of science that deal with the same or similar physical phenomena use different concepts and techniques in their theoretical approach to the physical reality. In brief, this sort of differences does not represent any contradiction since such differences are in the shape and form and not in the essence and content. Therefore, we can have more than one correct theory about the same phenomenon as long as they do not contradict each other since each theory can represent, describe, conceptualize and quantify the phenomenon from a certain perspective and hence it does not necessarily lead to a contradiction with the other theory.

3. Give some examples of attempts made by special relativists to defend the logical inconsistencies of special relativity.
4. Discuss the stand of some special relativists who tried to challenge the logic by experimental evidence in support of special relativity and its postulates.

11.2 Essential Elements of Potentially Acceptable Interpretation

Based on our analysis of the formalism of Lorentz mechanics, and considering the above logical and epistemological criteria, we think any acceptable interpretation of Lorentz mechanics and its formalism should consider incorporating the following elements and facts that have been drawn directly or indirectly from analyzing the formalism:

1. The distortion of spacetime by motion which is represented by the effects of length contraction and time dilation or what we call spacetime contraction.
2. The existence of a unique absolute frame to justify the difference between inertial and non-inertial frames and the reality of the characteristic Lorentzian effects like time dilation and length contraction. Accordingly, the unrestricted relativity principle that is based on the denial of this existence should be dismissed and replaced by the restricted relativity principle where the "relativity" in this restricted version is ultimately referred to this unique absolute frame. Implications and consequences of the Lorentzian formalism, like the relativity of simultaneity and the relativity of co-positionality, that apparently contradict the existence of absolute frame should be interpreted harmoniously to comply with this existence, as explained earlier.
3. The dependence of the velocity of light on the velocity of its source and hence the light signal should have a velocity component from the velocity of its source which may be interpreted as having a projectile-like model for light propagation where the characteristic speed of light is relative to its source.
4. The invariance of the speed of light across all inertial frames but in an apparent sense due to the contraction of spatial and temporal coordinates of spacetime by the same γ factor. This proportionate contraction, in addition to the added velocity component that compensates for the displacement of objects in the moving frame, is the cause of this invariance.
5. The fact that the whole theory of Lorentz mechanics is based on adopting the characteristic speed of light as a standard for calibrating the measurement of space and time. Hence, in principle a different physical theory with different transformations to those of Lorentz can be proposed if we change this choice of calibration.
6. The merge of space and time into spacetime manifold where this merge should be seen as a mathematical artifact that underlies the formal theoretical structure of Lorentz mechanics. This merge may be regarded as another demonstration for adopting a speed (i.e. the characteristic speed of light) for spacetime calibration because speed is a concept that links space and time since it is a ratio between space and time.

Solved Problems

11.2 Essential Elements of Potentially Acceptable Interpretation

1. Outline the conclusions that can be obtained from analyzing the formalism of Lorentz mechanics about the speed and velocity of light
   Answer: The formalism of Lorentz mechanics indicates that the speed of light is invariant across all inertial frames (refer for example to § 4.2.2 and § 4.2.3). However, the velocity of light is not independent of the velocity of its source since it has a velocity component from the velocity of the source. This indicates a projectile-like model for the propagation of light (see § 1.13.1 and § 1.14) and rules out the possibility of a classical wave propagation model and the need for a propagation medium like the hypothesized luminiferous ether (see § 1.13.2 and § 1.14). We note that the projectile model for propagation is consistent with the existence of an absolute frame although it does not require such a frame. The reader is referred to our previous analysis (see for example § 9.7) where it was found that the light signal behaves like a projectile.
2. Why the issue of the postulates of Lorentz mechanics is not discussed in this section as an essential element of potentially acceptable interpretation?
   Answer: We think this is a matter of choice and preference and belongs to the presentation and packaging more than the essence and content and hence it is not a crucial element for interpretation. However, a scientifically based set of postulates which is closely related to the formalism should be superior and advantageous. Our preferred approach, which was fully explained earlier (see for example § 5), is to consider the Lorentz spacetime coordinate transformations as the main postulates of Lorentz mechanics to be more connected to the formalism and to avoid being entrapped by interpretations and getting involved in controversial philosophical and epistemological issues from the beginning. However, if we should propose a different postulate then we can propose a single postulate for Lorenz mechanics from which the Lorentz spacetime coordinate transformations can be derived and hence other formulations are obtained subsequently. We can simply propose the contraction of space and time by motion as the main postulate for Lorentz mechanics where this postulate may be phrased like this: the spacetime coordinates contract by the Lorentz $\gamma(v)$ factor under the influence of motion through spacetime with speed v relative to the absolute frame (refer to § 12.4.2).

Exercises
1. Explain why the existence of absolute frame is needed to have a valid interpretation.
2. Try to explain why inertial frames are treated equally by the space in their validity for formulating the known laws of physics despite their possible motion with various speeds relative to the absolute frame while non-inertial frames are treated differently.
3. Demonstrate the invariance of the speed of light and the non-invariance of its velocity using a simple diagram.
4. Why the formalism of Lorentz mechanics indicates a projectile-like model for the propagation of light?
5. Outline a potential interpretation that naturally emerges from analyzing the formalism of Lorentz mechanics and incorporates the proposed elements in this section.
6. Inspect and analyze the logical and physical consequences of the interpretation that

11.2 Essential Elements of Potentially Acceptable Interpretation

was proposed in the previous question to assess its validity.

7. Explain length contraction and time dilation effects from the viewpoint of the proposed interpretation of Lorentz mechanics and outline their relation to the absolute frame and to relative frames.
8. Imagine two inertial observers in a state of standard setting. Following the instant $t = t' = 0$, each one of these observers will keep watching the time of the other observer. What will they see? Base your answer on the proposed interpretation.
9. Our observations, such as the prolongation of lifetime of elementary particles, should suggest, according to the proposed interpretation, that we (i.e. the inhabitants of the Earth) are in the absolute frame, which may be very unlikely, unless we accept the special relativistic view that effects like time dilation occur in each frame in compliance with the unrestricted relativity principle. What is your answer to this challenge?
10. Apply the proposed interpretation to a specific example, e.g. the prolongation of lifetime of elementary particles.
11. Can you provide more clarification about the relation between the velocity component of the light signal from its source and the invariance of the speed of light across all inertial frames?
12. Let assume that the experimental evidence is consistent with the special relativity interpretation (e.g. in correlating the characteristic Lorentzian effects to the relative motion between frames in accord with the unrestricted relativity principle) but not with other interpretations (such as the proposed interpretation). What should we conclude?

Chapter 12
Appendices

In this chapter, we provide additional materials about some topics related to Lorentz mechanics. As these materials are not essential to the understanding of Lorentz mechanics and may cause confusion or interruption if they are presented within the main body of the book they are provided as appendices.

12.1 Maxwell's Equations

In the 1860s, James Clerk Maxwell formulated his electromagnetic theory by proposing the following four partial differential equations as a complete and consistent set (in addition to the wave equation) for describing all classical electromagnetic phenomena:[271]

$$\nabla \cdot \mathbf{E} = \frac{\rho}{\varepsilon_0} \tag{344}$$

$$\nabla \cdot \mathbf{B} = 0 \tag{345}$$

$$\nabla \times \mathbf{E} = -\frac{\partial \mathbf{B}}{\partial t} \tag{346}$$

$$\nabla \times \mathbf{B} = \mu_0 \left(\varepsilon_0 \frac{\partial \mathbf{E}}{\partial t} + \mathbf{j} \right) \tag{347}$$

where ∇ is the nabla differential operator, $\mathbf{E}$ is the electric field vector, ρ is the electric charge density, ε_0 is the permittivity of free space, $\mathbf{B}$ is the magnetic field vector, t is the time, μ_0 is the permeability of free space, and $\mathbf{j}$ is the electric current density vector. We note that the above formulation is based on employing the SI system of units (Système international d'unités). Some of these equations take a slightly different form when using other unit systems. We also note that as well as the above differential form of Maxwell's equations, they can also be formulated in an integral form. Maxwell's equations may also be formulated in electromagnetic field variables other than $\mathbf{E}$ and $\mathbf{B}$ (namely electric displacement vector $\mathbf{D}$ and magnetic strength vector $\mathbf{H}$).

The first of these equations is Gauss law of electrostatics whose essence is that the divergence of the electric field is proportional to the charge density. The second of these equations is Gauss law of magnetism whose essence is that the divergence of the magnetic field vanishes identically and hence there is no magnetic monopole. The third of these equations is Faraday's law of induction whose essence is that the curl of the electric field is proportional in magnitude and opposite in sense to the temporal rate of change of magnetic

---

[271] These equations were already in existence since they have been formulated by other scholars except the last term of the fourth equation which was added by Maxwell himself using symmetry arguments, and hence the main credit to Maxwell is for merging them in a consistent system for describing electromagnetic phenomena and developing a wave propagation model for these phenomena.

field. The fourth of these equations is Ampere's circuital law of magnetic induction whose essence is that the curl of the magnetic field is proportional to the temporal rate of change of electric field with the second term on the right hand side (which is added by Maxwell himself) representing the contribution of electric current density to the total magnetic induction. As we will see in § 12.3.2, Maxwell's equations are invariant under the Lorentz transformations but they are not invariant under the Galilean transformations.

12.2 Michelson-Morley Experiment

As seen earlier, the purpose of the Michelson-Morley experiment is to detect the Earth movement through the ether and measure its speed relative to the ether which causes "ether wind". The logic of this experiment is to measure the light speed in two mutually-perpendicular directions where the measuring device is supposed to have different speeds relative to the ether in these directions. This hypothesized difference in the light speed in these directions can then be detected by interference between these two light rays when they combine and produce interference fringes. This experiment failed to detect the Earth motion through the ether and this led to questioning the existence of the ether.

The Michelson-Morley device consists of the following components (refer to Figure 24):
1. A monochromatic light source L_s with a collimator C to produce a parallel beam of light.
2. A mirror M_p which is partially silvered to be partially transparent and hence it splits the light beam into two sub-beams: transmitted sub-beam that follows the path PP_1 (horizontal) of length L_1 and reflected sub-beam that follows the path PP_2 (vertical) of length L_2 in the forward-backward journeys (i.e. PP_1P and PP_2P).
3. Mirrors M_1 and M_2 to reflect the light.
4. A detector D to detect and observe the interference pattern.
5. A compensating transparent glass plate C_p whose function is to compensate for the path difference between the transmitted and reflected sub-beams.

The working principle of this device is simple that is the light beam that originates from the light source will hit the mirror M_p and hence the beam will split into two sub-beams that propagate in the two perpendicular directions PP_1 and PP_2. In their return journey, these sub-beams will merge at point P, where they interfere constructively or destructively according to their phase shift, and hence they produce an interference pattern that can be detected and observed using the detector D.[272]

Now, let analyze this experiment to see how it is supposed to detect the ether wind. Let assume that the ether wind is in the horizontal direction and hence according to the Galilean velocity transformation (assuming a classical wave propagation model) the light speed in its forward and backward directions in the PP_1 arm should be $c \pm v$ where v is the speed of the ether wind which is caused by the Earth movement with respect to the ether (refer to Figure 25 a). Accordingly, the time for the light in its PP_1P journey will

---

[272] The reader is referred to standard textbooks on general physics to understand the details of interference and how the interference pattern is generated on the screen of the detector.

12.2 Michelson-Morley Experiment

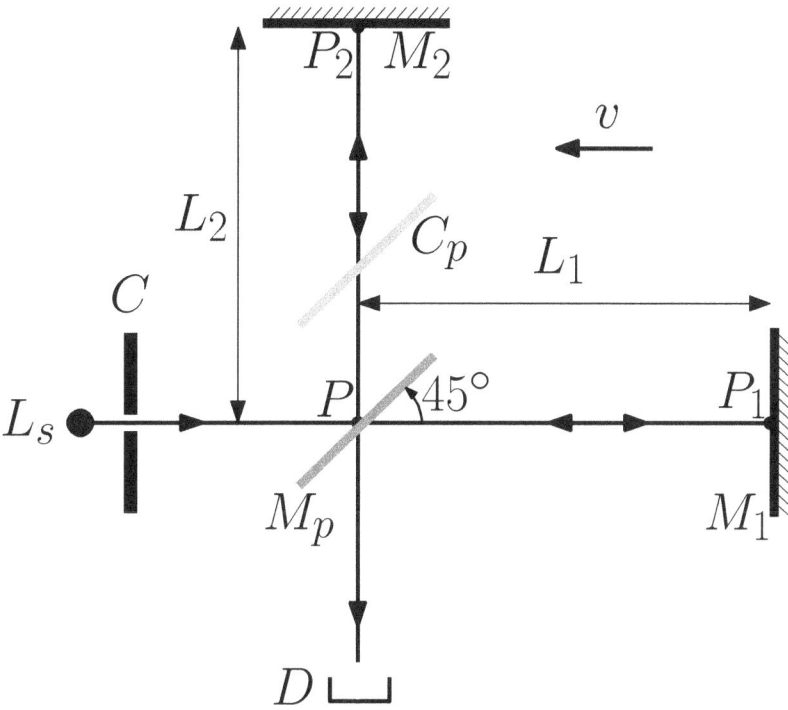

Figure 24: Schematic illustration of the Michelson-Morley apparatus to detect the ether wind.

be:

$$t_1 = \frac{L_1}{c-v} + \frac{L_1}{c+v} = \frac{L_1(c+v) + L_1(c-v)}{(c-v)(c+v)} = \frac{2cL_1}{c^2 - v^2} \tag{348}$$

Similarly, the time for the light in its PP_2P journey will be:

$$t_2 = \frac{2L_2}{\sqrt{c^2 - v^2}} \tag{349}$$

The denominator of this equation is based on the Pythagoras theorem where the light has speed c along the hypotenuse (representing its path in the ether) and hence the square of the horizontal speed v of the ether wind should be subtracted from c^2 to obtain the light speed along the vertical path PP_2P (refer to Figure 25 b). Accordingly, the difference between these two times will be:

$$\Delta t = t_2 - t_1 = \frac{2L_2}{\sqrt{c^2 - v^2}} - \frac{2cL_1}{c^2 - v^2} \tag{350}$$

Now, if we rotate the device 90° so that L_1 and L_2 exchange their roles, we obtain:

$$\Delta T = T_2 - T_1 = \frac{2cL_2}{c^2 - v^2} - \frac{2L_1}{\sqrt{c^2 - v^2}} \tag{351}$$

where the last equation is obtained from its predecessor by just exchanging the L_1 and L_2 labels (due to the exchange in their roles as parallel and perpendicular to the direction

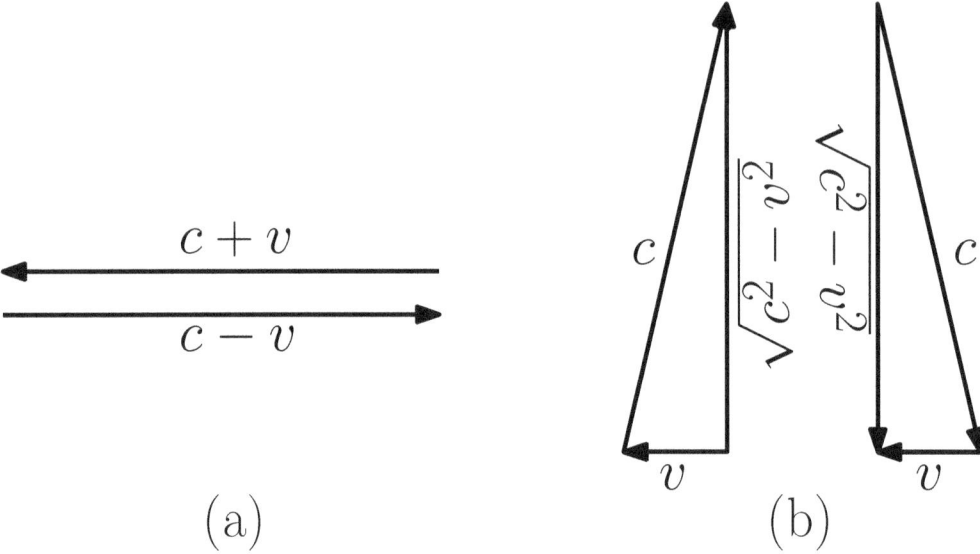

Figure 25: The speed of light in the Michelson-Morley apparatus (a) in the horizontal path PP_1P and (b) in the vertical path PP_2P according to the Galilean velocity transformation where v here represents the speed of the ether wind. We note that the speed along the diagonals in (b) is c because we are assuming a classical wave propagation model where the source and observer (which are in relative rest) are in relative motion with respect to the medium.

of the ether wind) and reversing the order of the subtraction terms (since this order is determined by their order as 1 and 2). Hence, the time difference between these two situations (i.e. before and after rotation) is:

$$\begin{align*}
\Delta T - \Delta t &= \frac{2cL_2}{c^2-v^2} - \frac{2L_1}{\sqrt{c^2-v^2}} - \left(\frac{2L_2}{\sqrt{c^2-v^2}} - \frac{2cL_1}{c^2-v^2}\right) \tag{352}\\
&= \frac{2cL_2}{c^2-v^2} - \frac{2L_1}{\sqrt{c^2-v^2}} - \frac{2L_2}{\sqrt{c^2-v^2}} + \frac{2cL_1}{c^2-v^2}\\
&= \frac{2cL_1}{c^2-v^2} + \frac{2cL_2}{c^2-v^2} - \frac{2L_1}{\sqrt{c^2-v^2}} - \frac{2L_2}{\sqrt{c^2-v^2}}\\
&= \frac{2c(L_1+L_2)}{c^2-v^2} - \frac{2(L_1+L_2)}{\sqrt{c^2-v^2}}\\
&= \frac{2cL}{c^2-v^2} - \frac{2L}{\sqrt{c^2-v^2}}\\
&= 2L\left[\frac{c}{c^2-v^2} - \frac{1}{\sqrt{c^2-v^2}}\right]\\
&= 2L\left[\frac{1}{c(1-v^2/c^2)} - \frac{1}{c\sqrt{1-v^2/c^2}}\right]
\end{align*}$$

12.2 Michelson-Morley Experiment

$$= \frac{2L}{c}\left(\frac{1}{1-v^2/c^2} - \frac{1}{\sqrt{1-v^2/c^2}}\right)$$

$$= \frac{2L}{c}\left(\gamma^2 - \gamma\right)$$

$$\simeq \frac{2L}{c}\frac{v^2}{2c^2} = \frac{Lv^2}{c^3}$$

where $L = L_1 + L_2$ and the last line is based on the approximation of Eq. 14.

This difference in time corresponds to a difference in the path length Δl of light between the two situations that can be obtained by multiplying the time difference by c, that is:

$$\Delta l = c\left(\Delta T - \Delta t\right) \simeq \frac{Lv^2}{c^2} \tag{353}$$

The phase shift N corresponding to this difference in the path length will then be obtained from the formula:[273]

$$N = \frac{\Delta l}{\lambda} \tag{354}$$

where λ is the wavelength of the monochromatic beam. Now, the visible part of the electromagnetic spectrum is between about 4×10^{-7} m and 7×10^{-7} m. So, if the wavelength λ of the monochromatic light beam is about 5×10^{-7} m, and we assume that the speed of the ether wind is the same as the orbital speed of the Earth around the Sun which is about 30000 m/s and we use a reasonable value for the length of the two arms $L_1 \simeq L_2 \simeq 10$ m and hence $L \simeq 20$ m, then we obtain:[274]

$$N = \frac{\Delta l}{\lambda} = \frac{Lv^2}{c^2\lambda} \simeq \frac{20 \times (3 \times 10^4)^2}{(3 \times 10^8)^2 \times 5 \times 10^{-7}} = 0.4 \tag{355}$$

which is a reasonably large phase shift that can be easily detected by the Michelson-Morley apparatus.

Now, let assume that the ether wind is in a given direction which is parallel to the PP_1 path and a certain interference pattern is seen. Then on rotating the apparatus through 90° we should observe a shift in the interference pattern because the difference in the path length that corresponds to this change in orientation is sufficient to produce an observable phase shift in the interference pattern, as shown above. The direction of the ether wind is unknown to the experimenter, but if he keeps rotating the apparatus he should observe a shift in the interference pattern by the above amount when the apparatus is rotated through 90° from a given initial orientation that corresponds to the parallel orientation according to the above logic.

---

[273] The phase shift N may be defined more clearly as the number of fringes that move past the cross-hair of the telescope of the interferometer. We note that this number is not necessarily an integer.

[274] We note that these numbers represent only reasonable order of magnitude values and hence they are not supposed to report the actual values used in the original Michelson-Morley experiment and calculation although they are generally close and representative of the actual values according to the available historical records.

Finally, we draw the attention of the reader to the following remarks about the Michelson-Morley device and experiment:

1. It is obvious that the working principle of the Michelson-Morley device is based on the assumption that the ether wind is in the plane of the apparatus, or at least it has a projected component on that plane. Hence, if the ether wind is in a perpendicular direction to the plane of the apparatus then the above device will not work. This assumption seems logical, especially when considering different times of the day and year where the orientation of the Earth motion relative to the ether rest frame should vary due to its continuous time-dependent motion (i.e. around its axis, in its orbit around the Sun, in its trajectory as part of the Galaxy, etc.), although there may still be place for an argument.

2. The function of the compensating glass plate is to offset the difference in the light path[275] since one of the sub-beams (either the transmitted or the reflected) should pass through the glass of M_p twice. If M_p is silvered on the front then C_p should be inserted in the path of the reflected beam (because reflection occurs at the front face and hence the transmitted beam will pass through M_p twice), while if M_p is silvered on the back then C_p should be inserted in the path of the transmitted beam (because reflection occurs at the back face and hence the reflected beam will pass through M_p twice). The compensating plate in Figure 24 is based on the former case. We note that the thickness of C_p should be equal to the thickness of M_p to have the correct compensation. We also assume that the thickness of the silver layer is negligible and the glass of M_p and C_p is of the same material so that they have the same refractive index.

3. It may be the case that at a certain time of the day or year (corresponding to a certain direction of the Earth motion in its spin around its axis and its orbit around the Sun) there may be no ether wind, i.e. there is no relative motion between the Earth and the ether. Hence, the experiment was repeated during different phases of the Earth orbital motion corresponding to different times of the day and year.

4. A number of possibilities about the ether wind have been excluded. These include the possibility that the Earth rest frame is always the same as the ether rest frame although this possibility was considered locally by the ether drag hypothesis as an explanation for the failure of the Michelson-Morley experiment to detect the ether wind (see § 2.6).

Exercises

1. What is the main limitation of the Michelson-Morley experiment and its analysis?

---

[275] The importance of offsetting the difference in the light path should arise from the variability of the speed of light as a function of direction (according to the assumptions of the Michelson-Morley experiment), and hence any potential variability due to the passage through glass should be eliminated. However, this depends on a number of issues and assumptions that cannot be discussed in detail here although some of which (related to the type of propagation through transparent media and the role of ether in this propagation) have been indicated earlier.

12.3 Invariance of Laws under Galilean and Lorentz Transformations

In this appendix, we discuss the issue of invariance of physical laws under the Galilean and Lorentz transformations within the subject of this book which is Lorentz mechanics. We will establish that under the Galilean transformations only the laws of classical mechanics are invariant, while under the Lorentz transformations Maxwell's equations are also invariant. In summary, we have the following:
1. The laws of classical mechanics are invariant under the Galilean transformations.
2. The laws of classical mechanics (with some modifications) are invariant under the Lorentz transformations.
3. Maxwell's equations are not invariant under the Galilean transformations. This also applies to the electromagnetic wave equation which is based on Maxwell's equations.
4. Maxwell's equations (and electromagnetic wave equation) are invariant under the Lorentz transformations.

In the following subsections we will investigate these invariance properties briefly with one detailed example from the electromagnetic wave equation.

Exercises
1. What we mean when we describe a law as being invariant under certain transformations, e.g. Lorentz transformations?
2. Summarize what is invariant and what is not invariant in classical and Lorentz mechanics.

12.3.1 Laws of Classical Mechanics

Newton's three laws of motion are invariant under the Galilean transformations. It was shown earlier (see the exercises of § 1.9.4) that the second law is invariant under these transformations. Because the first law is a special case of the second law, it should also be invariant. The third law involves only forces and hence it should also be invariant under the Galilean transformations since force transforms invariantly under these transformations. So, all Newton's three laws of motion are invariant under the Galilean transformations. A number of other subsidiary laws of classical mechanics have also been shown to be invariant under the Galilean transformations (e.g. Hooke's law in § 1.9.4). In fact, because all the laws of classical mechanics are based directly or indirectly on Newton's three laws of motion, we can generalize our claim by saying that all the laws of classical mechanics are invariant under the Galilean transformations.

It can also be shown that the laws of classical mechanics are invariant under the Lorentz transformations with some appropriate modifications to the definitions and formulations to comply with the formalism of Lorentz mechanics. Some of these invariant properties have been discussed previously within other contexts. The reader is also referred to the wider literature on this subject for some conditions and restrictions on the invariance of some of these laws.[276]

12.3.2 Maxwell's Equations

The invariance of Maxwell's equations under the Lorentz transformations can be easily established from the fact that these equations can be put in a tensorial form, as we did in § 6.5.7, and hence their invariance under the Lorentz transformations can be concluded from the invariance property of tensors. The non-invariance of these equations under the Galilean transformations may then be concluded from their invariance under the Lorentz transformations plus the fact that the Galilean and Lorentzian transformations are incompatible with each other and hence the invariance property cannot apply to both.[277] Although this "proof" is rather brief and may not be rigorous, it should be sufficient for the objectives of this book. The detailed proof of the invariance property of Maxwell's equations with respect to the Lorentzian, but not Galilean, transformations can be found in several textbooks and research papers.[278]

We should remark that in the research literature there are some question marks about the validity and generality of the available methods for proving the invariance property of Maxwell's equations under the Lorentz transformations which some readers (especially those who are interested in inspecting the foundations of Lorentz mechanics deeply and thoroughly) may find useful to follow and investigate further. If these question marks proved to be legitimate, then this may require a fundamental revision to the theory of Lorentz mechanics which was originally developed to address issues like the non-invariance of Maxwell's equations under the Galilean transformations.

12.3.3 Electromagnetic Wave Equation

Here, we use the electromagnetic wave equation as an example for demonstrating the invariance property in detail and how it can be proved. The electromagnetic wave equation is given by:[279]

$$\frac{\partial^2 h}{\partial x^2} + \frac{\partial^2 h}{\partial y^2} + \frac{\partial^2 h}{\partial z^2} - \frac{1}{c^2}\frac{\partial^2 h}{\partial t^2} = 0 \tag{356}$$

In the following we show that this equation is not invariant under the Galilean transformations but it is invariant under the Lorentz transformations.

---

[276] We neglect many details about these issues because they are not essential to the objectives of the book and not proportionate to its intended size.

[277] We should also refer the reader to § 2.3 where we argued in one of the exercises that the non-invariance of the Maxwell's equations under the Galilean transformations can be based on the fact that these equations contain the speed of light which according to the Galilean transformations should be frame dependent since the speed of light transforms like any other speed due to the additive nature of the Galilean velocity composition. However, this argument may not stand because Maxwell's equations contain the constant c which is the characteristic (not necessarily the observed) speed of light and hence the argument is not conclusive since it requires other assumptions and rectifications to stand.

[278] In fact, the invariance of Maxwell's equations under the Lorentz transformations has been shown by Lorentz in one of his papers. Such a proof has also been attributed to Larmor and Poincare.

[279] In this context, h represents a component of the electric or magnetic field.

12.3.3 Electromagnetic Wave Equation

Using the Galilean transformations (Eqs. 28-31), we have:

$$\frac{\partial x'}{\partial x} = 1, \qquad \frac{\partial x'}{\partial t} = -v, \qquad \frac{\partial y'}{\partial y} = \frac{\partial z'}{\partial z} = \frac{\partial t'}{\partial t} = 1 \qquad (357)$$

while all the other derivatives (e.g. $\frac{\partial x'}{\partial y}$) are zero. Now, on using the chain rule we obtain:

$$\frac{\partial h}{\partial x} = \frac{\partial h}{\partial x'}\frac{\partial x'}{\partial x} + \frac{\partial h}{\partial y'}\frac{\partial y'}{\partial x} + \frac{\partial h}{\partial z'}\frac{\partial z'}{\partial x} + \frac{\partial h}{\partial t'}\frac{\partial t'}{\partial x} = \frac{\partial h}{\partial x'} \qquad (358)$$

On repeating this we get:

$$\frac{\partial^2 h}{\partial x^2} = \frac{\partial^2 h}{\partial x'^2} \qquad (359)$$

This similarly applies to the y and z derivatives, that is:

$$\frac{\partial^2 h}{\partial y^2} = \frac{\partial^2 h}{\partial y'^2} \qquad \text{and} \qquad \frac{\partial^2 h}{\partial z^2} = \frac{\partial^2 h}{\partial z'^2} \qquad (360)$$

We also have:

$$\begin{aligned}
\frac{\partial h}{\partial t} &= \frac{\partial h}{\partial x'}\frac{\partial x'}{\partial t} + \frac{\partial h}{\partial y'}\frac{\partial y'}{\partial t} + \frac{\partial h}{\partial z'}\frac{\partial z'}{\partial t} + \frac{\partial h}{\partial t'}\frac{\partial t'}{\partial t} \\
&= -v\frac{\partial h}{\partial x'} + 0 + 0 + \frac{\partial h}{\partial t'} \\
&= -v\frac{\partial h}{\partial x'} + \frac{\partial h}{\partial t'}
\end{aligned} \qquad (361)$$

and

$$\begin{aligned}
\frac{\partial^2 h}{\partial t^2} &= \frac{\partial}{\partial t}\left(-v\frac{\partial h}{\partial x'} + \frac{\partial h}{\partial t'}\right) \\
&= -v\frac{\partial}{\partial x'}\left(-v\frac{\partial h}{\partial x'} + \frac{\partial h}{\partial t'}\right) + \frac{\partial}{\partial t'}\left(-v\frac{\partial h}{\partial x'} + \frac{\partial h}{\partial t'}\right) \\
&= v^2\frac{\partial^2 h}{\partial x'^2} - v\frac{\partial^2 h}{\partial x'\partial t'} - v\frac{\partial^2 h}{\partial t'\partial x'} + \frac{\partial^2 h}{\partial t'^2} \\
&= v^2\frac{\partial^2 h}{\partial x'^2} - 2v\frac{\partial^2 h}{\partial x'\partial t'} + \frac{\partial^2 h}{\partial t'^2}
\end{aligned} \qquad (362)$$

On substituting from Eqs. 359, 360 and 362 into Eq. 356, we obtain:

$$\frac{\partial^2 h}{\partial x'^2} + \frac{\partial^2 h}{\partial y'^2} + \frac{\partial^2 h}{\partial z'^2} - \frac{1}{c^2}\left(v^2\frac{\partial^2 h}{\partial x'^2} - 2v\frac{\partial^2 h}{\partial x'\partial t'} + \frac{\partial^2 h}{\partial t'^2}\right) = 0 \qquad (363)$$

that is:

$$\frac{\partial^2 h}{\partial x'^2} + \frac{\partial^2 h}{\partial y'^2} + \frac{\partial^2 h}{\partial z'^2} - \frac{1}{c^2}\frac{\partial^2 h}{\partial t'^2} - \frac{1}{c^2}\left(v^2\frac{\partial^2 h}{\partial x'^2} - 2v\frac{\partial^2 h}{\partial x'\partial t'}\right) = 0 \qquad (364)$$

12.3.3 Electromagnetic Wave Equation

On comparing Eq. 356 and Eq. 364, we see that the wave equation is not invariant under the Galilean transformations due to the presence of the last two terms in the primed frame.

On the other hand, the wave equation is invariant under the Lorentz transformations as shown in the following. Using the Lorentz transformations of Eqs. 90-93, we have:

$$\frac{\partial x'}{\partial x} = \gamma \tag{365}$$

$$\frac{\partial t'}{\partial t} = \gamma \tag{366}$$

$$\frac{\partial x'}{\partial t} = -\gamma v \tag{367}$$

$$\frac{\partial t'}{\partial x} = -\gamma \frac{v}{c^2} \tag{368}$$

$$\frac{\partial y'}{\partial y} = 1 \tag{369}$$

$$\frac{\partial z'}{\partial z} = 1 \tag{370}$$

while all the other derivatives (e.g. $\frac{\partial x'}{\partial y}$) are zero. Now, on using the chain rule we obtain:

$$\begin{aligned} \frac{\partial h}{\partial x} &= \frac{\partial h}{\partial x'}\frac{\partial x'}{\partial x} + \frac{\partial h}{\partial y'}\frac{\partial y'}{\partial x} + \frac{\partial h}{\partial z'}\frac{\partial z'}{\partial x} + \frac{\partial h}{\partial t'}\frac{\partial t'}{\partial x} \\ &= \frac{\partial h}{\partial x'}\gamma + 0 + 0 + \frac{\partial h}{\partial t'}\left(-\gamma\frac{v}{c^2}\right) \\ &= \gamma\frac{\partial h}{\partial x'} - \gamma\frac{v}{c^2}\frac{\partial h}{\partial t'} \end{aligned} \tag{371}$$

On repeating this we get:

$$\begin{aligned} \frac{\partial^2 h}{\partial x^2} &= \frac{\partial h}{\partial x}\left(\gamma\frac{\partial h}{\partial x'} - \gamma\frac{v}{c^2}\frac{\partial h}{\partial t'}\right) \\ &= \gamma\frac{\partial}{\partial x'}\left(\gamma\frac{\partial h}{\partial x'} - \gamma\frac{v}{c^2}\frac{\partial h}{\partial t'}\right) - \gamma\frac{v}{c^2}\frac{\partial}{\partial t'}\left(\gamma\frac{\partial h}{\partial x'} - \gamma\frac{v}{c^2}\frac{\partial h}{\partial t'}\right) \\ &= \gamma^2\frac{\partial^2 h}{\partial x'^2} - \gamma^2\frac{v}{c^2}\frac{\partial^2 h}{\partial x'\partial t'} - \gamma^2\frac{v}{c^2}\frac{\partial^2 h}{\partial t'\partial x'} + \gamma^2\frac{v^2}{c^4}\frac{\partial^2 h}{\partial t'^2} \\ &= \gamma^2\left(\frac{\partial^2 h}{\partial x'^2} - 2\frac{v}{c^2}\frac{\partial^2 h}{\partial x'\partial t'} + \frac{v^2}{c^4}\frac{\partial^2 h}{\partial t'^2}\right) \end{aligned} \tag{372}$$

Similarly, we have:

$$\frac{\partial^2 h}{\partial y^2} = \frac{\partial^2 h}{\partial y'^2} \tag{373}$$

$$\frac{\partial^2 h}{\partial z^2} = \frac{\partial^2 h}{\partial z'^2} \tag{374}$$

We also have:

$$\frac{\partial h}{\partial t} = \frac{\partial h}{\partial x'}\frac{\partial x'}{\partial t} + \frac{\partial h}{\partial y'}\frac{\partial y'}{\partial t} + \frac{\partial h}{\partial z'}\frac{\partial z'}{\partial t} + \frac{\partial h}{\partial t'}\frac{\partial t'}{\partial t} \qquad (375)$$

$$= \frac{\partial h}{\partial x'}(-\gamma v) + 0 + 0 + \frac{\partial h}{\partial t'}\gamma$$

$$= -\gamma v \frac{\partial h}{\partial x'} + \gamma \frac{\partial h}{\partial t'}$$

and

$$\frac{\partial^2 h}{\partial t^2} = \frac{\partial}{\partial t}\left(-\gamma v \frac{\partial h}{\partial x'} + \gamma \frac{\partial h}{\partial t'}\right) \qquad (376)$$

$$= -\gamma v \frac{\partial}{\partial x'}\left(-\gamma v \frac{\partial h}{\partial x'} + \gamma \frac{\partial h}{\partial t'}\right) + \gamma \frac{\partial}{\partial t'}\left(-\gamma v \frac{\partial h}{\partial x'} + \gamma \frac{\partial h}{\partial t'}\right)$$

$$= \gamma^2 v^2 \frac{\partial^2 h}{\partial x'^2} - \gamma^2 v \frac{\partial^2 h}{\partial x'\partial t'} - \gamma^2 v \frac{\partial^2 h}{\partial t'\partial x'} + \gamma^2 \frac{\partial^2 h}{\partial t'^2}$$

$$= \gamma^2 \left(v^2 \frac{\partial^2 h}{\partial x'^2} - 2v \frac{\partial^2 h}{\partial x'\partial t'} + \frac{\partial^2 h}{\partial t'^2}\right)$$

On substituting from Eqs. 372, 373, 374 and 376 into Eq. 356, we obtain:

$$\gamma^2 \left(\frac{\partial^2 h}{\partial x'^2} - 2\frac{v}{c^2}\frac{\partial^2 h}{\partial x'\partial t'} + \frac{v^2}{c^4}\frac{\partial^2 h}{\partial t'^2}\right) + \frac{\partial^2 h}{\partial y'^2} + \frac{\partial^2 h}{\partial z'^2} - \frac{\gamma^2}{c^2}\left(v^2 \frac{\partial^2 h}{\partial x'^2} - 2v \frac{\partial^2 h}{\partial x'\partial t'} + \frac{\partial^2 h}{\partial t'^2}\right) = 0 \qquad (377)$$

$$\gamma^2 \left(\frac{\partial^2 h}{\partial x'^2} + \frac{v^2}{c^4}\frac{\partial^2 h}{\partial t'^2}\right) + \frac{\partial^2 h}{\partial y'^2} + \frac{\partial^2 h}{\partial z'^2} - \frac{\gamma^2}{c^2}\left(v^2 \frac{\partial^2 h}{\partial x'^2} + \frac{\partial^2 h}{\partial t'^2}\right) = 0 \qquad (378)$$

$$\left(\gamma^2 - \gamma^2 \frac{v^2}{c^2}\right)\frac{\partial^2 h}{\partial x'^2} + \frac{\partial^2 h}{\partial y'^2} + \frac{\partial^2 h}{\partial z'^2} - \frac{1}{c^2}\left(\gamma^2 - \gamma^2 \frac{v^2}{c^2}\right)\frac{\partial^2 h}{\partial t'^2} = 0 \qquad (379)$$

$$\left(\gamma^2 - \gamma^2 \beta^2\right)\frac{\partial^2 h}{\partial x'^2} + \frac{\partial^2 h}{\partial y'^2} + \frac{\partial^2 h}{\partial z'^2} - \frac{1}{c^2}\left(\gamma^2 - \gamma^2 \beta^2\right)\frac{\partial^2 h}{\partial t'^2} = 0 \qquad (380)$$

Now, since $\gamma^2 - \gamma^2 \beta^2 = 1$ (see Eq. 7) we obtain:

$$\frac{\partial^2 h}{\partial x'^2} + \frac{\partial^2 h}{\partial y'^2} + \frac{\partial^2 h}{\partial z'^2} - \frac{1}{c^2}\frac{\partial^2 h}{\partial t'^2} = 0 \qquad (381)$$

i.e. the wave equation takes the same form in the primed frame (Eq. 381) as in the unprimed frame (Eq. 356) and hence it is invariant under the Lorentz transformations.

12.4 Derivation of Lorentz Spacetime Coordinate Transformations

As we explained earlier (see for example § 1.1 and § 5), we prefer to postulate the Lorentz spacetime coordinate transformations and derive the other transformations and formulations of Lorentz mechanics from these postulates. However, the general approach in the literature of Lorentz mechanics is to derive the Lorentz transformations from other postulates. So, for the purpose of completeness we present in this appendix two possible methods for deriving these transformations if they are not postulated. Although there are several methods for deriving the Lorentz transformations of spacetime coordinates, the circulating literature of Lorentz mechanics is overwhelmingly in favor of special relativity and hence the majority of the available derivations are based on the postulates, assumptions and principles of special relativity. Therefore, in the first subsection we will investigate one of the common special relativistic methods of derivation. This will be followed by a second subsection where we present our preferred method of derivation (which is not based on special relativity) if we choose to derive rather than postulate these transformations. As we will see, our postulate is physical in nature and hence it is not far from the essence and spirit of the formalism.

12.4.1 Special Relativity Method of Derivation

Although the Lorentz spacetime coordinate transformations have been suggested prior to the proposal of special relativity, they have been derived from the two postulates of special relativity, i.e. the equivalence of all inertial frames and the invariance of the observed speed of light. Let have two inertial frames, O and O', in relative motion with speed v and they are in a state of standard setting as described in § 1.5. When the two origins of these frames meet at $t = t' = 0$ a light signal is emitted from the common origin in all directions. Since the speed of light is the same in both frames according to the second postulate of special relativity, the wave front will have a spherical shape in both frames, where the spherical wave front will be represented by the following equations in O and O' respectively:

$$c^2 t^2 = x^2 + y^2 + z^2 \tag{382}$$
$$c^2 t'^2 = x'^2 + y'^2 + z'^2 \tag{383}$$

Because the speed of light is independent of the state of rest and motion of its source, the source can be in frame O or frame O' (or indeed in any other frame). As we noticed earlier, the Galilean transformations for the space coordinates and time (i.e. Eqs. 28-31) do not satisfy Eqs. 382 and 383 identically. This can be verified by substituting from Eqs. 28-31 into Eq. 383 (or Eqs. 35-38 into Eq. 382) where both equations will hold true only if $v = 0$ (i.e. Eqs. 382 and 383 are not invariant under the Galilean transformations). In fact, this is the basis of the invariance of spacetime interval under the Lorentz transformations but not under the Galilean transformations, as discussed and verified in 3.9.4.

Now, since the frames are in a state of relative motion only along the common x-x' direction while they are relatively at rest in the y-y' and z-z' directions, the wave fronts will have the same coordinates in the other two directions, that is:[280]

$$y = y' \tag{384}$$

---

[280] We note that we need here to assume that there is no contraction of coordinates (or other type of alteration caused by the motion) in the directions that are perpendicular to the direction of motion.

12.4.1 Special Relativity Method of Derivation

$$z = z' \tag{385}$$

Regarding the common x-x' axis, we can assume that the x-x' coordinates in the two frames are transformed linearly[281] and hence they are given by:[282]

$$x = \gamma'(x' + vt') \tag{386}$$
$$x' = \gamma(x - vt) \tag{387}$$

where γ and γ' are proportionality factors which are unknown yet but they will be determined in the process. Now, by the principle of relativity, which is the first postulate of special relativity, we should have:[283]

$$\gamma = \gamma' \tag{388}$$

and hence we have:

$$x = \gamma(x' + vt') \tag{389}$$
$$x' = \gamma(x - vt) \tag{390}$$

But since on the common x-x' axis we have: $y = y' = z = z' = 0$, Eqs. 382 and 383 for the coordinates of the wave front on the x-x' axis become:

$$ct = x \tag{391}$$
$$ct' = x' \tag{392}$$

and hence Eqs. 389 and 390 become:

$$ct = \gamma(ct' + vt') \tag{393}$$
$$ct' = \gamma(ct - vt) \tag{394}$$

Now, from Eq. 394 we have:

$$t' = \gamma\left(t - \frac{v}{c}t\right) \tag{395}$$

On substituting from the last equation into Eq. 393 for t', we obtain:

$$ct = \gamma\left[c\gamma\left(t - \frac{v}{c}t\right) + v\gamma\left(t - \frac{v}{c}t\right)\right] \tag{396}$$

---

[281] The condition of linearity may be justified by the validity of Newton's first law in inertial frames since in the presence of higher order terms in the transformations a uniformly moving object in one frame will be seen as accelerating in the other frame.

[282] This form of transformation should be based on an insight from the Galilean transformation and possibly a hindsight which may compromise the objectivity and independence of this derivation. We note that the difference in the sign of v can be easily explained by what we have indicated before that is the difference in the sign of v stands for the difference between the primed and unprimed versions of v and hence we can replace $+v$ with $-v'$ (where $v' = -v$) which gives identical form.

[283] This may be challenged by the dependence of this on the nature of γ-γ' and on which variables and parameters they actually depend which may need at least another hindsight.

12.4.1 Special Relativity Method of Derivation

$$t = \gamma^2 \left[\left(t - \frac{v}{c}t\right) + \frac{v}{c}\left(t - \frac{v}{c}t\right)\right] \tag{397}$$

$$t = \gamma^2 \left(t - \frac{v}{c}t\right)\left[1 + \frac{v}{c}\right] \tag{398}$$

$$t = \gamma^2 t \left(1 - \frac{v}{c}\right)\left(1 + \frac{v}{c}\right) \tag{399}$$

$$1 = \gamma^2 \left(1 - \frac{v^2}{c^2}\right) \tag{400}$$

that is:

$$\gamma = \frac{1}{\sqrt{1 - \frac{v^2}{c^2}}} = \frac{1}{\sqrt{1 - \beta^2}} \tag{401}$$

where only the positive root is maintained. This equation is the same as Eq. 5. Hence, according to Eqs. 389 and 390 we have the following transformations for the x-x' coordinates:

$$x = \frac{x' + vt'}{\sqrt{1 - \beta^2}} \tag{402}$$

$$x' = \frac{x - vt}{\sqrt{1 - \beta^2}} \tag{403}$$

Now, to obtain the time transformation we use Eq. 390, that is:

$$x' = \gamma(x - vt) = \gamma\left[\gamma(x' + vt') - vt\right] \tag{404}$$

where Eq. 389 is used in the last step. Hence, we have:

$$x' = \gamma\left[\gamma(x' + vt') - vt\right] \tag{405}$$

$$x' = \gamma^2(x' + vt') - \gamma vt \tag{406}$$

$$x' = \gamma^2 x' + \gamma^2 vt' - \gamma vt \tag{407}$$

$$\gamma vt = \gamma^2 x' + \gamma^2 vt' - x' \tag{408}$$

$$t = \gamma\frac{x'}{v} + \gamma t' - \frac{x'}{\gamma v} \tag{409}$$

$$t = \frac{x'}{v}\left(\gamma - \frac{1}{\gamma}\right) + \gamma t' \tag{410}$$

$$t = \frac{x'}{v}\gamma\left(\frac{v}{c}\right)^2 + \gamma t' \tag{411}$$

$$t = \gamma\left(t' + \frac{vx'}{c^2}\right) \tag{412}$$

where Eq. 11 is used in the seventh step. By a similar method, where Eq. 390 is substituted into Eq. 389, we obtain:

$$t' = \gamma\left(t - \frac{vx}{c^2}\right) \tag{413}$$

12.4.2 Our Method of Derivation 327

The above equations (i.e. Eqs. 402 and 403 for the x-x' coordinates, Eq. 384 for the y-y' coordinates, Eq. 385 for the z-z' coordinates, and Eqs. 412 and 413 for the t-t' coordinates) are the equations of Lorentz transformations for spacetime coordinates. The other formulae of Lorentz mechanics will then be derived from these basic transformations of spacetime coordinates, as we did in § 5. There are other forms of this method of derivation, such as using the differentials of x-x' and t-t' coordinates. However, they rely on the same principles and logic as the above method and hence they are not independent methods. Many other methods (whether based on the postulates of special relativity or not) can also be found in the circulating literature of Lorentz mechanics where most of these are just variants of others with minor modifications. We should remark that the above derivation method may be questioned in some of its steps which are based on certain assumptions or claims that may be challenged and hence the method may not be seen as sufficiently rigorous. Some of these question marks are indicated in the previous footnotes.

12.4.2 Our Method of Derivation

If we have to derive (rather than postulate) the Lorentz spacetime coordinate transformations from a postulate, then our choice is to base the derivation of these transformations on a single postulate which states: "the c-calibrated spacetime coordinates contract in the direction of motion[284] by the Lorentz $\gamma(v)$ factor under the influence of motion through spacetime with speed v relative to the absolute frame". The derivation of the Lorentz transformations from this postulate is given in the following.

Let have two inertial frames, O and O', which are in a state of standard setting where one of these frames (say O) is the absolute frame. The characteristic speed of light in O frame is given by:[285]

$$c = \frac{d}{t} \qquad (414)$$

where d is the c-calibrated distance and t is the c-calibrated time interval in the O frame. Hence, we should have:

$$c^2 = \frac{d^2}{t^2} = \frac{x^2 + y^2 + z^2}{t^2} \qquad (415)$$

that is:

$$x^2 + y^2 + z^2 - c^2 t^2 = 0 \qquad (416)$$

Similarly, in O' we should also have:

$$c = \frac{d'}{t'} \qquad (417)$$

where d' is the c-calibrated distance and t' is the c-calibrated time interval in the O' frame. Hence, we should have:

$$c^2 = \frac{d'^2}{t'^2} = \frac{x'^2 + y'^2 + z'^2}{t'^2} \qquad (418)$$

---

[284] For the temporal coordinate, the motion always occurs in this dimension since any motion should occur in time. Moreover, if we define the temporal spacetime coordinate as ct then the situation may be easier to comprehend.

[285] To simplify the notation, we use t, x, y, z in the following to represent $\Delta t, \Delta x, \Delta y, \Delta z$.

12.4.2 Our Method of Derivation

that is:
$$x'^2 + y'^2 + z'^2 - c^2 t'^2 = 0 \tag{419}$$

Therefore, from Eqs. 416 and 419 we get:[286]
$$x^2 + y^2 + z^2 - c^2 t^2 = x'^2 + y'^2 + z'^2 - c^2 t'^2 \tag{420}$$

Now, since we are using a standard setting then there is no relative motion between the frames along the y and z directions and hence according to the postulate there is no contraction in the y and z coordinates, that is:

$$y = y' \tag{421}$$
$$z = z' \tag{422}$$

Accordingly, Eq. 420 becomes:
$$x^2 - c^2 t^2 = x'^2 - c^2 t'^2 \tag{423}$$

Now, since the coordinates of O' frame are contracted then we should have:
$$x' = \frac{x}{\gamma} \quad \text{and} \quad t' = \frac{t}{\gamma} \tag{424}$$

and hence:
$$x = \gamma x' \quad \text{and} \quad t = \gamma t' \tag{425}$$

On substituting from Eqs. 424 and 425 into the right hand side of Eq. 423 we obtain:

$$
\begin{aligned}
x^2 - c^2 t^2 &= x'^2 - c^2 t'^2 \\
&= \frac{x^2}{\gamma^2} - c^2 \frac{t^2}{\gamma^2} \\
&= \frac{\gamma^2}{\gamma^2} x'^2 - c^2 \frac{\gamma^2}{\gamma^2} t'^2 \\
&= \frac{\gamma^2}{\gamma^2} x'^2 - \frac{\gamma^2}{\gamma^2} c^2 t'^2 \\
&= \gamma^2 \left(1 - \frac{v^2}{c^2}\right) x'^2 - \gamma^2 \left(1 - \frac{v^2}{c^2}\right) c^2 t'^2 \\
&= \left(\gamma^2 x'^2 - \gamma^2 \frac{v^2 x'^2}{c^2}\right) + \left(\gamma^2 v^2 t'^2 - c^2 \gamma^2 t'^2\right) \\
&= \gamma^2 x'^2 + \gamma^2 v^2 t'^2 - c^2 \gamma^2 t'^2 - \gamma^2 \frac{v^2 x'^2}{c^2} \\
&= \gamma^2 x'^2 + 2\gamma^2 x' vt' + \gamma^2 v^2 t'^2 - c^2 \gamma^2 t'^2 - 2\gamma^2 x' vt' - \gamma^2 \frac{v^2 x'^2}{c^2}
\end{aligned}
\tag{426}
$$

---

[286] In fact, this is no more than the invariance of spacetime interval and hence the link (which we indicated earlier) between this invariance and the use of c for calibrating the spacetime measurements becomes more obvious.

12.4.2 Our Method of Derivation

$$\begin{aligned}
&= \gamma^2 x'^2 + 2\gamma^2 x' vt' + \gamma^2 v^2 t'^2 - c^2\gamma^2 t'^2 - 2c^2\gamma^2 t'\frac{vx'}{c^2} - c^2\gamma^2 \frac{v^2 x'^2}{c^4} \\
&= \gamma^2 \left(x'^2 + 2x'vt' + v^2 t'^2\right) - c^2\gamma^2 \left(t'^2 + 2t'\frac{vx'}{c^2} + \frac{v^2 x'^2}{c^4}\right) \\
&= \gamma^2 (x' + vt')^2 - c^2\gamma^2 \left(t' + \frac{vx'}{c^2}\right)^2
\end{aligned}$$

that is:

$$x^2 - c^2 t^2 = \gamma^2 (x' + vt')^2 - c^2\gamma^2 \left(t' + \frac{vx'}{c^2}\right)^2 \tag{427}$$

Now, since the dimensions of spacetime are mutually independent,[287] then the above equation is identically valid *iff* we have:

$$x = \gamma(x' + vt') \tag{428}$$
$$t = \gamma\left(t' + \frac{vx'}{c^2}\right) \tag{429}$$

On combining these equations with Eqs. 421 and 422 we obtain the following set of Lorentz spacetime coordinate transformations from frame O' to frame O:

$$x = \gamma(x' + vt') \tag{430}$$
$$y = y' \tag{431}$$
$$z = z' \tag{432}$$
$$t = \gamma\left(t' + \frac{vx'}{c^2}\right) \tag{433}$$

Finally, we obtain the opposite transformations (from frame O to frame O') by exchanging the primed and unprimed symbols and reversing the sign of v, that is:[288]

$$x' = \gamma(x - vt) \tag{434}$$
$$y' = y \tag{435}$$
$$z' = z \tag{436}$$
$$t' = \gamma\left(t - \frac{vx}{c^2}\right) \tag{437}$$

We should remark that the above derivation shows that the invariance of spacetime interval is an equally valid postulate for deriving the Lorentz transformations, if these transformations are not postulated, due to the equivalence between this invariance and the use of c for calibrating spacetime measurements, as indicated earlier. Also, this invariance

---

[287] The independence of the dimensions of spacetime is not the same as the dependence of the coordinates in each frame.

[288] These opposite transformations are just formal results based on the labeling and hence they do not represent symmetrical observations according to our interpretation.

12.4.2 Our Method of Derivation

is equivalent to the invariance of the speed of light across inertial frames which is also linked to the use of the speed of light for calibrating space and time measurements in Lorentz mechanics across all inertial frames. We should also remark that the mutual independence of the dimensions of spacetime (which should be obvious for the 4D manifold) is consistent with the dependence of each one of the spatial and temporal coordinates in one frame on the spatial and temporal coordinates in the other frame which is based on the merge of space and time into spacetime.

Again, this method like the previous method may be questioned in some of its steps, assumptions and claims and hence the method may not be seen as sufficiently rigorous. In fact, these methods are more appropriate for the purpose of demonstration and highlighting certain physical principles on which the transformations are supposed to be based[289] than for the purpose of rigorously establishing these transformations and the theory of Lorentz mechanics. This is inline with our belief that the best and most natural, reliable and objective postulates for establishing Lorentz mechanics are the Lorentz transformations themselves where the established experimental evidence is the source and justification for these transformations and all their consequences.

Exercises

1. Discuss the issue of lack of clarity and rigor in most (if not all) the derivation methods of Lorentz transformations.

---

[289] Which is a matter related to the adopted interpretation more than to the formalism.

Epilogue

We may finalize this book with the following points which outline our main conclusions from this investigation:

• Although there is massive evidence in support of Lorentz mechanics, it should not be seen as a final theory. Hence, we should keep looking for a better and more general theory.

• We should clearly distinguish between Lorentz mechanics (as a scientific theory) and special relativity (as a philosophical and epistemological theory). Hence, all non-scientific attachments to Lorentz mechanics (which mostly come from special relativity) should be discarded. Also, although special relativity is a logically inconsistent theory, Lorentz mechanics is a logically consistent theory. Hence, any rejection to Lorentz mechanics that is based on logical inconsistency should be directed toward special relativity and not toward the formalism of Lorentz mechanics. This should put an end to many disputes about the validity and rationality of Lorentz mechanics. However, the logical consistency of Lorentz mechanics should not rule out potential limitations due to lack of experimental evidence in support of certain aspects or even potential clash with experimental evidence, similar to what have been witnessed in the past with classical mechanics which is logically consistent but experimentally limited or invalid in certain aspects.

• We should be fair in ascribing the credit for the development of Lorentz mechanics. In particular, the tasteless and baseless exaggeration of the role of Einstein by giving him all the credit for this development (and hence denying the real developers the credit that they deserve) should be stopped. This is not a minor issue about a trivial historical fact but it is an important ethical issue that touches the foundations and objectives of science as an honest and sacred enterprise to discover the truth rather than a corrupt project for making gains and getting prestige, fame and power.

References

M. Dalarsson; N. Dalarsson. *Tensors, Relativity, and Cosmology.* Academic Press, first edition, 2005.

R. D'Auria; M. Trigiante. *From Special Relativity to Feynman Diagrams.* Springer, second edition, 2016.

H. Dingle. *Science At the Crossroads.* Martin Brian & O'Keeffe, first edition, 1972.

R. Gautreau; W. Savin. *Schaum's Outline of Theory and Problems of Modern Physics.* McGraw-Hill, second edition, 1999.

D.C. Kay. *Schaum's Outline of Theory and Problems of Tensor Calculus.* McGraw-Hill, first edition, 1988.

D.F. Lawden. *An Introduction to Tensor Calculus and Relativity.* Methuen & Co Ltd, second edition, 1967.

A.J. McConnell. *Applications of Tensor Analysis.* Dover Publications, first edition, 1960.

P. Norrington. *AMA303: Tensor Field Theory.* The School of Mathematics and Physics Queens University Belfast, first edition, 2003.

Open University Team (R.J.A. Lambourne et al.). *Relativity, Gravitation and Cosmology.* Cambridge University Press, first edition, 2010.

T. Sochi. *Tensor Calculus Made Simple.* CreateSpace, first edition, 2016.

T. Sochi. *Introduction to Differential Geometry of Space Curves and Surfaces.* CreateSpace, first edition, 2017.

T. Sochi. *Principles of Tensor Calculus.* CreateSpace, first edition, 2017.

I.S. Sokolnikoff. *Tensor Analysis Theory and Applications.* John Wiley & Sons, Inc., first edition, 1951.

B. Spain. *Tensor Calculus: A Concise Course.* Dover Publications, third edition, 2003.

Note: as well as the above references, we also consulted during our work on the preparation of this book many other books, research papers and general articles about this subject.

Index

3-operator, 217, 219
3-tensor, 217
3-vector, 167, 217, 218, 229, 231, 233, 234, 236, 238, 239
3D space, 9–11, 28, 29, 85, 139, 142, 217, 218, 229, 233, 268
4-operator, 167, 217, 219–221
4-tensor, 217, 226, 227, 234, 235
4-vector, 167, 207, 217–219, 226–234
4D spacetime, 9–11, 15, 18, 114, 177, 217–219, 224, 226, 229, 230, 234, 268

Absolute
    derivative, 9, 218
    frame, 28, 34, 39, 40, 51, 53, 54, 58–60, 66–68, 73, 82, 90–96, 140, 241, 242, 244, 245, 248, 258, 259, 261–263, 268, 269, 273, 275, 276, 285, 296, 297, 299–305, 310–312, 327
    space, 34, 39, 60, 64–69, 73, 93–95, 245, 258, 259, 269, 270
    time, 34, 67, 68, 78, 242, 244, 245, 258, 259, 269–271, 275, 279, 304
Abstract device, 31, 102, 185, 277, 281, 282, 294, 299
Accelerated motion, 86, 91, 92, 297, 298, 301
Accelerating
    frame, 17, 66, 67, 81, 82, 94, 261, 268, 302
    observer, 42
Acceleration, 49–51, 125, 126, 152, 153, 180–183, 230
Ampere law, 236, 314
Anti-identical, 18, 246, 247
Arc length, 10, 101
Area, 118, 119
Aristotelian philosophy, 62, 64
Atomic clock experiment, 256, 300

Ballistic propagation model, 70, 75
Barn-pole paradox, 302–305
beta particle, 49, 142
Blue shift, 155–157, 198, 199, 206

Calibration, 87, 89, 186, 263, 265, 267, 278–280, 284, 285, 294, 310, 328–330
Cartesian coordinate system, 11, 18, 19, 27, 30, 85, 89, 218–225, 230
Causal relation, 94, 95, 99, 108–111, 113, 114, 217, 260

Causality, 95, 98
Cause, 53, 60, 66, 75, 86, 91, 94, 108, 110, 111, 113, 240, 241, 243, 310, 313
Center of mass, 201, 202, 211, 249
Characteristic speed, 9, 18, 20, 22, 31, 32, 35, 56–61, 70, 95, 96, 100, 260, 265, 310, 327
Charge density, 11, 159, 160, 167, 188, 189, 233, 313
Christoffel symbol, 11
Circular
    argument, 51, 90, 198
    shape, 108, 196, 197
Circularity, 193, 199, 203
Classical
    mechanics, 12, 17, 19, 25, 27, 30, 31, 34, 44–47, 49–54, 62–65, 67, 68, 170, 171, 319
    relativity principle, 65, 259
    view, 53, 59, 62–65, 95, 269
Clock paradox, 296
Co-positional, 18, 20, 94, 114, 121, 122, 136, 137, 184, 242, 244–248, 265
Communication signal, 94, 95, 108, 110, 111
Conservation
    laws, 161, 166, 167, 169, 209, 215
    of charge, 167, 209, 233
    of energy, 41, 47, 166, 169, 200, 231
    of energy-momentum, 231
    of kinetic energy, 214
    of Lorentzian mass, 169, 174, 187, 209–212, 214
    of mass, 166, 169
    of mass-energy, 166, 169
    of momentum, 41, 47, 166, 168, 187, 189, 193, 198, 199, 201, 209–212, 214, 215, 231
    of total energy, 167–169, 187, 204, 209–212, 214, 215
    principles, 169, 209, 215
Contravariant tensor, 9, 19, 218–220, 226–230
Coordinate system, 16, 20, 27, 29, 31, 41, 84, 85, 89, 102, 106, 114, 142, 145, 148, 214, 218–221, 223–226, 230, 233, 299
Coulomb law, 95
Covariant
    derivative, 218
    tensor, 9, 218–220, 224, 225, 227–229
Curl, 9, 313, 314
Current density, 10, 159, 160, 167, 188, 189, 233, 234, 236, 313, 314
Curvature of spacetime, 81, 82, 92

Curved space, 82, 92, 114, 260
Curvilinear coordinate system, 218, 220, 224, 230, 233
Cylindrical coordinate system, 11, 224, 226

d'Alembert operator, 219
d'Alembertian operator, 9, 219–221, 233, 240
Deceleration, 298–300
Determinism, 39, 62, 64
Diagonal, 9, 118, 119, 223, 225, 316
Diffraction, 72
Dirac, 15
Dispersion, 34
Displacement, 166, 203, 208, 227, 267, 278, 279, 285, 287, 289, 293, 310
    vector, 228, 229, 313
Distance, 10, 28, 46, 87, 94, 95, 100, 112, 114, 119, 120, 136, 201, 202, 242–245, 248, 254–256, 265, 274, 277 281, 287, 291, 293, 327
Divergence, 9, 221, 313
Doppler
    effect, 154–156
    shift, 154–156, 159, 187
Dragging coefficient, 10, 143
Dual field strength tensor, 10, 235
Dynamic, 12, 174, 231

Earth, 73–76, 92, 100, 122, 145, 157, 158, 256, 296, 299, 312, 314, 317, 318
Eddington, 14
Einstein, 1, 2, 13–16, 70, 77–79, 206, 251, 258, 268, 308, 331
Electric field, 9, 313, 314, 320
Electromagnetic
    scalar potential, 11, 233, 234
    vector potential, 9, 218, 233
Electromagnetism, 42, 52, 69, 71, 72, 75, 77, 80, 92, 134, 163, 233, 234
Electron, 10, 124, 131–133, 149, 150, 162–165, 169, 172
    volt, 132, 162
Elementary particle, 122, 253–256, 297, 300, 312
Elsewhere region, 109–111, 114
Energy, 128–133, 161–166, 194–209, 230, 249, 250, 253
Epistemological, 1, 2, 12, 13, 15, 19, 30, 34–37, 53, 55, 77, 111, 174, 175, 244, 258, 270, 279, 309–311, 331
Ether, 34, 39, 57, 58, 60, 68, 69, 72–76, 78, 95, 96, 256, 282, 311, 314, 315, 318
    drag, 74–76, 318
    wind, 73, 74, 76, 100, 314–318

Euclidean, 31, 62, 63, 100, 223, 226
Event, 10, 16–20, 26, 28, 35
    space, 18, 90, 102, 107, 108, 110–112
Extrinsic property, 40, 41, 56, 66, 93, 124, 132, 154

Faraday law, 238, 313
Field strength tensor, 10, 227, 234, 235, 238, 239
First postulate of special relativity, 260, 262, 297, 325
FitzGerald, 14, 15, 76, 78, 133
FitzGerald-Lorentz length contraction, 14, 75–78, 240
Fizeau, 143
    experiment, 76
    formula, 143
Flat
    metric, 85
    space, 82, 85, 92, 113, 114, 223, 225, 260
Force, 49–52, 127, 128, 193, 194, 231–233
Form invariance, 19, 53, 70, 77, 189, 202
Frame of reference, 16, 17, 20, 27–35, 89–94
Free
    particle, 16, 107, 226
    space, 11, 18, 27, 30, 35, 57, 59, 69, 72, 95–97, 143, 233, 259, 313
Frequency, 11, 27, 34, 131, 154–159, 187, 188, 200
    shift, 154–156, 159, 200
Fresnel, 143
    drag, 15
    drift effect, 76
Future, 108, 109, 111, 114

Galilean
    invariance, 68
    invariant, 20, 220
    relativity, 52, 65, 67, 68, 92, 260
    transformations, 19, 42–48, 50, 51, 53, 54, 61, 68–72, 75–78, 92, 100, 101, 112, 115, 116, 133–135, 178, 220, 222, 224, 243, 314, 319, 320, 322, 324
Galileo, 52, 65–68, 274
Gauss
    law of electrostatics, 235, 313
    law of magnetism, 237, 313
General
    relativistic mechanics, 12, 14, 81
    relativity, 14, 63, 78, 81, 82, 84, 92, 260, 297, 301, 302
Geodesic, 113, 226, 230
Global observer, 16, 32, 243, 274, 276
Gradient, 9, 221
Gravitation, 12, 18, 78, 81, 82, 92, 95, 256

Gravity, 12, 14, 27, 34, 51, 52, 81–83

Hasenohrl, 206
Homogeneity, 62, 63, 87, 88, 231, 260, 279
Homogeneous coordinates, 223, 225
Hooke, 52
    law, 51, 53, 319
Huygens, 59, 72

Identical, 18, 94, 242, 244–248
Imaginary, 9, 19, 83, 96, 97, 112, 113, 134, 172, 223, 224
Improper
    distance, 255
    frame, 153, 186, 204
    length, 30, 118, 153, 183, 241
    mass, 123, 124, 126, 228
    time, 190
    time interval, 184, 186, 241
Inertia, 17, 42, 52, 54, 67, 90, 91, 94
Inertial
    frame, 10, 12, 17, 20, 27–29, 32, 35, 40, 42, 43, 45, 46, 48–51, 53, 60, 64–67, 81–83, 85, 88, 90–97, 104, 106–108, 112, 114, 116, 118, 120, 122, 125, 126, 133–137, 140–142, 144, 152, 153, 156–160, 172, 174–178, 183, 184, 186–190, 194, 209–212, 214, 215, 217, 226, 227, 229, 231, 232, 241, 242, 244, 245, 247, 248, 259–264, 268, 278, 285, 294, 298, 299, 301, 302, 310–312, 324, 325, 327, 330
    observer, 29, 33, 42, 44, 46, 48–50, 60, 67, 75, 96, 100, 102, 114, 117–119, 122, 126–128, 134, 136–140, 143, 148, 152–158, 183, 190–192, 198, 200, 222, 242–245, 248, 258, 259, 263, 267, 268, 277, 278, 282, 297, 302, 312
Inner product, 167, 218–220, 229
Instantaneous rest frame, 17, 190
Interference, 72, 74, 314
    fringes, 314
    pattern, 74, 77, 314, 317
Intrinsic
    derivative, 9, 218
    property, 40, 41, 47, 50, 56, 64, 66, 69, 94, 117, 154
Invariance of
    charge, 209, 233
    conservation of kinetic energy, 214
    conservation of momentum, 199, 212, 214, 215
    conservation of total energy, 214, 215
    conservation principles, 209
    d'Alembertian operator, 220
    electric charge, 174, 188
    kinetic energy, 215
    laws of mechanics, 319
    length, 193
    line element, 114, 222, 223
    mass, 54, 123, 186, 187
    Maxwell equations, 70, 75, 92, 133, 259, 320
    momentum, 190, 191, 193
    Newton's second law, 50
    physical laws, 19, 41–43, 77, 189, 259, 319
    physical properties, 41
    proper time, 223
    quadratic form, 224, 226, 229
    rest energy, 211
    rest mass, 174, 187, 209–211, 214
    space interval, 102, 114, 116, 193
    spacetime interval, 101, 102, 114, 116, 134, 185, 186, 222–224, 226, 229, 263, 268, 324, 328, 329
    speed of light, 88, 106, 140, 142, 156, 176, 186, 258, 259, 263, 267–269, 274, 279, 281, 294, 310–312, 324, 330
    time interval, 102, 116
    type of spacetime interval, 113
    velocity of light, 311
Inverse
    Lorentz matrix, 10, 226, 227
    Lorentz tensor, 10, 227, 228
    Lorentz transformations, 135, 138
Isotropy, 62, 63, 87, 88, 99, 260, 278, 279

Jupiter, 274

Kinematic, 12, 91
Kinetic energy, 41, 47, 48, 124, 128–133, 162, 164–167, 170, 171, 187, 194, 195, 200, 201, 203, 204, 208–211, 214, 231, 249

Ladder paradox, 302
Laplacian operator, 9, 219–221, 233, 240
Larmor, 14, 15, 76, 78, 206, 253, 320
Lateral speed, 99, 306
Laue, 14, 15
Law of inertia, 17, 42, 52, 67, 90, 91, 94
Lenard, 15, 206, 253
Length, 117, 118, 153, 183, 184, 240, 241
    contraction, 14, 33, 37–39, 75–78, 116, 120, 121, 133, 153, 170, 183–186, 240–243, 246, 249, 250, 253–256, 261, 263–268, 271, 279, 280, 284, 287–289, 291, 297, 302–305, 310, 312
    dilation, 184, 186
Light
    clock, 13, 185, 203, 277–285, 287–290, 292–295, 298

cone, 102, 107–111, 113, 114, 299
Lightlike, 108, 109, 112–114
Line element, 9, 101, 114, 116, 222–225, 240
Local
    observer, 16, 32, 155, 274, 276
    time, 78
Logically
    consistent, 66, 95, 302, 309, 331
    inconsistent, 299, 331
Lorentz, 14–16, 69, 76–79, 92, 133, 258, 280, 310, 320
    covariant, 19
    factor, 11, 20, 21, 25, 26, 33, 83, 96, 97, 116, 123, 129, 133–135, 167, 170, 172, 177, 178, 243, 245, 254, 263, 264, 297, 310, 311
    group, 14
    interpretation, 67, 70, 308, 309
    invariant, 19, 211, 212, 214, 217, 232
    matrix, 10, 226, 227
    mechanics, 12–16, 30, 62, 77–84, 100–102, 117, 216, 240, 251, 308
    metric, 221
    tensor, 10, 217, 227, 228
    transformations, 76–78, 135, 176–178, 226, 253, 319, 323, 324, 327, 329, 330
Lorentzian mass, 166, 167, 169, 174, 187, 193, 209–212, 214, 228, 249
Luminiferous
    ether, 72, 73, 311
    medium, 73, 74, 77, 259

Magnetic field, 9, 196, 234, 238, 239, 313, 320
Manifold, 17, 18, 30, 31, 82, 90, 100–102, 108, 114, 177, 178, 218, 219, 225, 226, 240, 260, 310, 330
Mass, 52, 123, 124, 154, 161–163, 186, 187, 197–206, 228, 249, 250, 253
    energy, 167, 249
Mass-energy
    equivalence, 53, 123, 128, 161, 164, 167–169, 197–201, 205, 249, 253
    relation, 132, 161, 162, 197, 198, 203, 205, 206, 249
Massive, 16, 20, 29, 31, 48, 52, 56, 59, 91, 99, 107, 111, 123–125, 129, 130, 132, 133, 141, 149, 154, 163–165, 171, 172, 194, 199, 201, 208, 210, 231, 233, 249, 258
Massless, 16, 20, 31, 96, 97, 130, 163, 165, 172, 175, 198, 207, 258, 307
Maxwell, 15, 52, 69, 72, 235, 236, 313, 314
    equations, 42, 44, 52, 59, 68–72, 74–76, 92, 133, 134, 163, 198, 201, 202, 205, 235–238, 259, 313, 314, 319, 320
Metric tensor, 9, 220–226, 228
Michelson, 15, 73, 74, 76, 77
Michelson-Morley
    analysis, 100
    apparatus, 77, 133, 315–317
    device, 314, 318
    experiment, 15, 70, 73, 75–78, 100, 133, 279, 314, 317, 318
Milky Way, 92
Minkowski, 14, 15, 101, 102, 270
    diagram, 102
    metric tensor, 221, 227
    space, 18, 20, 240
    spacetime, 9, 15, 18, 106, 107, 222–226, 230, 233, 268
Mixed tensor, 217, 226, 227
Modern convention about mass, 123, 129, 167, 169, 186, 187, 197, 249
Momentum, 52, 126, 127, 160, 163–165, 189–193, 196, 206–208, 230, 231
Momentum-energy relation, 163–165, 198, 203, 206–208, 231
Morley, 15, 73, 74, 76, 77
Mutually exclusive, 18, 246, 247

nabla operator, 9, 167, 219, 220, 313
Neutron, 10, 123, 127, 132, 133, 162, 163, 165
Newton, 51, 52, 64, 72
    first law, 17, 53, 54, 90, 93, 260, 325
    law of gravity, 51, 95
    laws, 17, 19, 20, 42, 44, 51–54, 66, 67, 90, 91, 93, 161, 319
    second law, 41, 50, 51, 54, 127, 128, 160, 161, 189, 193, 194, 196, 197, 211, 231–233
    third law, 54
Newtonian principle of relativity, 68
Non-
    accelerating, 66, 67, 91, 299, 300
    inertial frame, 17, 32, 41, 64, 66, 68, 90–94, 97, 98, 256, 261, 262, 298, 310, 311
    invariance of Maxwell equations, 75, 320
    invariance of quadratic form, 224
    invariance of speed of light, 274
    invariance of velocity of light, 311
    kinetic energy, 123, 167, 187, 194, 210, 249
    local reality, 306, 307
    rest frame, 159
    rest mass, 124, 154, 167, 249
    simultaneous, 272, 273, 276, 304
    singular, 172
Null

cone, 107
geodesic, 107, 110, 113

Observed speed, 15, 22, 31, 35, 56–61, 64, 70, 88, 95, 96, 100, 258–260, 263, 267, 268, 276, 324
Observer, 16, 17, 19, 29, 32, 33
Old convention about mass, 53, 123, 167, 169, 186, 187, 193, 197, 204, 228, 249
Olinto De Pretto, 206

Partial derivative, 9, 218, 219
Past, 108, 109, 111, 114
Pauli, 15, 206, 253
Permeability of free space, 11, 30, 69, 313
Permittivity of free space, 11, 30, 69, 233, 313
Phase shift, 74, 75, 77, 314, 317
Philosophical, 1, 2, 12, 13, 15, 30, 34–37, 51–53, 62, 64, 66, 67, 111, 174, 175, 244, 258, 270, 279, 309, 311, 331
Physical
reality and truth, 33, 35, 36, 39–41, 63, 150, 242, 261, 296, 309
relations, 160, 196
transformations, 133, 176
Planck, 14, 15, 206
constant, 131
Poincare, 1, 13–16, 69, 76, 78, 87, 101, 162, 206, 258, 308, 320
interpretation, 13, 15, 258
mass-energy relation, 129, 132, 161, 162, 166, 168, 197, 198, 203, 205, 206, 249
synchronization procedure, 87–90
Polarization, 72
Pole-barn paradox, 302
Position vector, 10, 227, 228
Positive definite, 112, 113, 224
Postulate of
constancy of speed of light, 176
invariance of speed of light, 258
relativity, 176, 258
Postulates of special relativity, 1, 15, 32, 98, 174, 259–261, 268, 276, 277, 324, 327
Preciseness, 39, 62, 64
Present, 108, 109, 111, 114
Principles of reality and truth, 39, 40, 43, 55, 262, 282, 296, 297, 304, 309
Projectile propagation model, 60, 70, 75, 88, 140, 273, 276, 311
Prolongation of lifetime, 253–255, 297, 298, 300, 312
Propagation medium, 57, 59, 60, 64, 95, 140, 273, 276, 282, 311

Proper
area, 118
charge density, 11, 159, 188
distance, 255
energy, 17
frame, 16, 188, 204
frequency, 11, 155–157, 159
length, 9, 10, 16, 30, 117, 118, 153, 217, 241, 255, 303
mass, 17, 53, 123, 210, 228
time, 10, 17, 190, 217–219, 223, 255
time interval, 119, 138, 186, 218, 241
time parameter, 11, 19, 212, 216, 217, 220, 221, 230
volume, 10, 188
wavelength, 11, 156, 157
Proper-improper perspective, 120, 122, 185
Proton, 10, 127, 130, 132, 162, 163, 165
Pseudo Euclidean, 223
Pythagoras theorem, 202, 278, 284, 286, 288, 290, 291, 315

Quadratic form, 112, 113, 218, 222–226, 229

Rank of tensor, 217, 226, 227, 234, 235, 239
Red shift, 156–158, 198, 199, 206
Reflection, 16, 110
Relativity
of co-positionality, 137, 243, 245, 246, 248, 269, 310
of simultaneity, 242–246, 248, 269, 271, 272, 274–276, 280, 303, 304, 310
Rest
energy, 17, 125, 128–132, 161–163, 166, 168, 169, 197, 204, 207, 210, 211, 231, 249
frame, 16, 20, 32–34, 37–39, 48, 49, 56–60, 64–66, 69, 72, 88, 92, 93, 119, 122, 143, 148, 149, 151, 154, 159, 160, 184, 185, 187, 188, 190, 191, 200, 218, 241, 254, 273, 276, 278, 281, 282, 286, 290, 294, 318
mass, 10, 17, 123, 154, 167, 169, 174, 186, 187, 192–194, 204, 207, 209–211, 214, 249
Restricted
relativity principle, 261, 262, 310
speed, 31, 84, 96–99, 104, 175, 258
Romer, 274
Rotation, 16, 28, 92, 106, 114, 142, 316

Scaling, 16, 243, 267, 285, 294
Second postulate of special relativity, 22, 88, 98, 99, 111, 114, 176, 260, 268, 273, 276, 278, 281–283, 324

Shearing, 16
Simhony, 253
Simultaneity
　　of observation, 32, 243, 244, 272, 274, 276
　　of occurrence, 32, 243, 244, 274–276
Simultaneous, 18, 20, 95, 114, 121, 122, 183, 242–248, 265, 272–274, 276, 304
Singular, 134, 172
Singularity, 83, 96, 97
Solar eclipse expedition, 14, 78
Sommerfeld, 14, 15
Space, 44–46, 84, 85, 89, 240
　　coordination, 84, 85
　　interval, 10, 18–20, 114, 121, 134, 224
Spacelike, 108, 109, 112–114, 217
Spacetime, 100–104, 106–108, 110–116, 135, 136, 176–178, 228, 229, 240, 323, 324
　　contraction, 40, 140, 184, 185, 253, 258, 261, 267, 268, 285, 289, 294, 295, 310
　　diagram, 101–106, 108, 110, 247, 248, 299
　　distortion, 240, 243, 285, 286, 310
　　interval, 11, 19, 20, 101, 107, 112–116, 134, 185, 216, 218, 222, 224–226, 229, 263, 268, 324, 328, 329
Spatial separation, 113–115, 137, 138, 177, 178, 245, 248
Special
　　relativistic mechanics, 1, 14
　　relativity, 1, 3, 12–16, 22, 30, 31, 33, 34, 54, 55, 58, 59, 65, 67–70, 75, 78, 79, 81, 82, 84, 86–88, 92, 95–99, 111, 114, 174–176, 243, 244, 251, 256, 258–263, 268–285, 293, 294, 296–310, 312, 324, 325, 327, 331
Speed, 56–58, 170–173, 306
　　of light, 9, 31, 32, 59–61, 69, 70, 95–100, 263–268, 306
　　of projectile, 56, 58, 69, 95
　　of wave, 57, 58
　　ratio, 11, 20, 21, 25, 26, 83, 135, 170
Spherical coordinate system, 10, 224, 226
Standard
　　configuration, 27
　　setting, 27–30, 33, 35, 44–49, 51, 85, 102, 104, 106, 115, 118, 121, 134–137, 139, 140, 142–144, 153, 172, 175–178, 183–185, 187, 190–192, 212, 214, 226–228, 231, 232, 240, 244, 245, 248, 264, 267, 268, 297–299, 312, 324, 327, 328
Stellar aberration, 74, 256
Summation convention, 19, 26, 216, 219, 223, 229, 235
Sun, 14, 75, 78, 92, 317, 318

Sweep speed, 99

Temporal separation, 113–115, 248
Thought experiment, 13, 54, 55, 185, 200, 271–277, 283, 305
Time, 44–46, 85–89, 100–102, 119–123, 153, 154, 184–186, 240–242
　　contraction, 184–186, 264
　　dilation, 37, 86, 116, 119–122, 153, 156, 170, 184–186, 191, 202, 203, 207, 220, 241–243, 246, 249, 253–256, 261–268, 271, 277–280, 282–285, 287–289, 292–294, 296–301, 303, 304, 310, 312
　　dilation triangle, 202, 203, 207
　　interval, 28, 46, 63, 100, 103, 119, 121, 122, 134, 136–138, 153, 177, 178, 184–186, 189, 201, 241, 244, 245, 248, 264–266, 277, 278, 305, 327
　　synchronization, 86, 88
Timelike, 108, 109, 112–114, 217, 221
Total
　　derivative, 218
　　energy, 9, 125, 128–132, 163–165, 167–169, 187, 195, 197, 202, 204, 206, 208–215, 231
Train thought experiment, 271–277
Trajectory, 17, 18, 112, 113, 125, 126, 128, 157, 226, 318
Transformation of
　　acceleration, 43, 152, 180, 183
　　charge density, 159, 188
　　current density, 159, 188
　　frequency, 154, 187
　　length, 153, 183–186
　　mass, 154, 186, 190, 193
　　space, 29, 34, 43, 44, 48, 51, 77, 92, 121, 153, 224, 240, 259
　　spacetime, 1, 12, 13, 15, 20, 42, 77, 78, 81, 89, 98, 101, 112, 114, 115, 121, 122, 134–138, 174, 176, 178, 180, 185, 187, 209, 212, 214, 217, 224, 226, 240, 242, 245, 253, 258, 259, 261, 266, 267, 276, 280, 283, 284, 311, 323, 324, 327, 329
　　time, 29, 34, 43, 44, 48, 51, 67, 77, 83, 92, 121, 177, 184, 224, 241, 259, 326
　　time interval, 153, 184–186
　　velocity, 46–49, 83, 88, 96, 136, 138, 139, 141, 143, 144, 150–152, 170, 172, 178, 180, 191, 259, 268, 284, 314, 316
Translation, 16, 27, 29, 43, 44, 51–54, 243, 260, 261
Twin paradox, 256, 296, 297, 300–304

Ultimate speed, 31, 84, 96–99, 104, 108, 110, 111, 175, 258
Uniform
    acceleration, 81, 105
    motion, 16, 20, 27, 32, 46, 64–68, 86, 90–93, 117, 119, 133, 138, 222, 259–261, 263, 277, 296, 298, 302
Unrestricted relativity principle, 261, 262, 269, 273, 302, 310, 312

Vacuum, 9, 20, 31, 249, 306
Value invariance, 19, 54, 190
Velocity, 46–49, 124, 125, 138–140, 142–145, 148–152, 178–180, 229, 230
    composition, 48, 49, 69, 96, 98, 100, 144, 145, 148–152, 172, 285, 289, 294, 320
Voigt, 14, 15, 76, 78
Volume, 10, 119, 159, 188

Wave propagation model, 60, 70, 75, 88, 100, 140, 273, 276, 282, 311, 313, 314, 316
Wavelength, 11, 27, 131, 156–158, 187, 317
Work, 133, 165, 166, 189, 194, 203, 208
Work-energy relation, 165, 166, 208, 209
World line, 18, 20, 102–104, 106, 107, 111, 233

Author Notes

- All copyrights of this book are held by the author.
- This book, like any other academic document, is protected by the terms and conditions of the universally recognized intellectual property rights. Hence, any quotation or use of any part of the book should be acknowledged and cited according to the scholarly approved traditions.